the wankel engine

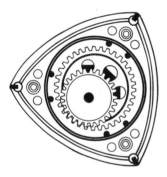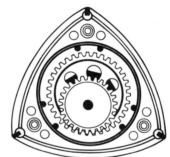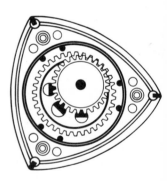

CHILTON BOOK COMPANY

the wankel engine

**DESIGN
DEVELOPMENT
APPLICATIONS**

JAN P. NORBYE

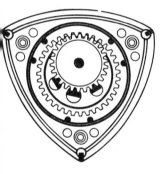

PHILADELPHIA NEW YORK LONDON

Copyright © 1971 by Jan P. Norbye
First Edition

Second Printing, August 1972

All rights reserved

Published in Philadelphia by Chilton Book Company
and simultaneously in Ontario, Canada,
by Thomas Nelson & Sons, Ltd.

ISBN 0-8019-5591-2

Library of Congress Catalog Card Number 73-161624

Designed by Cypher Associates, Inc.

Manufactured in the United States of America

Introduction

IT WAS A LONG TIME from the day I first read about the Wankel rotary engine to the day its potential became clear to me. And it would not have become clear at all if it had not been for the articles published by the technical press in the meantime. For my own understanding of the Wankel engine, I am indebted to the work of many, notably R. F. Ansdale, Harry Mundy and Karl E. Ludvigsen.

It has been my good fortune to see and drive Wankel-powered cars of all origins, some even in the prototype or experimental stage, and my archivist instincts led me to collect notes on the subject.

Reflecting on the material in these files led me to start writing this book. It is not a definitive work. Things are happening much too fast, and new chapters would have to be added every month just to keep it up-to-date. What I have tried to do is to present the full background of the Wankel engine itself and the first and second generation of cars powered by it. Of course, the Wankel engine is still too young to be assessed in proper historical perspective.

The full importance of a new development is not normally realized at the moment of its invention. John Ambrose Fleming could have had little idea of the value of his twin-electrode thermionic valve when he invented it in 1904, but it was later to revolutionize the entire technique of radiocommunication. The situation must have been similar when Valdemar Poulsen invented the magnetic tape recorder in 1900, and when Chester F. Carlson developed electrophotography (xerography) in 1942.

Frank Whittle probably did not even suspect, when taking out his basic gas turbine patents in 1930, that his invention was to replace the piston-type aviation engine for practically all commercial and military aircraft in the space of only 30 years!

A discussion of the Wankel engine today, in terms of its importance to the future of transportation, can hardly be more accurate than speculation on the subject of jet-powered aeroplanes would have been in 1941. In both cases, feasibility had been established. Beyond that, the rest is guesswork. Let me now come to the defense of guesswork. Nations, armies, corporations, individuals—all are guided by guesswork. The most momentous decisions taken by world leaders depend on educated guesses, and I believe that the more educated the guesswork, the higher the potential quality of the decisions. If we are to try and make up our minds about the role of the Wankel engine in tomorrow's transportation picture, the first thing we need is information about the Wankel engine. We must take full account of its rivals, and of all conditions (legal, environmental, economical, and supply-related) that affect the choice of power generation equipment for future vehicular transport.

What will be the requirements for future automotive powerplants? Briefly, it looks like this:

Minimum pollution
Minimum depletion of resources
Minimum noise level
Minimum cost
Maximum reliability
Minimum maintenance and wear
Maximum efficiency

Maximum freedom for the automotive designer
Maximum adaptability to existing traffic conditions
Maximum ease of handling and operation

Future requirements are no different from those of the past, except for the new emphasis on environmental issues (pollution and resources). Otherwise, automobile engineers have always had the same goals, and the same priorities. Try to think back to the year 1900, and you'll realize that proponents of different power sources were then engaged in as hard a struggle for domination of the automotive field as they are now.

Then the struggle raged between gasoline engines, battery-electric systems, and steam power. The winner was the gasoline engine, for its higher efficiency, lower cost, superior power-to-weight ratio, greater operational range, and minimum of fuss. The gasoline-burning internal combustion engine that has dominated the automobile industry for half a century operates on the four-stroke cycle. Despite its 80-year development history, it has many drawbacks: mechanical complexity, weight, wear, and a need for maintenance.

The Wankel engine is also a four-stroke internal combustion engine, but it is only on the threshold of its career as an automotive powerplant, whereas the reciprocating piston engine has a long development history behind it.

A multitude of minor problems with the piston engine have been solved over the course of the years, and there is a knowledge bank concerning the design and production of such engines that does not exist for the Wankel engine. The challenger, being new, lacks the advantage of pre-tried solutions. It will take time before it can reach the same level of perfection as is currently to be found in the typical Detroit V8.

In all comparisons between piston engines and Wankel engines, we are comparing a type of power unit barely beyond the experimental stage with an engine type developed at the cost of many millions of dollars. It can safely be assumed that the Wankel engine can be improved substantially as we gain more experience with it and try out new ideas that seem promising in one area or another.

In other words, it would be presumptuous to attempt to judge the Wankel engine in its present stage of development. We can safely accept its proven advantages. Whatever drawbacks it contains at the moment may be overcome by continued research. But back to basics. Why should the Wankel engine be considered as a potential source of motive power for tomorrow's automobile? For the answer, look at the alternatives:

Fuel cells

Atomic power

Electricity

Gas turbines

Steam engines

Diesel engines

Stirling engines

Free-piston engines

Hybrid-electric systems

All the above are energy-conversion systems, and it is axiomatic that you cannot get something for nothing by converting energy from one form to another. On the contrary, there is always a loss of energy. For the anti-pollution league, it should be a sobering thought, too, that whenever energy is converted, there is a risk of the process not being complete, which in turn means waste, sometimes of toxic matter, sometimes not. These "unconventional" power plants are not necessarily emission-free. Nor must it be assumed that internal combustion engines will produce exhaust gases with pollutants.

1. FUEL CELLS

A fuel cell cannot actually drive an engine. Rather, it *is* an engine, but only in the sense that it produces electric power. Fuel cells have been successfully developed (at extremely high cost) for spacecraft and sub-marines, and many organizations are now investigating the possibilities of fuel cell power for use in road vehicles. The fuel cell produces electric energy without moving parts and with very low emissions. It has many other claimed advantages, such as highest efficiency at low loads, quiet operation and no loss of energy at idle. It promises close to 100 percent efficiency. On the other hand, there may be starting difficulties, and the power-to-weight ratio is low. High utilization time will be necessary to offset the high cost.

For a fuel cell to be able to propel a car, it would also have to be bulky. Its fuels, oxidants and catalysts can be both dangerous and ex-pensive. The automotive fuel cell is strictly experimental. Future ad-vances in fuel cell technology are needed before we can consider even specialized vehicle applications. In any event, research and development costs would run into millions of dollars.

2. Atomic Power

We are not concerned with atomic power generated in stationary nuclear reactors (such as now in use by some utility companies) and made available to the consumer as electricity. At the vehicle end, such power sources would be dependent on electric vehicle technology which is dealt with below. Here we are talking about an on-board nuclear reactor, feeding electric batteries to drive an electric motor. Possible heat sources are plutonium, curium, thalium, promethium, and uranium. Of these, only uranium U235 is mass-produced. The others are in very short supply.

Development is so far advanced that a reactor could be made as light as 800 pounds, which would be considered "portable" in the automotive industry. However, such a reactor would require massive concrete shield-ing to keep radiation below hazardous levels, in the order of 15,000 to 25,000 pounds. The size and weight of such a package remove the on-board nuclear reactor from consideration for passenger cars.

3. Electric Vehicles

The electric car promises silent operation as well as absence of emis-sions. Electric motors deliver full torque at all speeds, giving a higher rate of acceleration from a dead stop. At standstill, no fuel is used. Because the electric motor is such a simple piece of machinery, main-tenance and repair costs can be expected to stay minimal. The electric car has no need for the costly automatic transmission of today's pas-senger car and needs no differential, engine cooling system, generator, exhaust system or muffler. What threatens to exclude the electric car from tomorrow's traffic picture is that the best batteries available today are inadequate to meet present performance requirements. Electric prototypes I have driven have low top speed and restricted range. Batteries are heavy, bulky, and costly. The bases for judging batteries are energy density (watt-hours per pound of battery); power density (watts per pound of battery); and life (measured in time and cycles).

High energy density is required for high mileage per battery charge, and high power density is needed for rapid acceleration. Different types of battery may be combined within the same system, so as to complement each other's characteristics. It applies to all of them that they tend to run hot under sustained periods of discharge, so a cooling system will be needed. One obvious way to stretch the interval between stops for recharging or change of batteries is to add a regenerative device which charges the batteries when the vehicle is coasting downhill or decelerating. The most direct method would be a generator driven by the road wheels.

There are no batteries available today that fulfill the requirements of a practical electric vehicle. They are either prohibitively heavy, costly, or inefficient. The common lead-acid battery has low energy density. The sodium-sulphur battery runs hot. Zinc-air batteries require a compressor to feed oxygen to the cells, and a spent-air separator to remove the gas components that are pumped through the cathode without participating in the reaction. The lithium-halide battery has a long charging time and short life. Short life also excludes the lithium-chlorine battery. The nickel-cadmium battery has low energy density and high cost. Silver-zinc batteries have short life and expensive components.

The creation of a modern electric car is contingent upon a breakthrough in battery technology. Given an adequate storage battery, the car would still require a new type of motor. Present motors are not suitable for automobile propulsion. Experiments with homopolar motors, alternating current induction motors and permanent magnet motors have so far proved inconclusive. Spokesmen for some of the companies that have been most active in the electric vehicle field have admitted that the electric car is, at best, a distant-future possibility.

4. Gas Turbines

The gas turbine is attractive to automobile engineers because of its low emission levels, vibrationless operation, excellent torque characteristics, high power-to-weight ratio, multi-fuel capabilities, and non-stalling characteristics. Cost has been the key problem about developing gas turbines for cars. But today we are approaching the point where the gas turbine can compete effectively on a cost-per-horsepower basis with the largest truck diesel engines now in production. The gas turbine, during some 20 years of development for road vehicles, has overcome most of its problems. The worst operational drawbacks were excessive fuel consumption at part load, lagging throttle response, a low pressure ratio ceiling, and lack of retardation on closed throttle.

Concurrently with new advances in fiberglass, glass ceramic, and other materials, new heat exchangers have been designed, giving over 70 percent effectiveness combined with a pressure drop of only 8 percent. The pressure ratio of a gas turbine is limited by the heat-resisting properties of the turbine material. New alloys have been developed to withstand gas temperatures up to 1,800°F., and the pressure ratio is no longer felt to be a limiting factor in raising gas turbine output.

The acceleration lag was caused by the necessity of speeding up the compressor before the power turbine could deliver the required torque. Throttle response in the best gas turbines is still not instantaneous as in gasoline-driven piston engines, but the acceleration lag has been reduced to an acceptable 1.0 second (against 0.3 second in the modern V8) by greater precision of fuel control and reducing rotor inertia to a minimum. The objections regarding lack of retardation on the overrun have been conquered by installing a variable-pitch nozzle system in the power turbine. This feature also improves part-load efficiency. Automotive gas turbines down to 360 horsepower are on the way. Truck and bus applications are under study, but no concrete plans for turbine-driven passenger cars are in process.

5. STEAM ENGINES

Although the basic steam engine is less expensive than a modern V8 a complete steam power system would cost three times more! However, since the steam engine produces maximum torque at zero r.p.m. it can use a cheaper and less complex transmission. Emission controls are simpler and the exhaust system for a steam engine would cost about half of that on a V8 gasoline engine. The steam engine does not require a cooling system with a radiator, thermostat, fan and fan drive, and water pump. It has multi-fuel capability and offers reverse torque operation for retardation.

In contrast with the gasoline engine, the steam engine does not suffer an efficiency loss under part load. No steam is wasted at standstill, while the V8 continues to use gasoline while idling. Impressive advantages? Yes, but there's the other side of the coin. The steam engine requires an expensive burner assembly, with an electronic brain, a steam generator, plus complex sensors and controls. The condenser costs about twice as much as a radiator in a conventional car. The steam car needs a reservoir for the working fluid (water or Freon) in addition to the fuel tank for the burner. Accessory drives are more complicated than on a V8, and the steam engine needs an oil separator to reduce the risk of water contamination. The manual steam cutoff is more expensive than the throttle linkage in today's car.

It is clear that the steam car has disadvantages in both cost and weight compared with present V8s. It may have the same ease of driving and offer simplified maintenance, however. Even if modern steam technology can overcome the traditional drawbacks, such as a starting delay, the freezing risk and the explosion risk, the thermal efficiency of the steam engine remains inferior to the gasoline engine's. The steam turbine is a promising alternative to the reciprocating steam engine. It is more compact, with higher power-to-weight ratios, but also considerably more costly.

Like the gas turbine, it requires a torque converter and reduction gearing. It should be easier to maintain than conventional steam engines, and has proved its excellent reliability in ships and locomotives.

6. DIESEL ENGINES

Common in trucks and buses, with a bad reputation for producing dark, dense, and smelly exhaust gas, the diesel engine may turn out to be a real sleeper in the long-term clean air car race. The diesel engine is the main power plant throughout the world today for railway, marine, heavy truck and industrial uses, and there are some passenger car applications (Mercedes-Benz, Peugeot, Perkins and Austin). Keep in mind that the Wankel engine does have diesel-fuel capability.

A compression-ignition engine, with high air/fuel ratios and extremely high compression ratios, the diesel's exhaust is actually healthier (if this can be said) than that of the spark-ignition gasoline engine. The diesel exhaust is very low in carbon monoxide content. The diesel engine is more expensive, has a poor power-to-weight ratio, is not free of vibration and gives off loud combustion noise. Great progress has been made in the past ten years, but a total solution is not yet in sight.

7. STIRLING ENGINES

The Stirling engine is a "hot air" engine, an external combustion engine running on kerosene fuel, with hydrogen as its working fluid. It has multi-fuel capability, very low noise levels, and is practically vibrationless. It is usually thought of as a power source in hybrid-electric cars. It is best suited for constant-speed operation, although it could be provided with gearing to drive the wheels direct. In the hybrid application, the Stirling engine plus the electric batteries, alternator and motor, controls and reservoir, form an extremely bulky installation. It is also heavy and costly. But thermal efficiency is high (39 percent) and its specific output beats both the diesel and the gasoline engine.

Only two companies are devoting any great amount of attention to the Stirling engine: Philips Gloeilampenfabriek in Eindhoven, Holland, and General Motors. Judging by reports on their experimental units, we can only conclude that a hybrid car with Stirling engine is not a practical automobile.

8. Free-Piston Engines

The free-piston engine is not capable of running a mechanical drive train. It is an external combustion engine and a gas mover. The best proposals for its use in automobiles involve its application as a feeder for a gas turbine. Combining a free piston engine with a gas turbine gives the same low emission levels that the gas turbine delivers, with even greater multi-fuel capabilities. With a gas thermal efficiency of 44 percent and a turbine efficiency of 83 percent, the overall thermal efficiency at the shaft is 36.5 percent, which is comparable to a modern diesel engine. The free-piston engine is inherently balanced and virtually vibrationless. Torque characteristics, thanks to the turbine, are excellent.

What are the drawbacks? Since the free-piston engine has no revolving shaft, outside units are required to provide accessory drives, for starting, oil pumps, coolant pumps, and fuel pumps. Starting is usually accomplished by compressed air fed into the bounce chambers at 400 psi. The normal automobile has no high pressure air source, but it is claimed that a standard truck-type brake system air compressor can do the job. The turbine can conveniently be equipped to drive pumps and other accessories.

The added weight and complexity of running both a free-piston engine and a gas turbine in the same vehicle has led engineers to the logical conclusion that they are better off with the gas turbine alone, built as a complete unit to undertake all operating functions for itself rather than receiving exhaust gas flow from a separate engine.

Conclusions

There are many indications that the best short- and medium-term solution lies with the internal combustion engine (piston-type or Wankel). With the use of positive crankcase ventilation, pre-heated intake air, transmission-controlled spark advance, afterburners with or without supplementary air injection, exhaust gas recirculation, and catalytic converters, it is possible to meet current standards with a comfortable margin. This is true of both Wankel and piston-type engines.

Independently conducted tests show the following results:

Engine	CH	CO	NO$_x$
NSU Ro-80	1.54	19.7	—
Mazda R-110	1.4	11.9	—
Curtiss-Wright	1.11	17.8	1.44
1971 limits	2.2	23.0	4.0
1974 limits	1.5	23.0	1.3

The emission characteristics of the Wankel engine are fully discussed in Section II, Chapter 15.

The advantages of the Wankel engine over the reciprocating piston engine are dealt with in Section I, Chapter 7. There is just one aspect of the Wankel, for use in passenger cars, that I would like to enlarge on here. It concerns the greater design freedom offered to product planners, engineers and stylists. In other words, when Detroit starts manufacturing Wankel engines, we should look not for detail modifications in transmission systems or chassis engineering, but for basic changes in vehicle architecture.

How will adoption of the Wankel engine affect future passenger car design? It will have immediate effects in three areas: safety, comfort and cost.

SAFETY

The Wankel engine will enable significant gains to be made in two areas of accident avoidance: visibility, and handling precision. Visibility can be improved because the small engine will allow a lower hood line and a sizeable gain in glass area. A driver who can see more is a safer driver. Handling precision will be improved because the Wankel engine, being smaller, will have a lower center of gravity, and its lightness means a reduced engine mass, which translates into improved weight distribution. A lower center of gravity in combination with improved weight distribution means increased controllability, more predictable steering response, and reduced risk of rollover.

The Wankel engine will have little influence on the effects of a collision. But there is this point: future cars are expected to have bumpers good for 15-mph impacts without damage to the car. Some form of side impact protection is also likely to become a legal requirement. All this will add to the weight of the car. With retention of the V8, the car would gain weight. The Wankel engine will allow these safety features to be installed without making the car heavier than present cars.

Comfort

How can the Wankel engine possibly contribute toward greater creature comfort? Since it is practically free of vibration, all occupants will be more comfortable. Next, let us suppose that the designer of a new medium-size car had to work on two parallel versions of the same basic vehicle, one with a six-cylinder reciprocating piston engine and the other with a twin-rotor Wankel engine. He would soon find that the smallness of the Wankel engine would allow a roomier interior. A net increase in legroom, for instance, promises more comfort. A roomier interior also contains more air, which means more efficient ventilation. It also plays a part in safety, since efficient ventilation can help combat driver fatigue.

Cost

Certain savings can be had in the Wankel-powered car as a direct result of the engine's compact size and light weight. With a front-engined car of medium size, the need for power steering would be reduced. Since the engine's space requirement is smaller, it will take less material to package it. The space saving therefore also indicates that a saving can be made in the frame, the engine cradle, and the sheet metal around it. The Wankel engine's compact size invites transverse installation for front-wheel-drive or midships engine installation, thereby eliminating the propeller shaft and avoiding other costly transmission complications.

For some examples of how this works out in practice, take a look at the cars I have described in Section III of the book.

No doubt a brief study of these automobiles will raise a number of questions in your mind. It is my hope that I have provided the answers elsewhere in the book. If I haven't, I would like to know about it, and invite your correspondence.

Jan P. Norbye

Contents

Section I

DESIGN

1

Principles of
Operation

THE WANKEL ENGINE is an internal
combustion engine operating on the four-stroke cycle: intake, compression, combustion and exhaust. These four strokes serve to draw in the
air-fuel mixture, compress it to intensify the charge so that a greater force
is released during combustion, and clean the combustion chamber in
preparation for the cycle to begin again. Keeping this in mind, an internal combustion engine can be thought of as a device for harnessing an
expansion force and converting it into motion. In this sense, the Wankel
engine can be thought of as a halfway point between the conventional
automotive piston engine and the ground transport gas turbine (as opposed to aircraft turbines for jet planes).

All operational cycles occur in the same area in the piston engine,
while each phase of the cycle occurs in an area specifically designed for
it in the Wankel and turbine engines. The Wankel and turbine engine
functions are spaced out according to the *gas flow path*, while in the
piston engine, everything takes place in the same area and the various
phases are separated only by time. But, even though the Wankel engine
appears to be more closely related to the turbine than to the piston engine (because of their common rotary motion), there is a great difference
—the Wankel engine (as well as the piston engine) is an internal combustion engine while the turbine works by external combustion.

A better understanding of the Wankel may be gained by reviewing
briefly the operating principles of both the four-stroke cycle piston engine and the gas turbine. A single-cylinder piston engine takes four
strokes (two crankshaft revolutions) to complete its operating cycle.
Starting at top dead center, between cycles, the first stroke is a down

stroke—intake. Fresh gas mixture is drawn into the cylinder by the vacuum created by the downward motion of the piston. The piston then moves upward in the compression stroke and squeezes the mixture into a small space at the top of the cylinder (combustion chamber). The combustion stroke (often called the power stroke) is the downward working stroke that produces the power. The compressed gas is ignited by a spark plug, and the force of expansion created by the burning of the air-fuel mixture drives the piston downward under very high pressure. The exhaust stroke is an up stroke, following the power stroke. The motion of the piston helps push the burned gases out of the cylinder. At the end of the exhaust stroke, the cylinder is emptied of burned gases and ready for another intake of air-fuel mixture. Then the cycle is repeated.

The intake phase of the gas turbine takes place through an annular chamber which leads to a compressor. A compressor is an air pump which, for gas turbines, consists usually of radial-flow or axial-flow vane wheels working inside closed chambers to build up air pressure. There is continuous flow throughout the compressor and turbine; therefore, instead of having a compression ratio, the gas turbine has a pressure ratio. This pressure ratio (4.5:1 is considered high) is far lower than the compression ratio of an internal combustion engine (where 8:1 is normal and 11:1 is high). Diesel engines, which rely on compression pressure rather than spark for ignition, have compression ratios as high as 22.5:1.

The combustion phase of the turbine takes place in the burner section. Ignition is performed by a continuously glowing plug which functions as a spark plug. Gas turbines do not have carburetors to mix air and fuel; they compress clean air and inject fuel into the combustion section. The same principle is used in piston engines and Wankel engines with fuel injection, but the point is that gas turbines *depend* on fuel injection for their efficient operation while internal combustion engines work well with *either* carburetors or fuel injection.

Just as the piston engine harnesses an expansion force, so does the gas turbine. A turbine is like a compressor in reverse—a vane wheel gets its momentum from gas flow and converts the gas thrust into mechanical torque. The automotive gas turbine actually is two turbines which have no mechanical connection—the first drives the compressor and the engine accessories and the second drives the output shaft. The main difference between the gas turbine and the Wankel engine is that the turbine's operational functions are spaced out axially along the turbine shafts, while the Wankel's areas of activity are spaced out radially from the rotor.

Mechanically, the Wankel engine differs significantly from the piston engine. The cylinders and pistons are replaced by working chambers and

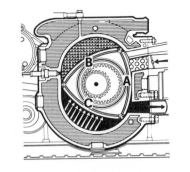

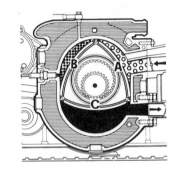

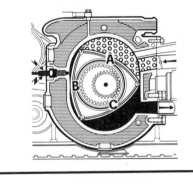

In the four-stroke Wankel, all three working chambers are in continuous action. While rotor face A is sweeping out the remaining exhaust gases and preparing to begin a new intake phase, chamber B is beginning compression and chamber C is about to complete its expansion phase. In the second sketch, chamber A goes ahead with its intake, while chamber B is approaching maximum compression. Chamber C has just started its exhaust phase. In the third sketch, ignition takes place in chamber B, chamber A is about to complete its intake phase, and chamber C is in the middle of the exhaust phase. The fourth sketch shows expansion in chamber B, completion of intake in chamber A, and continued scavenging in chamber C. (Rotor revolves CCW.)

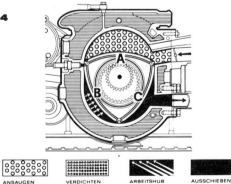

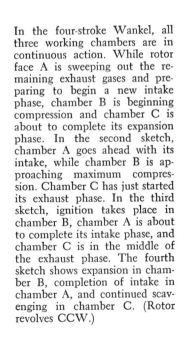

ANSAUGEN	VERDICHTEN	ARBEITSHUB	AUSSCHIEBEN

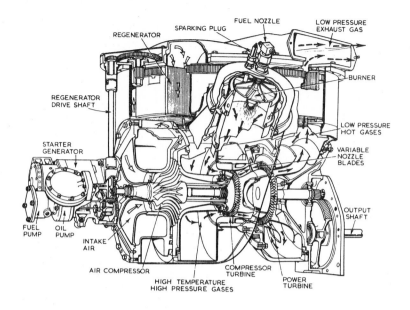

Chrysler CR2A gas turbine shows how a simple principle of operation nevertheless results in a lot of vital accessories and a sinuous gas flow path.

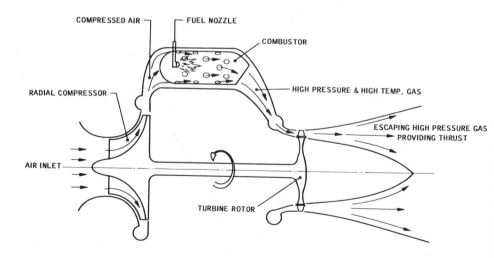

The jet aircraft gas turbine does not deliver shaft horsepower but thrust. The turbine merely drives the compressor, and the exhaust gas provides the thrust.

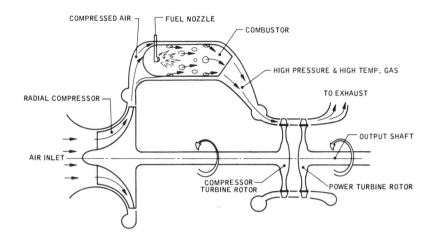

COMPRESSED AIR FUEL NOZZLE

COMBUSTOR

HIGH PRESSURE & HIGH TEMP. GAS

RADIAL COMPRESSOR

TO EXHAUST

OUTPUT SHAFT

AIR INLET

COMPRESSOR TURBINE ROTOR

POWER TURBINE ROTOR

The automotive gas turbine does deliver shaft horsepower because a second turbine is added behind the first one that drives the compressor. The second turbine is integral with the output shaft and can be coupled to a normal transmission.

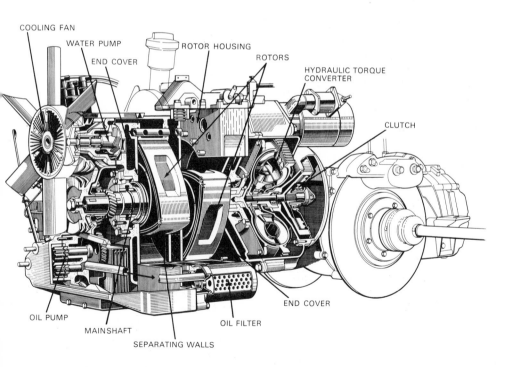

COOLING FAN

WATER PUMP ROTOR HOUSING

END COVER ROTORS

HYDRAULIC TORQUE CONVERTER

CLUTCH

END COVER

OIL PUMP

MAINSHAFT OIL FILTER

SEPARATING WALLS

This is the twin-rotor Wankel engine that powers the NSU Ro-80. Its simplicity is immediately apparent when you compare it with the other power units described and illustrated here.

rotors, encased in a stationary housing. As a result, phases replace strokes, rotor seals do the job of piston rings, and ports take the place of valves. Instead of a crankshaft, there is a mainshaft, the job of the crankpins being performed by eccentrics.

The housing is not made in one piece, but is built up around the working chamber. This chamber, shaped like a figure-eight with a fat waistline, contains a three-cornered rotor which moves around the working surface in an orbiting motion, all three corners always being in contact at some point. The working chamber is provided with ports for intake and exhaust and a spark plug for ignition of the mixture.

Wankel volumetric efficiency is largely determined by the position, size, and shape of the intake port. The rotor, along its directed travel, moves into positions which provide each of the four operational phases

The cross-section of the three-rotor Mercedes-Benz C-111 engine shows very clearly the rotor phasing gears, with 24 teeth on the stationary reaction gear and 36 teeth on the inner ring gear in the rotor.

(strokes) with its own area. In other words, each area along the working surface corresponds to a part of the familiar four-stroke cycle.

The rotor leaves a certain amount of free space between its faces and the working surface throughout its movement. This space is continuously changing in size, shape, and position. These volumetric changes provide the pumping action for gas intake, compression, combustion and exhaust. Minimum volume in the working chambers is achieved at the minor axis, which corresponds to the end of the exhaust phase and the beginning of the intake phase on one side of the housing. On the other side, minimum volume corresponds to the end of compression and the start of combustion. Conversely, maximum volume is achieved along the major axis. At the top end, this corresponds to the end of intake and the beginning of compression, while at the bottom it signals the end of the expansion resulting from combustion and the beginning of the exhaust phase.

This four-cylinder, four-stroke Citroën DS-21 engine typifies design trends of the Sixties. It has a multitude of moving parts that are lacking in the Wankel engine.

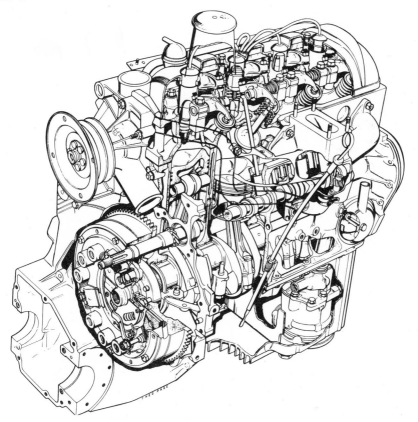

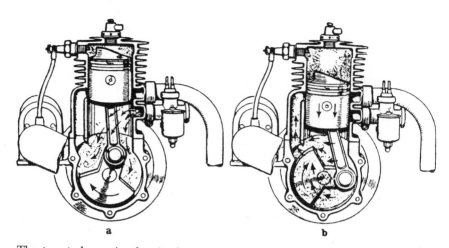

a   b

The two-stroke engine has intake, compression, combustion and exhaust periods, but the four stages are compressed into one crankshaft revolution, or two piston strokes. The four-stroke engine takes two revolutions. In a one-cylinder, two-stroke engine, as shown, every revolution brings a power stroke. The plug fires every time the piston approaches top dead center. The explosion of the burned gases and the intake of fresh mixture is squeezed into the short time between the end of the power stroke and the beginning of the next compression stroke. Naturally, the power and compression strokes are only partial strokes. The two-stroke engine has no valves. The pistons open and close ports as they move up and down. This obviates the need for the complicated valve gear and eliminates a potential source of wear and noise. At the start of the two-stroke cycle, the rising piston uncovers the intake port and draws a fresh charge of mixture into the crankcase, below the piston. On top, at the same time, the piston is compressing the charge from the previous cycle. The compressed charge is ignited and the combustion pressure forces the piston down. During the power stroke, the piston uncovers the exhaust port, allowing burned gases to escape. A transfer port in the block connects the crankcase with a port placed below the exhaust port.

When the piston uncovers the transfer port, the partly compressed mixture in the crankcase rushes to the combustion chamber. The piston then starts its upward travel and compresses the charge, while fresh mixture is drawn into the crankcase. Each main bearing is sealed so that each cylinder in a two-stroke engine is fed

This picture is considerably complicated by the fact that all three faces of the rotor are going through part of the operational cycle simultaneously, whereas in a piston engine all the action takes place on top—one "stroke" at a time. The result is that when one chamber is in the middle of the compression phase, the next is beginning the exhaust phase, and the one behind is well into the intake phase. This triple action is possible because of the radial location of the various functions. For example, fresh mixture enters at "10 o'clock"; compression begins before "12 o'clock," and reaches maximum at "3 o'clock"; from "3 o'clock" to between "7 and 8 o'clock" is pure combustion, until the rotor uncovers the exhaust port and the exhaust phase begins. The exhaust phase overlaps both combustion and intake phases. In fact, there

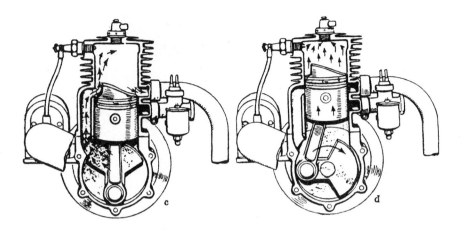

from a separate crankcase compartment. Otherwise, all pre-compression would be lost.

The two-stroke engine has twice as many power strokes in a given number of revolutions as the four-stroke engine. It would seem logical that a two-stroke engine with an equal number of cylinders would produce twice the power of its four-stroke counterpart. In practice, that is not so. Because the intake and exhaust phases of the two-stroke are not clearly defined and overlap with each other and with the compression and combustion phases, the two-stroke engine has poorer volumetric efficiency. Its specific power output is not significantly higher than in four-stroke engines. The power of a two-stroke engine is limited by the brief opening of the intake ports. It is almost impossible for the mixture to be fed in through the ports for more than 150 degrees crankshaft rotation, whereas the four-stroke engine permits intake durations of 250 degrees without difficulty. Two-stroke engines can make up this deficiency to some extent by using greater port area. A certain port overlap is inevitable. The fresh charge tends to mix with the exhaust fumes, with the result that some of the fresh charge disappears via the exhaust port. The ports can be so designed that practically all burned gases are evacuated from the cylinder. If this is done, some portion of the fresh mixture ready to be compressed follows the exhaust gases and is wasted, raising fuel consumption. A two-stroke engine cannot move a given car with better fuel economy than a four-stroke engine of the same cylinder displacement.

is considerable overlap all around, but without detrimental effects because gas flow is controlled by rotor motion. The rotor opens and closes the ports and compresses and scavenges the mixture.

The rotor can be better understood by comparing it, again, to the piston engine. Picture the rotor as being the pistons and connecting rods all rolled into one—if mounted on the crankpin, it would revolve with the crankpin. If it were also to *rotate* on the crankpin, its center would still describe a circle, but its corners would describe other types of curves. If the rotor rotation were carefully timed to the crankpin rotation in a specific ratio, the curve could be predicted and, if the ratio were kept constant, the curve would be repeated for every orbit and every revolution.

This cutaway shows the high-performance version of the three-cylinder, two-stroke Saab engine. Although it has far fewer moving parts than a four-stroke engine, it is less efficient.

In actual operation, the pressure of expanding gases acts on one lobe of the rotor to produce rotary motion. That explains why the rotor revolves around itself. Gas pressure on one face turns the rotor, which brings up another face, and the process is repeated. For the Wankel's rotating piston to exert leverage on the mainshaft, it must act on a point away from the shaft centerline. The rotor is thus mounted on eccentric bearings to provide this leverage. Rotor rotation is kept in phase with the rotation of the eccentric bearing by use of phasing gears. One part of the phasing gears is a stationary reaction gear fixed to the end cover plate but mounted concentrically with the mainshaft. This reaction gear meshes with an inner ring gear on the rotor. If the reaction gear has 36 teeth, the rotor's inner ring gear must have 54 teeth, providing a 3:2 ratio. The remaining one-third of the motion is performed by the output shaft. The relationship is as follows:

Rotor	54 teeth	3 revolutions
Gear	36 teeth	0 revolutions
Shaft	0 teeth	1 revolution

When the rotor advances 90 degrees, the eccentric advances 30 degrees. Each time a rotor apex passes the intake port, the mainshaft starts another complete revolution, thus there is a power impulse for every one-third turn of the rotor, giving one power phase for each complete mainshaft revolution.

2

Geometry

THE WANKEL ENGINE has no direct counterparts to the piston engine's bore and stroke. The *dimensions* of the working chamber are dictated by rotor width, rotor radius, and rotor eccentricity. Rotor width is a straight line, drawn radially across the rotor face, rotor radius is a straight line drawn from the rotor center to the rotor apex, and eccentricity is the distance from the mainshaft center to the rotor center (corresponding to crank throw in a piston engine). The *shape* of the working chamber is determined by the rotor radius and the rotor eccentricity.

The working surface can be compared to the cylinder wall in a piston engine so far as function is concerned—it is the inner surface of the housing. As was mentioned in the previous chapter, the rotor, mounted on eccentrics, both rotates on its axis and orbits around the mainshaft center while the three rotor apices maintain permanent contact with the chamber working surface. The shape of the working surface is defined as an epitrochoid, but, in order to understand the generative process, the evolution of the epitrochoid begins with the trochoid. A trochoidal curve is the path described by a chosen point on or within a circle as this circle rolls, without slip, around the periphery of another circle until it returns to its starting position. During this rotation, the center of the circle merely describes another circle, but all other points describe other curves. It is these other curves that make the Wankel engine possible from a theoretical point of view.

An epitrochoid is the curve described by the chosen point when the rolling (generating) circle is positioned *outside* the stationary (base) circle. If the generating circle rolls around the *inside* of the base circle,

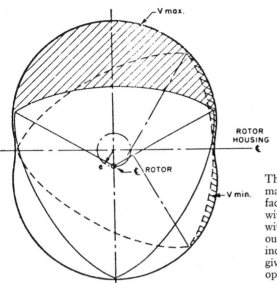

The epitrochoidal chamber has maximum volume on one rotor face when one apex coincides with the major axis (white rotor with firm outline). The dotted-outline rotor with one apex coinciding with the minor axis gives minimum volume on the opposing face.

the resultant shape is called a hypotrochoid. If the point is actually located *on the periphery* of the generating circle, the trochoid thereby created is called a cycloid. A cycloid is the curved path traced by a point on the periphery of the generating circle as it rolls along a straight line. If the base is changed from a straight line to a circle and the generating circle rolls around the outside of the base circle, the curve generated is called an epicycloid. The base circle must remain stationary and its center must not be displaced—no rotation can take place. If, on the other hand, the generating circle rolls around the inside of the base circle, the curve generated is called a hypocycloid. But, if the point chosen for a curve is moved from the periphery of the generating circle closer to its center, the curves generated will be quite different. They will be, in this case, trochoids instead of cycloids. A variation of the trochoid, the epitrochoid, is formed if the point is not on the circumference and it never meets the periphery of the stationary base circle while the generating circle rolls around the base.

Throughout these generative processes it must be kept in mind that the relative size of the circles does not matter except for one condition—if the circles are not of equal size, one must have a radius that is a whole multiple of the other's radius. In other words, fractional variations between the base circle and generating circle radii cannot exist. Any attempt to use a ratio of 1:1.5 or 1:2.2, for example, would result in an irregular curve in which the chosen point would fail to return to its point of departure after one complete orbit.

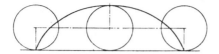

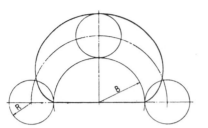

The drawing on top shows the curve described by a chosen point on the periphery of a circle as it rolls along a flat surface. Remove the flat surface and substitute another circle (the base circle) and roll the generating circle around it, without slipping. The chosen point describes a pincer movement.

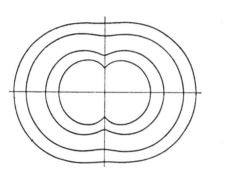

This family of epitrochoids has been obtained by changing the position of the chosen point on the generating circle. The innermost epitrochoid was created with the chosen point at the periphery of the generating circle. The outermost epitrochoid was created by a point close to the center of the generating circle. The distance between the chosen point and the center of the generating circle determines rotor eccentricity for the resulting Wankel engine.

If the radii of the generating and base circles are equal, and the point chosen to describe the curve is located on the periphery, a kidney-shaped epicycloid is generated. A straight line drawn through the narrowest part of the kidney is called the minor axis, while a similar line drawn through the widest part is called the major axis. In this case, both circles being the same size, a 3:1 relationship exists between the minor axis and the center of the base circle. The chosen point on the circumference of the generating circle touches the periphery of the base circle at only one point, and the rest of the way it moves farther and farther away from the center of the base circle until it reaches its apogee at three times the radius. The place where this occurs is diametrically opposed to the spot where the chosen point coincided with the base circle periphery.

If the two circles are the same size and the chosen point is placed at the halfway mark on the radius of the generating circle, the resultant curve is an epicycloid, kidney-shaped but with a longer minor axis than when the point was on the periphery of the generating circle. This is because, in this instance, the chosen point never touches the base circle.

But, in order to create the geometry of a true Wankel engine, a ratio between the two circles must be established.

Make the generating circle radius one-half that of the base circle, and choose a point on its periphery. Roll the generating circle around the base circle, and the path of the chosen point describes a figure-eight shaped epitrochoid. The minor axis, in this case, is exactly twice the base circle radius because the point touches the periphery twice on its tour; the major axis is four times the inner circle radius. The ratios resulting are as follows:

Base circle radius	1.0 unit
Generating circle radius	0.5 unit
Half major axis	2.0 units
Full major axis	4.0 units

If the chosen point is moved from the periphery of the generating circle to a position at one-half the radius, and the generating circle is rolled around the base circle, the curve generated is once again a figure-eight epitrochoid, but this time it has a fatter "waistline." The minor axis is 2.5 times the base circle radius, and the major axis is 3.5 times the base circle radius. The chosen point never gets closer to the base circle periphery than one-quarter of the base circle radius, and never gets farther away than three-quarters of the base circle radius. It is this basic two-lobe epitrochoidal shape that represents the shape of the working chamber in a practical Wankel engine. Of course, the number of usable trochoid shapes created using a generating circle radius half as long as the base circle radius is practically unlimited. The chosen point on the generating circle can be placed any distance from the center, and moving it toward the center of the generating circle reduces the eccentricity of the rotor motion. Briefly stated, eccentricity in the generating circle corresponds to rotor eccentricity.

The ratio between the circle diameters is analogous to rotor gearing. A two-lobe chamber demands a three-lobe rotor, which, in turn, necessitates a 3:2 gearing relationship between the rotor annular gear and the reaction gear. If the inner gear in the rotor has 72 teeth, the reaction gear must have 48 teeth. Without this phasing the engine cannot run.

Variations in the geometry for the two-lobe working chamber can be created by shortening the radius of the generating circle to one-third of the base circle radius. Choosing a point on the periphery of the generating circle, roll the generating circle around outside of the base circle. The chosen point will touch the base circle periphery, not twice but three times, on its tour and the resultant epitrochoid will resemble

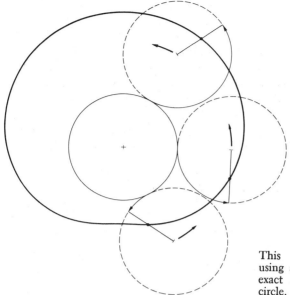

This epitrochoid results from using a generating circle of the exact same radius as the base circle.

a three-leafed clover. A Wankel engine could be designed with a working surface of this shape, but it would require a four-lobe rotor.

For a three-lobe chamber with a four-lobe rotor, a gear ratio of 4:3 is necessary (i.e., the reaction gear must have 25% fewer teeth than the rotor gear). If the inner gear in the rotor has 64 teeth, the reaction gear must have 48 teeth to keep the rotor in phase. To continue this geometric experimentation, set the radius of the generating circle at one-quarter the base circle radius. Choose a point on its circumference and the shape generated will be an epicycloid resembling a four-leaf clover. Placing the chosen point closer to the center of the generating circle will bring the four leaves together nearer their tips, and the tips will be brought closer to the stem. A Wankel engine designed in this manner would need a five-lobe rotor. As the generating circle radius is shortened with relation to the base circle radius, the number of rotor lobes required to produce a workable Wankel engine becomes greater. Unless the rotor has one more apex than the working surface has chambers, the rotor cannot be made to assume the duties of gas flow direction and control.

The limits of practicality are soon reached, however. All companies that are now engaged in development of the Wankel engine on an industrial scale have adopted the two-lobe epitrochoid.

The efficiency of the Wankel engine depends on compression ratio, just as in a conventional piston engine. The compression ratio in a piston engine, however, is virtually unlimited—to raise it, the designer

This two-lobe epitrochoid results
from using a generating circle
with one-half the radius of the
base circle.

This three-lobe epitrochoid is created by a generating circle having
one-third the radius of the base
circle.

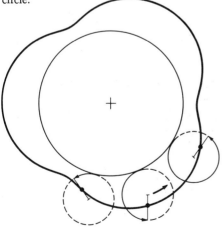

Definition of terms: 1, 2 and
3 = Rotor apex. A, B, and C =
Working chamber. a = Rotor
face. b = Trochoidal surface. c
and d = Phasing gears. e, f and
g = Rotor normal. h = Eccentric shaft. i = Mainshaft center.
M = Center for the rotor nor-
mals and phasing gear contact
point.

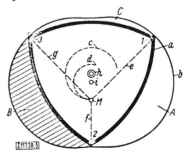

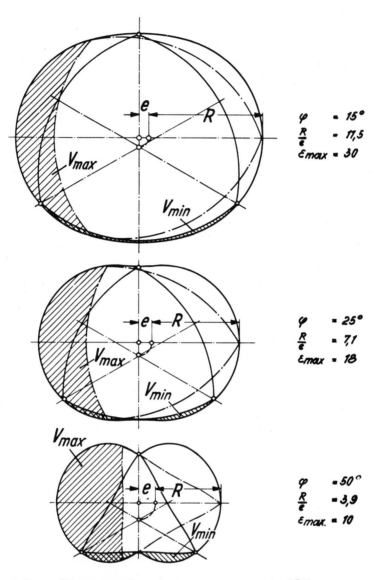

$$\varphi = 15°$$
$$\frac{R}{e} = 17{,}5$$
$$\mathcal{E}_{max} = 30$$

$$\varphi = 25°$$
$$\frac{R}{e} = 7{,}1$$
$$\mathcal{E}_{max} = 18$$

$$\varphi = 50°$$
$$\frac{R}{e} = 3{,}9$$
$$\mathcal{E}_{max} = 10$$

2-LOBE EPITROCHOID WITH INNER ENVELOPE
EFFECT OF TROCHOID-SHAPE ON
OVERALL DIMENSIONS OF ENGINES
COMPRISING IDENTICAL SWEPT VOLUME

Variations in R/e ratios produce enormous differences in maximum compression ratio potential and dictate the apex seal sliding angles. V *max* means maximum volume, and V *min* means minimum volume. R means rotor radius and e means eccentricity. The sliding angle is denoted by γ. E *max* means highest potential compression ratio.

has only to build up the piston crown. But, in the Wankel engine, the compression ratio is limited by the rotor radius and eccentricity. Once these two dimensions are selected, the maximum possible compression ratio for the engine is fixed. More specifically, the compression ratio is not restricted by either the radius or the eccentricity, but by the radius-to-eccentricity *ratio*. This is usually expressed as the R/e ratio (engineers often refer to it as the K factor). Because rotor size and orbital path determine the shape of the working surface, it follows that the spaces between the rotor faces and the working surface are dictated by the same considerations. The combustion chamber is formed by the trochoidal surface, the rotor face, and the end covers. This means that the maximum and minimum volumes of these spaces are limited once the basic design parameters have been established. Rotor width has no theoretical limit, but if an extremely wide rotor was adopted, the result would be slow and incomplete combustion. On the other hand, an extremely narrow rotor could bring the maximum volume of the working chamber so low that the engine would not run. Rotor width can be

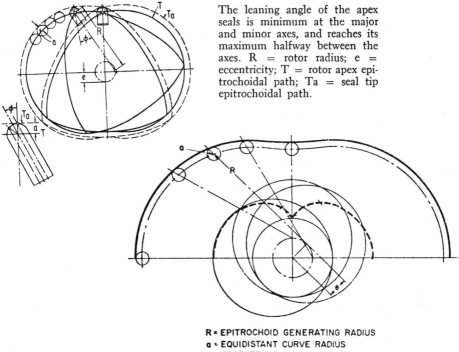

The leaning angle of the apex seals is minimum at the major and minor axes, and reaches its maximum halfway between the axes. R = rotor radius; e = eccentricity; T = rotor apex epitrochoidal path; Ta = seal tip epitrochoidal path.

R = EPITROCHOID GENERATING RADIUS
a = EQUIDISTANT CURVE RADIUS
e = ECCENTRICITY

Apex seal leaning angle is determined by rotor eccentricity and the equidistant curve radius.

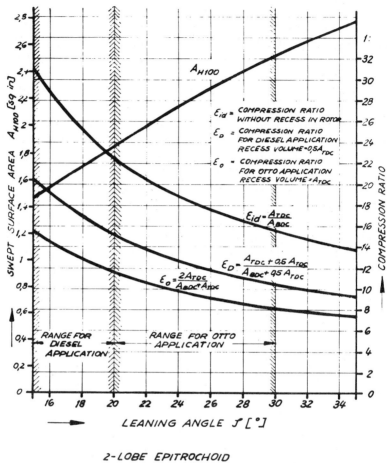

A_{H100}

ε_{id} = COMPRESSION RATIO WITHOUT RECESS IN ROTOR

ε_D = COMPRESSION RATIO FOR DIESEL APPLICATION RECESS VOLUME = $0.5 A_{TDC}$

ε_0 = COMPRESSION RATIO FOR OTTO APPLICATION RECESS VOLUME = A_{TDC}

$\varepsilon_{id} = \dfrac{A_{TDC}}{A_{BDC}}$

$\varepsilon_D = \dfrac{A_{TDC} + 0.5\, A_{TDC}}{A_{BDC} + 0.5\, A_{TDC}}$

$\varepsilon_0 = \dfrac{2 A_{TDC}}{A_{BDC} + A_{TDC}}$

RANGE FOR DIESEL APPLICATION

RANGE FOR OTTO APPLICATION

SWEPT SURFACE AREA A_{H100} [sq in]

COMPRESSION RATIO

LEANING ANGLE $\mathcal{T}$ [°]

2-LOBE EPITROCHOID
RELATION BETWEEN MAX. LEANING ANGLE [$\mathcal{T}$°]
AND SWEPT SURFACE AREA, RESPECTIVELY
MAXIMAL COMPRESSION RATIO
BASED ON AN ENGINE SIZE OF 100 mm FOR THE
MAJOR AXIS

This graph shows the relationship between maximum leaning angles, R/e ratios and highest potential compression ratios.

compared to cylinder bore—it is manipulated to gain the most reasonable combustion chamber. Most designers choose rotor width approximately equal to one-half the rotor radius.

A low R/e ratio (or K factor) gives the highest compression ratio, which means a small rotor radius and high eccentricity. The lower the K factor, the shorter the minor axis of the epitrochoid. A longer radius means a larger rotor, and bigger overall dimensions for the whole engine. With higher K factors, Wankel engines can be designed with compression ratios higher than is permissible with present-day fuels. For instance, a K factor of 11.5 allows a maximum compression ratio of 30:1, a 7.1 K factor allows 18:1, and a 3.9 K factor 10:1. The engine does not always run, of course, with the full compression ratio permitted by its design (because of combustion characteristics), and the actual compression ratio depends on design of the cavity in the rotor face. The designer aims for the highest compression ratio compatible with fuel anti-knock properties.

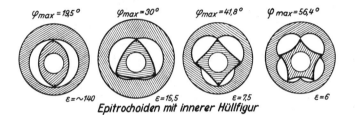

Epitrochoiden mit innerer Hüllfigur

Felix Wankel has made a complete study of possible rotary engine configurations. These four are described as epitrochoids with inner envelopes. In each case, the rotor has one lobe more than the envelope. The 1:2 configuration (left) permits compression ratios up to about 140:1 with apex seal leaning angles up to 19.5°. The 2:3 configuration (second from left) allows a maximum compression ratio of 15.5:1 and leaning angles are limited to 30°. The 3:4 configuration (third from left) cannot reach higher compression ratios than 7.5:1 and leaning angles are very high: 41.8°. The 4:5 configuration (right) has very low compression potential (6.0:1) and unacceptably high leaning angles with a maximum of 56.4°.

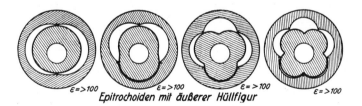

Epitrochoiden mit äußerer Hüllfigur

These configurations are characterized as epitrochoids with outer envelopes. That means the rotor is an epitrochoid as well as the envelope. All of them permit compression ratios in excess of 100:1. The 1:2 configuration (left) has a low maximum leaning angle of 19.5°. The others, reading from left to right, have leaning angles of 30°, 41.8° and 56.4°.

Just as the R/e ratio sets a high compression ratio limit, it also sets a low limit. Building the engine down to this limit will give the most compact housing dimensions. But, low R/e ratios give rise to another problem—because the "waistline" of the working chamber will be sharply marked and the overall dimensions small, the apex seals will be subject to far greater changes in angularity (relative to the working surface) than would be the case with higher R/e ratios. This is det-

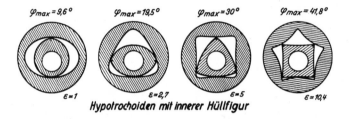

Hypotrochoiden mit innerer Hüllfigur

These configurations are hypotrochoids with inner envelopes. The hypotrochoid curve is traced by a generating circle that rolls, without slipping, inside the base circle instead of outside. All have poor compression potential. The 1:2 configuration (left) has a maximum compression ratio of exactly 0, while the leaning angle is only 9.6° maximum. The 2:3 configuration (second from left) allows compression ratios up to 2.7:1 while leaning angles reach 19.5°. The 3:4 configuration (third from left) gives a compression ratio as high as 5:1, and leaning angle restricted to 30°. The 4:5 configuration (right) allows a realistic compression ratio of 10.4:1, but its leaning angle goes up to 41.6°.

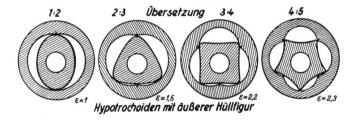

Hypotrochoiden mit äußerer Hüllfigur

These shapes are hypotrochoids with outer envelopes. The rotors used in these four examples correspond exactly to the envelopes of the four hypotrochoids with inner envelopes shown above. Compression potential is hopelessly low in all of them. The 1:2 configuration (left) has zero compression and leaning angle limited to 9.6°. The 2:3 configuration (second from left) has a 1.5:1 maximum compression ratio and maximum leaning angle of 19.5°. The 3:4 configuration (third from left) has a compression ratio limited to 2.2:1 and leaning angle can go as high as 30°. The 4:5 configuration (right) has a maximum compression ratio of 2.3:1 and maximum leaning angle of 41.6°.

rimental to sealing effectiveness as well as to seal durability. A low K factor does bring certain advantages, however, such as low bearing loads, a low surface-to-volume ratio, and large bearing surfaces in proportion to rotor dimensions. High K factors also have their strong points, among which are greater rotor cooling capacity, more available space for side seals and oil seals, lower oil seal rubbing speeds, smaller apex seal swing angles, and reduced sensitivity to exhaust/intake phasing overlap. With radially disposed apex seals, the maximum swing angularity of the seal on the working surface is 15 degrees in an engine with an 11.5 K factor, 25 degrees in an engine with a 7.1 K factor, and as high as 50 degrees in an engine with a 3.9 K factor.

3

Displacement

THERE ARE THREE PRINCIPAL REASONS why it is important to calculate the displacement of an engine:

1. Displacement provides a basis of comparison with other engine types (specific output). Better methods probably exist, such as rating horsepower against weight or horsepower against fuel consumption per hour, but the industry is conditioned to its own definition of *specific power output:* horsepower per liter (or per cubic inch) displacement.

2. International motor racing rules include engine displacement limits, therefore an equivalency formula for the Wankel engine is necessary to enable Wankel-powered cars to compete. This has long traditions, although formulae based on weight, piston area, and fuel consumption have been tried at various times.

3. Several countries have imposed a tax on motor vehicles with engine displacement as its basis.

When NSU started production of a Wankel-powered car, the German authorities made an attempt to tax it under the existing cylinder displacement system. Because each combustion chamber held 500 cc., NSU said the engine should be rated as a 500 cc. engine. The German tax authorities said the engine had to be taxed as a 1,500 cc. unit, because the rotor has three faces and there are three power impulses for each rotor revolution. The same thing happened in the racing world. In 1962, NSU submitted proposals to the FIA and FIM arguing in favor of rating the Wankel engine on the basis of the volume of a single working chamber. They received the support of the German and British delegates but were opposed by the representatives of France and Italy who maintained that the unit volume of one chamber must be multiplied by the number of faces on each rotor because each face goes through the same full operational cycle that takes place in each cylinder of a reciprocating piston engine. This led NSU to compare the Wankel engine with a single-cylinder, two-stroke piston engine, which has the same power-stroke frequency in relation to crankshaft revolutions. Both have one power stroke for each revolution of the output shaft.

Indeed, a single-rotor Wankel engine does share certain operational characteristics with the single-cylinder, two-stroke piston engine. These include mean effective pressure, torque, horsepower and r.p.m. However, it is wrong to rate them on a similar displacement basis, because the Wankel engine operates on the four-stroke cycle. It is fairer to compare the Wankel with a two-cylinder, four-stroke piston engine, which also has one power stroke per crankshaft revolution. A single-rotor Wankel engine goes through two complete operational cycles in two mainshaft revolutions. A single-cylinder piston engine requires two crankshaft revolutions to go through the four strokes that make up its operational cycle. Two-cylinder engines do not complete their cycles any faster, but do produce twice as many power strokes, therefore two-cylinder engines deliver one power impulse for every revolution

Graphic comparison of displacement in reciprocating piston engines and Wankel engines. The cylinder displacement in the piston engine at left is 250 cc. Two such cylinders in line would give a 500 cc. engine with comparable firing and torque-producing characteristics to the 500 cc. single-rotor Wankel engine shown in the center. At right, the configuration that would result if the Wankel engine were to be designed as a twin-rotor unit with 250 cc. displacement per chamber. In both Wankel engines shown, the R/e ratio is 7.15:1, and the maximum seal leaning angle is 25.5°.

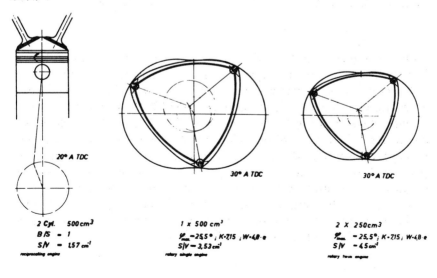

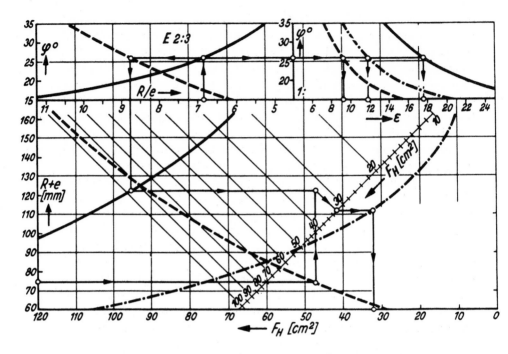

This chart is a nomogram that can be used for calculating the displacement of a Wankel engine once the R/e ratio, radius, and rotor width are known. Using the KKM-125 as an example: Radius = 65 mm. Eccentricity = 9.5 mm. R/e ratio = 6.85:1. R + e = 74.5 mm. Follow the arrow from the R/e point to the point of intersection with the eccentricity curve. That shows a maximum leaning angle of 26°. Follow the arrow left and down until it intersects with the R + e curve, then right to the surface line for the rotor and the surface curve for the epitrochoidal housing. Draw an arrow straight down to the surface scale, and the result is 32 cc. Multiply the surface figure by the rotor width (39 mm.) and chamber displacement equals 125 cc.

Others argued that the Wankel engine represents a three-cylinder, four-stroke engine, maintaining that positive torque is produced over 240 degrees of mainshaft rotation, just as a three-cylinder engine produces torque over two-thirds of each crankshaft revolution (it goes through three cycles in two revolutions). This argument was effectively put to rest when the leading independent expert on Wankel engines, Richard F. Ansdale, delivered a paper entitled "Rotary Engine Development and its Effect on Transport" to the Society of Automotive Engineers of Australasia (October 15, 1968), in which he proved conclusively that the torque characteristics of the single-rotor Wankel engine correspond to those of a two-cylinder, four-stroke piston engine and not to those of a three-cylinder engine.

A comparison between the single-rotor Wankel engine and the two-cylinder piston engine is as follows: a power stroke in cylinder number one corresponds to a power impulse against number one rotor face. At the same time, cylinder number two is taking in a fresh charge and rotor face number two is compressing fresh gas while face number three has just completed an exhaust phase. When face number two enters the power phase, cylinder number two in the piston engine begins its power stroke. When firing occurs on face number three, firing again occurs in cylinder number one. The torque fluctuations and volume variations in the two engines are identical, and completely in phase with regard to degrees of crankshaft and mainshaft rotation. Even with these questions settled, an equitable displacement racing formula for Wankel engines eluded the sanctioning bodies.

From time to time, moves were made in racing circles to circumvent the problem by abolishing the displacement limits. It has been suggested that a combination formula, stipulating both minimum weight and maximum fuel tank capacity, could be the solution. This would allow cars powered by piston engines, Wankel engines, gas turbines, and steam engines to compete on realistic terms, with variations in fuel tank capacity allowed according to the type of fuel used.

Gas turbine cars have completed in the 24-hour race at Le Mans and the Indy 500 race, and equivalency formulae, based on air consumption, have been worked out to produce a relative cylinder displacement rating for them. The Commission Sportive Internationale of the Federation Internationale de l'Automobile (the international governing body for motor racing) then looked into ways of applying the same rule to Wankel-powered cars. With a single rotor and a chamber volume of 500 cc., the intake air volume of a Wankel engine is only 500 cc. per mainshaft revolution. That means it is equivalent to a one-liter (1,000 cc.) four-stroke piston engine. A two-cylinder, one-liter engine will complete an intake stroke during one crankshaft revolution or, an equivalent of 500 cc. A four-cylinder engine, with 250 cc. per cylinder, has only two intake strokes during one crankshaft revolution, and again there is 500 cc. intake air volume. These facts influenced the CSI to adopt a formula which rated the Wankel engine's displacement at *twice the combustion chamber volume multiplied by the number of rotors*. This formula is still in effect.

4

Sealing

THE MOST CRUCIAL of all the problems that beset the Wankel engine during the early stages of its development concerned sealing. To obtain efficient operation, the working chambers must have seals that prevent gas leakage between them. There are several paths that must be blocked—across the apices between the rotor faces, and around the sides of the rotor faces.

The sealing performance of the Wankel engine is determined by many variables of engine design, such as basic configuration, material and dimensions of sealing elements, lubricating conditions, rotor and housing cooling conditions, and precision of machining and finish. When compared with the reciprocating piston engine, the Wankel engine has handicaps in its gas sealing mechanism. The length of the gas leakage paths, the size of the oil seal, the number of clearances, the configuration of the seals and their operating conditions constitute serious disadvantages. It is also undeniable that the gas sealing efficiency of the Wankel engine is unfavorable, especially at low speed because the angle of each phase of the Wankel engine is 1.5 times larger than that of the four-stroke reciprocating piston engine.

Sealing consists of providing uninterrupted contact over the whole primary and secondary sealing areas. The primary areas are between a seal and a stationary surface, the seal being carried in a moving component. Secondary sealing areas are between the seal and its slot, groove, or bore. Sufficient force must be applied to the sealing element to maintain contact. Broadly speaking, the rotor seals in a Wankel engine perform the same duties as the piston rings in a reciprocating engine because they provide a seal for the combustion gases.

Piston rings also aid heat flow from the piston and prevent excess lubricating oil from reaching the combustion chamber. All rings contact the cylinder wall around their entire periphery. They also have circumferential side contact with the piston grooves they are in (except under transient conditions). Most modern pistons have three compression rings and one oil control ring. The rings act as radially expanding springs against the cylinder bore. The most common piston ring material is cast-iron—it's springy and wears well. Chrome-plated rings reduce bore wear, but chrome plating is expensive. However, many engines use rings with a certain chrome content because chrome has two important properties: it's immune to corrosion, and far harder than iron—chrome melts at 3,407°F. It also has a beneficial burnishing action on the cylinder walls, because it smoothes down peaks and high spots instead of tearing them off. Chrome is itself long lasting and also gives longer bore life.

Most present-day engines have piston rings with a high molybdenum content. Molybdenum is a tough, slippery, silver-white metal that has a high melting point (4,370°F.) and offers high resistance to scuffing. In addition, it resists corrosion and has a high break-in or seating potential. Its abrasive-wear resistance is moderate (between iron and chrome), it has good scuff resistance (its greatest advantage), and is also somewhat porous and capable of maintaining an oil film better than chrome-plated or plain cast-iron rings.

The sealing line of the reciprocating piston engine is considerably shorter than that of the Wankel engine. At left, the sealing line of a two-cylinder one-liter piston engine. Center, the sealing line of a single-rotor 500 cc. Wankel engine. With more rotors, sealing problems are aggravated. At right, the sealing line of a twin-rotor 500 cc. Wankel engine.

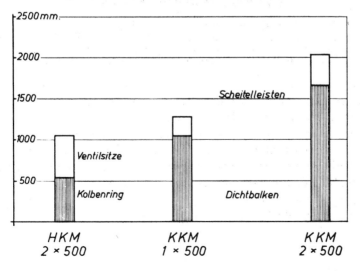

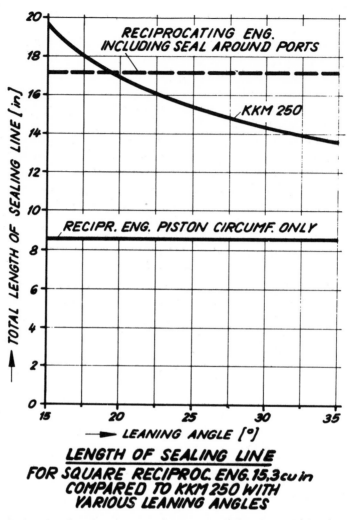

**LENGTH OF SEALING LINE
FOR SQUARE RECIPROC. ENG. 15.3 cu in
COMPARED TO KKM 250 WITH
VARIOUS LEANING ANGLES**

Sealing line length comparison between a single-cylinder 15.3 cubic inch piston engine and a single-rotor Wankel engine KKM-250. The dotted line represents the sealing line for the piston engine including the valve seats; the solid line represents the piston ring circumference only.

Because of the triangular shape of the Wankel engine rotor, it cannot use seals of the ring type. Rotor seals are of the strip type, some straight and some curved. Apex seals are straight strips inserted in radial slots at each rotor apex. Side seals are curved, to follow the curvature of the rotor flanks, and are inserted in segmental grooves in the rotor sides, as close to the edge as possible. However, each seal strip has two ends, facing perpendicularly from the direction of seal loading, and special corner seals to block the leakage path around the sides of the apex seals were eventually devised. Sealing the side seal corners was easier. Since three of them are carried on the same surface, they intersect near each apex. The problem was solved in different ways by the various manufacturers, but all solutions were based on the same principle of blocking the gas leakage path by interlocking the side seals together at all corners.

Both piston rings and Wankel engine rotor seals must be lubricated to prevent sticking. Both rely on gas pressure to form a seal with the opposing surface, and both are also spring-loaded to ensure contact at all times. If the rotor apex seals were not spring-loaded, resistance to cranking would be so high that a special high-capacity starter motor would be necessary. At the operational speeds of the Wankel engine, the spring-loading is of no value—it is only to help start the engine.

The top compression ring in a piston engine is backed up by another ring which is intended to trap blowby gases that manage to leak past. Even the oil control ring makes a contribution to gas sealing. As a result, the piston engine has the benefit of a three-stage gas sealing system, which is not feasible in the Wankel engine. The Wankel can have only one seal; neither apex seals nor side seals can be backed up by a second line of seals. This is one of the key reasons why the sealing problem was so serious, and why the solution had to be found in specific areas—at or near the apices and edges of the rotor. However, even assuming satisfactory sealing systems have been devised, there always will be the risk of gas leakage across the exhaust and intake ports, as in a two-stroke piston engine. (This corresponds to valve overlap in a four-stroke piston engine.) At very high rotor speeds, a gas

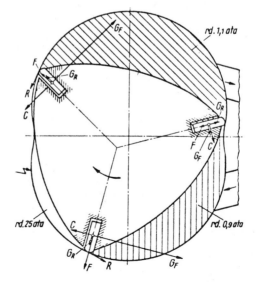

These are the forces that act on the apex seals in a Wankel engine. C = coriolis forces. F = centrifugal force. G = gas pressure. G_R = radial gas pressure. G_F = lateral gas pressure. R = friction. rd = gas pressure in percent of atmospheric.

leakage path also may be created at the constriction on the working surface due to reversed centrifugal loads on the apex seals. This form of leakage is equivalent to the effect of valve bounce in a four-stroke piston engine.

The accelerative forces imposed on the seals are light compared to those on piston rings, even though rotor seals have higher average sliding velocities. The reason is that the rotor seals have the advantage of unidirectional motion, while piston rings suffer frequent reversals of travel and are subject to very high acceleration. The sliding velocity of the apex seals varies with rotor position, even at constant mainshaft r.p.m., because of the duplex rotation of the rotor—it rotates on its own axis while orbiting around the mainshaft. For example, at 5,000

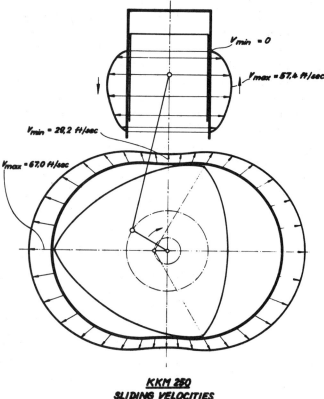

KKM 250
SLIDING VELOCITIES
COMPARING ROTATING COMB. ENG. AND
RECIPROCATING ENG. DISPLACEMENT 15.3cu in 5000RPM

A simple comparison between sliding velocities in a reciprocating piston engine and a Wankel engine.

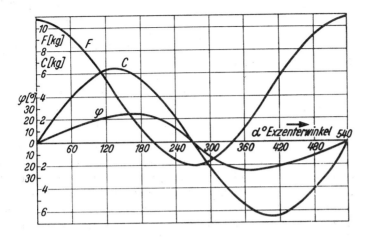

This graph shows the magnitude of the forces acting on the apex seals. F = centrifugal force. C = tangential acceleration (coriolis force). yo = leaning angle. (One apex seal strip in this example weighs only 2 grams. Rotor radius is fixed at 65 mm. and eccentricity is 9.5 mm. Rotational speed is a steady 10,000 r.p.m.)

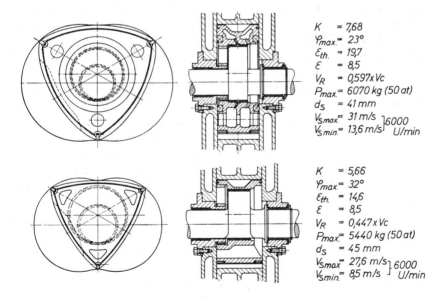

This comparison illustrates the influence of the K factor (R/e ratio), leaning angle, compression ratio, eccentricity and main bearing diameter on the sliding velocity of the apex seals. K = R/e ratio. R = rotor radius. e = rotor eccentricity. G = compression ratio. G_{th} = potential compression ratio. yo_{max} = maximum leaning angle. V_R = maximum chamber displacement. P_{max} = combustion pressure. d_s = main bearing diameter. V_s = apex seal sliding velocity (min and max).

r.p.m., the apex seals in Curtiss-Wright's RC2-60 have sliding velocities within a band stretching from 42.5–108 ft./sec. Compare this with the figures for some piston engines, all at 5,000 r.p.m.

Engine	Minimum	Maximum
Honda C-92	0	66.8 ft./sec.
Chrysler V8 (318)	0	63.5 ft./sec.
Mercedes-Benz 220	0	62.6 ft./sec.

The forces that work on the apex seals are centrifugal (positive and negative), gas pressure (both directions), and friction against the working surface. Negative centrifugal force is formed because during the approach of the rotor apex toward the minor axis (the waistline of the figure-eight), the apex is moving closer to the mainshaft center—not radially, but in an easy curve. This motion is opposed to the normal centrifugal loads provided by rotor rotation, and thus constitutes a negative force.

The apex seal is perpendicular to the working surface only when at the major and minor axes; at all other rotor positions, it is at an angle. This "leaning" angle of the apex seals against the working surface varies during the motion of the rotor. On its way from minor to major axis the seal adopts an increasing leading angle up to the halfway point, then the angle is gradually reduced until the apex seal returns to its perpendicular position, relative to the working surface, at the major axis. On its way from the major axis to the minor axis, the apex seal runs at a trailing angle throughout. Maximum angularity is reached at the halfway point between the axes. The maximum "leaning" angle in a particular engine is dependent on the radius/eccentricity ratio. The angle gets higher with lower R/e ratios. Extreme angles (from perpendicular) are detrimental to proper sealing, and therefore are to be avoided. It is generally held that 30 degrees is a practical limit.

The process of lubricating the gas sealing elements of the Wankel engine is fundamentally different from lubricating the piston rings in a reciprocating engine, and is considerably more complicated. During the initial stages of development, both NSU in Germany and Toyo Kogyo in Japan tested Wankel engines using oil mixed with the fuel in proportions of 50:1 or 100:1. Actually, there was no need to mix oil with the fuel, because the rotors were cooled by internal oil flow and there was considerable oil leakage from the rotor. This amount of oil leakage seemed to be sufficient for lubricating the seals and the working surface. Despite this presence of lubricating oil, asymmetrical wear of the apex seal tips and chatter marks on the chamber working surface were problems. Seal tip wear is, first and foremost, a matter of material compatibility—with a hard working surface and soft seals, the seals wear out quickly. Harder seals, however, wear the working surface. Both NSU and Toyo Kogyo were confident they were using compatible materials for surfaces in rubbing contact—materials that required a minimum of lubrication.

It was clear to the test engineers in both Germany and Japan that the chatter marks occurred because, under certain conditions, the apex seal configuration, movement, and construction broke down the oil film on the rubbing surface, resulting in direct metal-to-metal contact. Obviously, reasonable seal life cannot be expected without lubrication of the high-speed metal-to-metal interface. NSU attacked the problem in three ways: first they reduced the side clearance in the apex seal grooves, to dampen the rocking motion of the seals without causing seizure; then they inserted flat springs of beryllium bronze under the seals to brace them against the working surface; finally they chrome plated the working surface to reduce the friction that was a partial cause of the problem.

These are three rotor oil seal versions developed by NSU, presented in consecutive order. Their features are described in the text.

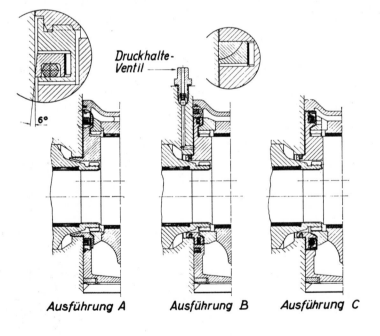

Ausführung A   Ausführung B   Ausführung C

The rocking motion of the apex seals is a phenomenon similar to piston ring flutter under extreme piston acceleration. It is brought about by a combination of seal tilting during reversals of gas pressure and radial movements of the seals caused by the difference in radial expansion of the working surface. Freedom of radial movement is, of course, necessary to prevent the formation of deposits. The apex seals have a clearance of about 0.001 inch in their slots. When the pressure in the working chamber is highest, gas pressure on the apex seals works to increase the contact pressure of the seals against the working surface, while friction tends to produce a tilted position of the apex seal under low-pressure conditions. The apex seal tip is held back by its rubbing on the working surface, while the seal base is carried ahead by the slot in the rotor.

The apex seal has three consecutive positions during the pressure reversal that occurs when combustion pressure from the chamber leading it overcomes the compression pressure from the chamber trailing it. On its way towards the minor axis, the apex seal is standing up against the leading wall in its groove, and is held in that position by compression pressure. When combustion begins ahead of the apex seal, gas pressure acting against the exposed part of the seal tilts it, forcing the tip back towards the trailing edge of the groove. Continued combustion pressure brings the bottom of the seal over to the trailing wall of the

Evolution of the sealing system. These examples are based on NSU practice. In the 1959 version (left), the apex seals had fragile corners because the number of components was excessive. In the 1963 version (center), the apex seal itself had adequate durability but did not provide proper corner sealing. In the 1966 version (right), the floating triangle corner of the apex seal provided requisite corner sealing without involving a degree of complexity that would create other problems.

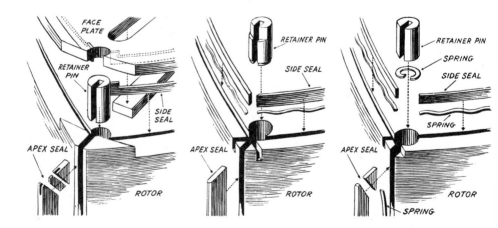

Assembled version of the 1966 sealing system developed by NSU.

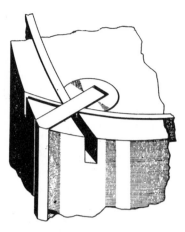

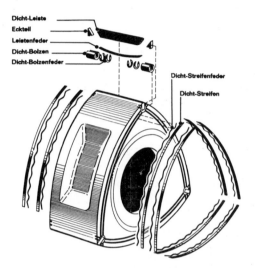

Sealing system finally adopted for the production version of the KKM-612 (Ro-80). Dicht-Leiste = Sealing strip. Eckteil = Corner triangle. Leistenfeder = Seal spring. Dicht-Bolzen = Trunnion. Dicht-Bolzenfeder = Trunnion spring. Dicht-Streifen-feder = Side seal spring. Dicht-Streifen = Side seal.

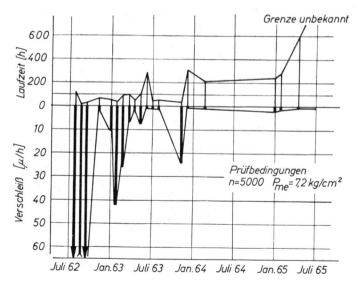

Life of NSU carbon seals improved in sensational leaps over a three-year span. Wear is measured in microns per hour and test duration time is in hours. The engines were running at a constant 5,000 r.p.m. at a mean effective pressure of 102.1 psi.

groove and holds it there until the exhaust phase begins in the leading chamber and compression pressure again drives the seal to the leading wall of the groove, where it stays throughout the combustion process. Control of this rocking motion of the apex seals was achieved by cutting gas pressure relief slots on the lower edge of the apex seal to reduce the lag before the higher pressure can reach the base of the groove and assist in the sealing. This modification was also effective against a phenomenon called "spitback" that occurred across the apex seals under conditions of very rapid pressure rise in the working chamber. Spitback

In addition to gas sealing, there are oil sealing duties. Because the rotor is oil-cooled in most Wankel engines, the oil has to be sealed inside the rotor and prevented from escaping into the working chambers. The seal has to be positioned at the bearing edges because the rotor bearing itself needs an oil film for lubrication purposes.

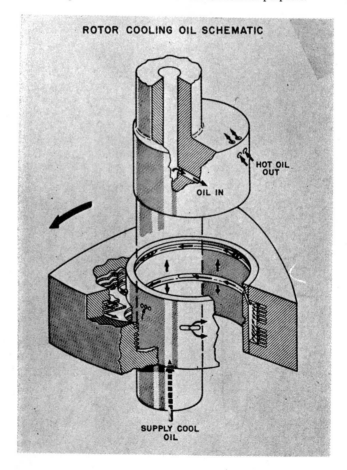

ROTOR COOLING OIL SCHEMATIC

HOT OIL OUT

OIL IN

SUPPLY COOL OIL

usually showed up as sharp pressure fluctuations, followed by low-frequency pressure waves within the intake ports.

Toyo Kogyo decided against a metal-to-metal interface, for which lubrication would be a vital requirement. They adopted a specially developed carbon compound as a new material for the apex seal because carbon has self-lubricating properties and the presence of an oil film is not absolutely necessary. However, because the lubricant in the internal combustion engine also serves as a gas sealing agent, it became necessary to study the lubrication of the apex seal in terms of gas sealing efficiency in addition to the problem of friction and wear.

NSU pointed out that carbon seals must be at least 5 mm. (0.197 inch) thick to have sufficient strength. Metallic apex seals need be only 1.6 mm. (0.063 inch) thick to be effective. Tests showed that the presence of a lubricant still had a favorable effect on the gas sealing performance, although the carbon apex seals formed a geometrically linear contact with the working surface.

There are three methods of supplying lubricating oil to the apex seals of the Wankel engine: one is to mix oil with the fuel mixture (as was done in early tests); the second is automatic metering of lubricating oil from the rotor side; and the third is an independent feed system by which oil is introduced into the intake ports. The oil-fuel mixture method would have worked if the necessary amount of oil had not been in constant proportion to the fuel flow. This method has been rejected as unsuitable for modern automobile engines. The second method also was rejected, because it is almost impossible to accurately control the amount of oil leakage from the side surfaces of the rotor under all engine operating conditions. Independent feed systems have been successfully developed, however, and they supply the necessary amount of oil to the intake port according to engine operating conditions.

5

The Combustion Process

THE WANKEL ENGINE RUNS on the same gasoline that is used in most reciprocating piston engines, and its combustion process is fundamentally the same. Air/fuel ratios are similar in both Wankel and piston engines, and compression ratios are closely comparable. To understand the differences, a comparison of the basic combustion process as applicable to both types of engine is necessary.

The air-fuel mixture of the internal combustion engine is ignited in the combustion chamber by a spark plug firing at a pre-set point, starting a wave of flame spreading out from the spark plug. This flame front continues to move across the combustion chamber until it reaches the other side and the compressed mixture is burned smoothly and evenly. Flame front speed varies from 20 ft./sec. to over 150 ft./sec. This speed depends mainly on air/fuel ratio, compression ratio, mixture turbulence, and combustion chamber design. Flame front travel is quite slow when the mixture is too rich or too lean. When the same amount of fresh mixture is compressed into a smaller space combustion will be quicker and more of the heat value in the fuel charge will be utilized, thus the compression ratio is a direct indication of efficiency. The pressure gain with a 5:1 compression ratio is about 300 psi. The pressure gain with a 10:1 compression ratio is about 700 psi. More work must be done to compress the mixture when the compression ratio is high, but there is a large gain in thermal efficiency and power because less heat is lost to the cooling system and more of the fuel energy is put to useful work. If the initial pressure at the moment of firing is 160 psi, the final pressure will be almost 600 psi (normal pressure rise is 3.5 to

4 times the initial pressure). Combustion pressures are high at wide open throttle and low during low-speed cruising or idling conditions. Under high pressure, combustion is speeded up. A controlled swirl or turbulence in the combustion chamber will also speed up the flame front. All types of combustion chambers can be given some amount of turbulence, but too much turbulence is undesirable because it increases the rate of heat loss to the cylinder walls. But the faster the flame front travels, the smaller the risk of abnormal combustion (knock, ping and rumble).

Knock occurs when end gas ignites spontaneously towards the end of the power stroke. The flame front spreads out from the plug and the mixture burns properly and evenly. Then, heat and pressure from the beginning combustion work together to heat and compress the last portion of mixture before the flame front reaches it. When temperature and pressure become too great for the unburned mixture, it ignites but doesn't burn smoothly. This explosive burning may increase pressure 200–300 psi above the pressure existing in the rest of the combustion chamber. The sudden pressure rise spreads throughout the chamber, and causes high-frequency pressure fluctuations which produce a knocking noise. The intensity of the knock causes the combustion chamber walls to go into sympathetic vibration, and this gives rise to the familiar "ping" sound.

Knock is often referred to as "detonation." It is unlikely that even the highest intensity of knock occurring in an engine in service reaches the level of a true detonation wave as found in laboratory tube-type apparatus, so knock is technically a more correct term. The pressure fluctuations also cause a heat loss from the combustion chamber to the surrounding engine parts which may result in a power loss. When they combine with surface ignition, the pressure fluctuations produce sharp cracking noises and the phenomenon is known as "wild ping." In addition to noise and power loss, knock leads to overheating of valves, spark plugs and pistons. Overheating promotes further knock and may also lead to pre-ignition—the two phenomena are mutually provocative. Overheating also severely shortens the life expectancy of valves and spark plugs.

Surface ignition is caused by one or more hot spots in the combustion chamber that ignite a portion of the mixture before the gas is properly ignited by the flame front. Such hot spots may be glowing combustion chamber deposits, poorly cooled areas, sharp edges on the head or block, an overheated spark plug electrode or an exposed spark plug thread. Surface ignition can occur either before or after the plug fires. If it occurs before, it is called pre-ignition and corresponds to random over-advance of the ignition timing. Pre-ignition opposes piston travel on the compression stroke. The force of too-early combustion may burn through the piston crown, melt the cylinder walls, sever valve heads or crush valve springs. When surface ignition occurs after the firing of the spark plug, total combustion time is reduced and the result is power loss. This form of combustion is called post-ignition, which often leads to "rumble." Rumble (or "pounding") is unique to reciprocating engines and probably has no counterpart in Wankel engines. It's a deep rattle that emanates from the crankshaft. The noise is due to vibration of connecting rods and crankshaft at a cycle time approaching bottom dead center, set off by abnormal combustion. Rumble occurs with unclean engines of medium compression ratios as a result of secondary flame fronts during combustion. Once begun, rumble tends to go on. If caused by surface ignition, it can disappear on full-throttle acceleration because the additional gas flow and pressure will eliminate glow and speed up flame front travel. Rumble is not confined to any one type of combustion chamber—it can exist with all current designs.

These graphs compare the volumetric changes in the Wankel engine with those of a four-stroke reciprocating piston engine. The solid line represents the Wankel engine; the dotted line represents the piston engine. Zündung = ignition. Auslass = exhaust. Einlass = intake. Entspannen = expansion. Ausschieben = exhaust. Ansaugen = intake. Verdichten = compression. ein Arbeitsspiel = one working cycle. Steuerquerschnitt = port and valve timing. Kammervolum V_k = chamber volume. Zylindervolum V_z = cylinder displacement. oT = top dead center. uT = bottom dead center. Zeit = time. V_c = compression chamber volume.

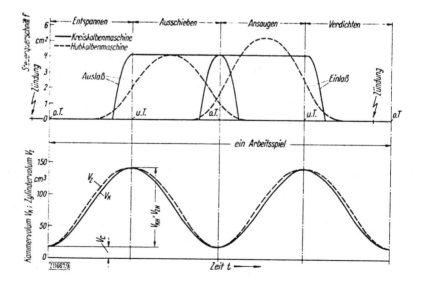

The top graph shows gas velocity in a working chamber of symmetrical design at 6,000 r.p.m., measured in meters per second. (One m. sec. = 3.06 ft. sec.) The line code indicates rotor position. 80° v OT = 80° BTC (before top dead center). 20° n OT = 20° ATC (after top dead center). The lower graph shows the same curves for a working chamber of non-symmetrical design developed to increase gas velocities.

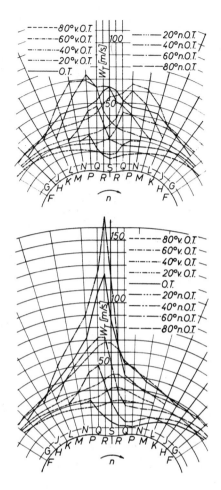

When an engine continues to run after the ignition has been turned off, the engine is "running-on" or "dieseling." Running-on is an erratic process which may lead to sudden reversals of rotation and consequent mechanical damage. It can be caused by the idling speed being too high, air/fuel mixture too lean, ignition timing too late, or fuel octane rating too low. Some sources believe running-on is a case of surface ignition, others maintain it is a case of compression-ignition. The condition cannot be cured by changing to a spark plug with a colder heat range.

Gas transfer should help speed up the flame front travel. One would expect that the gases ignited first would ride the crest of the gas transfer wave, and that the flame front would travel ahead of the gas transfer. Papers released by Curtiss-Wright, however, indicate that instead of being additive to the high gas transfer velocities, the flame front never quite catches up with the gas transfer wave, and never reaches all the combustible mixture. There is, therefore, an "end gas" problem

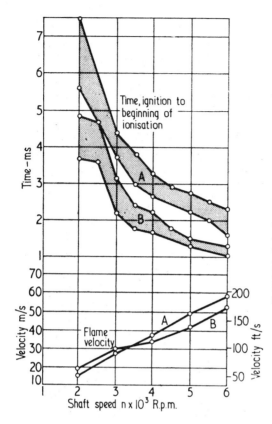

This chart shows the time taken from the moment of ignition to the beginning of ionization measured in milliseconds, measured over a 2,000 to 6,000 r.p.m. range. The dark "A" field shows the spread obtained between positions 2 and 4, i.e., just before the minor axis and about 60 degrees after the minor axis. The dark "B" field shows the spread obtained in the tight area between positions 2 and 3, i.e., just before the minor axis and about 25 degrees after the minor axis. The lower lines represent flame velocity. The "A" line represents the distance from position 2 to position 4, while the "B" line represents the distance from position 2 to position 3—a much shorter way.

with Wankel engines just as with many piston engines. Gas near the rotor's trailing apex is quenched between the rotor trailing and working surface, preventing the advance of the flame front towards the trailing apex. There is also some flame front travel in the direction opposite to the gas transfer, but it is very slow. However, the spark is spread "backwards" to some extent—especially in the early portion of the power phase. The flame front facing the rotor rotation just sits and waits for the fresh mixture to be brought to it, rather than going out to meet it. This produces relatively slow combustion because the entire combustion chamber is being displaced during the combustion phase.

Another factor adds to the complexity of the combustion process. In the vicinity of the minor axis, the combustion chamber is divided into front and rear sections. The trailing section undergoes compression at the same time expansion is taking place in the leading section. While the gases in a reciprocating piston engine always are contained within a cylinder area between the valves and spark plug on one side and the piston crown on the other, the gas flow in the Wankel engine is

continuous. Because of the rotor's motion and the ever-changing shape
and volume of the combustion chamber, the combustion process is
quite different.

Wankel engine research has given rise to a new term—*gas transfer
velocity*. The combustion chamber itself travels along the working surface,
pushed from behind by the rotor. The fresh charge travels in "pockets"
between the rotor face and the working surface. When a pocket widens
during the power phase, the gases, in full process of combustion, spread
towards the next apex with incredible speed and gas transfer velocity is
extremely high. Combustion spreads just as in a piston engine, with the
same kind of flame front. In a low-turbulence piston-engine combustion
chamber, the flame front advances from the spark—like ripples spread
from a stone dropped in a pool of water. In the Wankel engine, the
flame front cannot keep up with the gas transfer. It is this phenomenon
that makes the placement and location of the spark plugs very critical
factors in achieving effective combustion. The spark plug must ignite
the bulk of the mixture contained in front of the rotor face, not in the
narrow sections at either end of the working chamber. A spark plug
positioned well beyond the minor axis is presumed to be in a better position for successful flame propagation because the gas transfer is more
violent on the leading side than on the trailing side, and because the
leading side increases its volume as the rotor rotates. Flame propagation
to the trailing side is difficult because of violent gas flow and the non-
uniformity of the air-fuel mixture. On the other hand, a spark plug positioned at, or slightly ahead of, the minor axis is in a better position with
respect to flame propagation and air-fuel mixture uniformity; but the

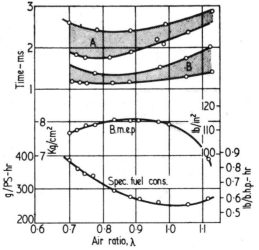

This graph indicates results of
experimentation with flame propagation on different air-fuel ratios
in an NSU engine. The "A" field
above shows the time taken by
the flame front to travel from
position 2 to position 4, while
the "B" field represents the distance between position 2 and
position 3.

flame front becomes stationary owing to the quenching action occurring on the trailing side, which leaves a certain volume of unburned end gas, swept into the peripheral exhaust port by the trailing apex. By adopting dual spark plugs, it is possible to maintain positive ignition and minimize the combustion duration. Dual plugs also ensure more *complete* combustion, which results in higher power output and reduced exhaust emissions.

As in piston engines, both centrifugal and vacuum spark advance can help acceleration. With single-plug engines, more spark advance is advisable because it provides more reliable ignition with highly variable air/fuel ratios. More power could be developed by later ignition (after the minor axis), although some loss of smoothness might result. Curtiss-Wright found that a spark advance of 4–5 degrees was needed (that corresponds to 27–30 degrees in a piston engine) for maximum power at 5,000 r.p.m. In a dual-plug ignition system with the ignition advance of the upper spark plug set at the minor axis and that of the lower spark plug variable, ignition advance of the lower spark plug has a great influence on flame front travel on the leading side, but has hardly any bearing on the flame propagation on the trailing side. Combustion on the trailing end is delayed, and, consequently, this is the area most prone to cause knocking.

Combustion problems arise any time rotor speed and gas velocity are increased. Combustion is affected by the manifold and port configuration, combustion chamber shape, gas transfer velocity during the minor axis transition, and spark plug location.

Flame front position at 10°, 20° and 40° after minor axis in an experimental NSU engine. These sketches show very graphically the problems of a traveling combustion chamber.

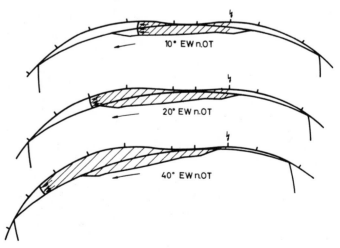

Various types of ports have been developed. The NSU-type Wankel engine has an intake port on the periphery of the rotor housing and Curtiss-Wright uses the side port type, in which the intake port is located on the sidewall. At Toyo Kogyo, only the peripheral port was tested at first, but later the side port and the combi-port (which is a combination of the two) were tested extensively. Later, Toyo Kogyo developed a system with dual side ports.

Each port arrangement has its merits and drawbacks. The peripheral port meets the requirements of high speed and high output. With side ports, combustion is stable under low-load conditions at low speed, including idling, and it is easy to determine the port timing most suitable for raising volumetric efficiency at low speed. However,

Variations in mean effective pressure (top chart), air-fuel ratio, specific fuel consumption caused by changes in spark plug position. The - - - - line represents spark plug position 20 mm. before the minor axis. The solid line represents a spark plug position 28 mm. before the minor axis. The - . - . - . line represents dual ignition with spark plug positions 56 mm. before, and 20 mm. after, the minor axis. P_{me} = Mean effective pressure (kg/cm²). 1.0 kg/cm² = 14.2235 psi (pounds per square inch). Luftzahl = air-fuel ratio. g/PSh = specific fuel consumption, grams per horsepower-hour. Drehzahl U/Min = r.p.m.

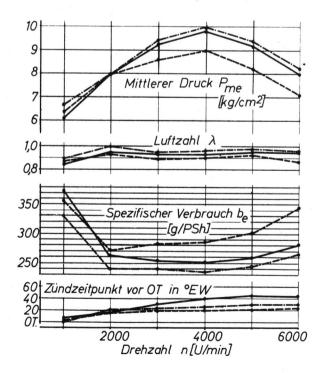

even with side ports an adequate time-port area for high-speed intake requirements is attainable. This is a key feature of the Wankel engine, and a major difference between it and the reciprocating engine (which has a mechanical restriction between the valve timing and valve lift). This advantage for the Wankel engine is due to the fact that its rotor moves through a larger angle during each operational phase than the crankshaft in a reciprocating engine moves during one stroke. Wankel engines with side ports have not only the best timing for low-speed operation, but also can provide the required intake port time-area to raise the useable r.p.m. range. Another distinction between the side port and the peripheral port lies in the turbulence of intake mixture. Intake gas turbulence obtained with side ports is more violent than that obtained with peripheral ports. This contributes to better atomization and air-fuel mixture uniformity. The pattern of the movement of intake mixture also varies with the shape of the rotor face cavity. The double side port engine makes it possible to use the two ports separately, one as a low-speed intake port and the other as a high-speed intake port. In this case, the intake gas velocity can be kept at optimum level at all times, depending on the operating conditions.

6

Housing and Rotor Cooling

THE WANKEL ENGINE can be adapted to either air or liquid cooling, just as a reciprocating piston engine, but the cooling problems are strikingly different. The basic duty of the housing cooling system is to lower the temperature of the areas exposed to the highest heat input and to minimize temperature differences throughout the housing.

The same stationary area on the working surface is always exposed to the same phase of the operational cycle, therefore the heat distribution in the housing is uneven. A cooling problem exists only in the area where combustion and expansion take place—the area immediately surrounding the spark plugs; the rest of the working surface requires no cooling. This uneven heating can cause housing distortion which, in turn, can prevent the proper functioning of the gas and oil sealing elements. The time during which the combustion chamber is cooled by fresh gas is fairly short, therefore the wall temperature of the combustion chamber is high and sensitive to changes in load.

The maximum temperature of the working surface is much higher than that of the sidewalls; local overheating can destroy the oil film on the surface. Sudden acceleration with a cold engine, especially in winter or when auto ignition occurs during high-speed driving, exposes the rotor housing combustion chamber wall to a repeated sudden and very large thermal load. As a result, thermal fatigue or thermal shock cracks can occur around the spark plug holes. In general, cracks occur most frequently across the spark plug holes in the axial direction, where there is a high stress concentration, although in extreme cases cracks can even reach the water jacket. There is also the danger that thermal distortion

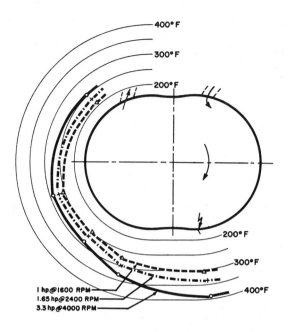

This chart shows rotor housing temperatures as measured with thermocouples under varied speed and power output conditions. The - - - - - line represents an output of 1 horsepower at 1,600 r.p.m. The - . - . - . line represents an output of 1.65 horsepower at 2,400 r.p.m. The solid line represents an output of 3.3 horsepower at 4,000 r.p.m. The spark plug is positioned about 20° before the minor axis, and it is in the area behind the plug that the highest temperatures were recorded.

The intake portion of the housing remains relatively cool while temperatures rise to great heights in the expansion portion. Housing material had a great influence in this connection. The solid line represents temperatures recorded in a cast-iron housing, while the dotted line represents temperature distribution with an aluminum housing, recorded under identical conditions.

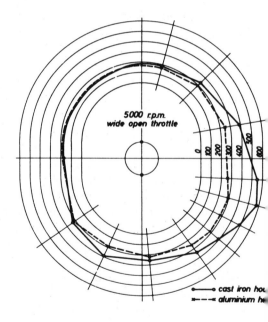

of the side covers will cause deterioration in the gas and oil sealing of the rotor side surfaces.

The housing assembly must be designed to limit this tendency toward thermal distortion, and the relationships between the cooling system,

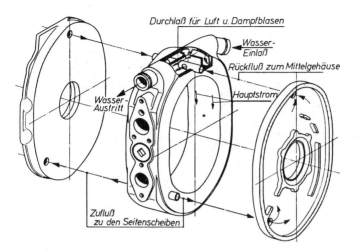

The water circulation system developed by NSU for the single-rotor KKM-400. Durchlass für Luft u. Dampfblasen = Passage for steam and air. Wasser-Einlass = Water entry. Rückfluss zum Mittelgehäuse = return line to center casing. Hauptstrom = Main coolant flow. Zufluss zu den Seitenscheiben = Coolant flow to the end covers. Wasser-Austritt = Water exit.

housing, and rotor seals must be kept in mind. New high-conductivity materials for the rotor housing have been developed in an effort to reduce the maximum temperature of the trochoidal surface, and coolant circulation methods that match the cooling capacity of each location with the distribution of heat in the housing have been devised. An example of this is the multi-pass forced-flow cooling system developed by Curtiss-Wright. The coolant flows back and forth through the rotor housing, from one end cover to the other, and back again. The flow passages are parallel to the mainshaft axis and internal ribbing of the end covers redirects the coolant flow. The flow area and wetted perimeter (and thus the hydraulic diameter) vary from pass to pass, in accordance with the heat input from the combustion chamber. In hot areas of the housing, coolant flow velocity is high and vice-versa. This system assures even heat dissipation and permits balanced cooling with large, clog-free, and easily manufactured passages which prevent vapor accumulation. Thermal distortions are small because the design allows uniformly thin heat transfer walls—the outer walls run cool and maintain their design profile. The zones of maximum heat input correspond to maximum stress zones, so that the addition of ribs serves to improve structural strength as well as heat transfer.

Cooling the rotor in the Wankel engine cannot be compared to cooling a piston in a reciprocating piston engine. The piston is largely air-cooled, having a voluminous chamber (cylinders plus crankcase) to draw

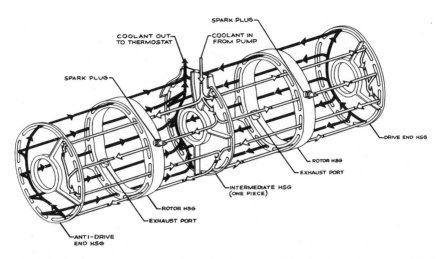

This is the cooling system developed by Curtiss-Wright for the twin-rotor RC2-60 U5 engine.

The cooling system of the NSU KKM-502 engine. The oil pump draws oil from the sump, pushes it through a filter and then into a heat exchanger. Still under pressure, the oil continues straight into the hub of the rotor, circulates inside it, and leaks out from the hub sides into the sump, where the pump can pick it up again. WASSER = Water; OEL = Oil.

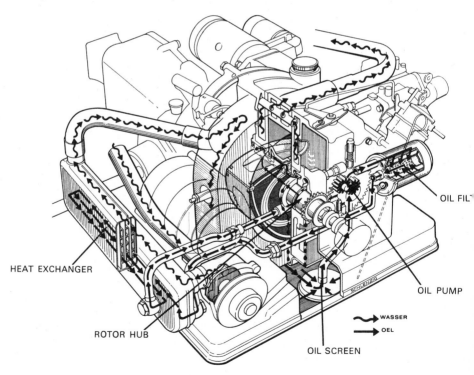

on for circulating air, while the rotor, encapsulated inside the housing, requires higher-intensity cooling. The very fact that the rotor moves at one-third mainshaft speed magnifies the cooling problem. If rotor cooling is not sufficient, various troubles will occur, such as sticking of the sealing elements to the sealing grooves or to the sliding surface of the housing, and pre-ignition resulting from local overheating on the combustion face.

There is an important relationship between the cooling characteristics of the rotor and its material, whether cast-iron or aluminum. Both cast-iron and aluminum rotors have their merits and drawbacks. Most manufacturers use cast-iron because their test results indicate that such a rotor is safer in practical use. An aluminum rotor has the benefit of low inertia and high thermal conductivity, but it demands more cooling. Air-cooling is hardly adequate for an automotive Wankel rotor and water cooling is feasible but would cause enormous mechanical complications. The Wankel rotor, however, has a very convenient structure for being internally cooled by oil circulation.

Lubricating oil of the same type that lubricates the main and eccentric bearings circulates inside the rotor and, after cooling the rotor efficiently, all the oil flows out of the rotor and fresh oil is pumped in.

This schematic shows the oil circulation system of the Curtiss-Wright RC2-60 U5 engine.

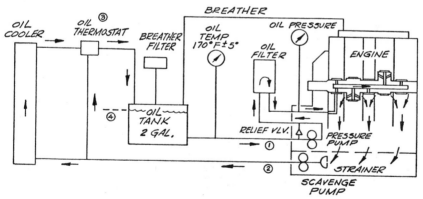

① INLET LINE CAPACITY IN ACCORDANCE WITH OIL FLOW CURVE.
② OUTLET LINE MUST HANDLE TOTAL OIL FLOW PLUS ENTRAINED AIR OF 2 CU. FT./MIN. MAXIMUM.
③ OIL THERMOSTAT MUST CONTROL OIL IN TEMPERATURE TO 170°F±5° AND LIMIT BACK PRESSURE ON SCAVENGE PUMP TO 30 PSI MAXIMUM.
④ OIL LEVEL IN EXTERNAL SYSTEM MUST BE AT LEAST 4" BELOW SHAFT CENTERLINE UNLESS VALVING IS PROVIDED TO ASSURE AGAINST DRAINBACK THROUGH PUMPS ON SHUTDOWN.

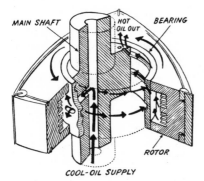

MAIN SHAFT   HOT OIL OUT   BEARING

ROTOR

COOL-OIL SUPPLY

NSU's first rotor cooling system used pressure-fed oil with an intricate circulation path inside the rotor. It was given up in favor of using rotor motion to force oil to the hottest areas.

Graphic illustration of the vector field that determines acceleration forces in the rotor.

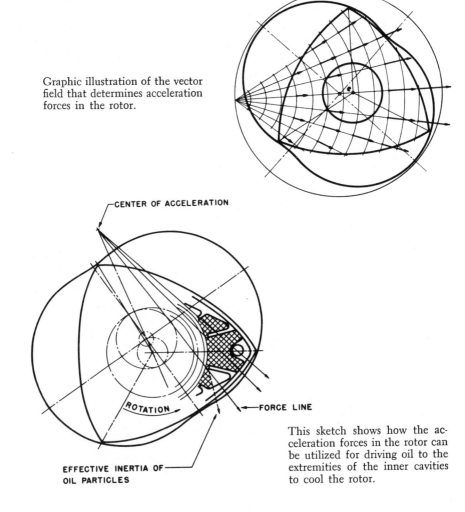

CENTER OF ACCELERATION

ROTATION   FORCE LINE

EFFECTIVE INERTIA OF OIL PARTICLES

This sketch shows how the acceleration forces in the rotor can be utilized for driving oil to the extremities of the inner cavities to cool the rotor.

Detail design of the inside of the rotor is important and, at the same time, the amount of the oil jetted from the oil hole in the eccentric bearing to the inside of the rotor must be carefully controlled. The design of the rotor cooling system affects lubrication as well as engine balance.

Oil entry temperature, temperature rise, and period of contact with the metal are controlled to avoid coking or other lubricant deterioration. Oil flow is continuous, but velocity and oil pressure are dependent on the engine's rotational speed. Shaft and rotor inertia pumping superimpose a roughly sinusoidal pressure increment on the oil feed pressure. The rotor cavities are maintained full or near-full at all times, oil being metered at the pump exit port.

The rotor bearings have oil seals that permit a certain controlled leakage so as to assure lubrication of any high-friction areas on the rotor flanks. In contrast with the metered apex seal oil, the lubricant for the side seals on the rotor flanks serves not only as gas sealing agent but also as antifriction material. The metering oil pump in the induction system supplies oil for the sole purpose of gas sealing. Maximum flow characteristics of this pump are determined through experiments, according to gas sealing performance requirements at full throttle. Just as the top ring of the reciprocating piston engine is adequately lubricated by a very small quantity of oil film which is left after being scraped by the second ring and the oil control ring, the side seals of the rotary combustion engine are continuously lubricated by the oil film which is left on the sidewalls being scraped by the eccentric bearing oil seals.

While the piston rings of the reciprocating engine must stop at top and bottom dead center and withstand extremely high acceleration

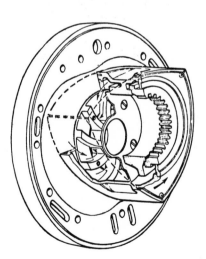

NSU developed a system of guide vanes to expel the hot oil from the rotor.

loads, the side seals of the rotary piston engine have the advantage of sliding on the sidewalls continuously—in one direction only. However, the movement of the side seals is not necessarily simple, therefore the design, machining and material of the side seals had to be studied in combination with the sliding surface of the side housing. When oil is used for cooling the rotor, and the rotor bearing is lubricated under pressure, considerable amounts of oil would be ejected onto the sliding surfaces of the sidewalls unless effective oil seals for preventing oil leakage into the combustion chamber were provided. During the early stages of development, the oil seals were relatively primitive and oil consumption was far from satisfactory. From the point of view of mechanical engineering, the oil sealing provided on the side surfaces of the rotor seemed to be simple at a glance, but the actual development of the oil seal turned out to be an unexpected difficulty.

In addition to the mechanical and thermal distortions inevitably occurring on the sliding surfaces of the side housing, the operational conditions and other requirements of the oil seal itself were by no means simple. During the rotor's rotation, one part of the oil seal rides on the oil film and separates it from the sidewall surface. At the same time, the other part of the seal is in contact with the surface of the sidewall. Obviously, the rotor side seals cannot also serve as oil seals. That would mean a presence of oil on the full rotor sides throughout the rotor's travel—except for the very small areas between the side seals and the rotor edges. It is not practical to keep such a large volume of oil on the rotor flanks during the rotor's complete cycle, nor can the existing side seals do an acceptable job of preventing leakage—the oil seal must be between the eccentric bearing and the rotor. All constructors now build their Wankel engines in accordance with this concept.

7

Advantages of the Wankel Engine

ONE OF THE MAIN ADVANTAGES of the Wankel engine is that it is about half the size and weight of a piston engine of comparable power output. Curtiss-Wright made a direct comparison between their RC2-60 U5 Wankel engine and a Chevrolet 283 cubic inch V8 piston engine. The Wankel engine was rated at 185 horsepower; the Chevrolet at 195. The Wankel weighed only 237 pounds while the V8 was over twice as heavy at 607 pounds. Regardless of maximum power output, the Wankel engine has a far superior weight per horsepower ratio when compared with reciprocating engines.

POUNDS PER HORSEPOWER

Horsepower	Wankel Engine	Piston Engine
10	3	11–15
20	3	12.5
30	3	11.1
100	1–1.8	4–9.5
250	0.8–1.5	2.1–4
500	0.6–1.2	1.5–3

Dimensional differences are approximately to the same scale:

	RC2-60 U5	283 V8
Height	21.5 inches	31.5 inches
Width	22.1 inches	28.0 inches
Length	18.0 inches	29.5 inches

The Chevrolet V8, with its essential accessories, occupies 23.2 cubic feet of volume. The Wankel engine requires only 5.1 cubic feet for its installation. The next major advantage is a reduction of parts.

The Wankel engine is far less complicated than a reciprocating piston engine. The block is replaced by a housing and two end covers. The housing is built up of sections containing trochoidal chambers, separated by sidewalls, and the cylinder head is eliminated. No valve train is necessary, because valves are replaced by ports that open and close automatically by rotor motion. The difference in the total number of parts in a piston engine versus a Wankel engine is enormous. The 283 cubic inch Chevrolet V8 has a total of 1,029 parts, the Curtiss-Wright RC2-60 U5 has only 633. There are 388 *moving* parts in the Chevrolet V8, while the RC2-60 U5 has only 154.

A reduction in the number of parts provides greater simplicity and lower cost. Studies have shown that a typical American V8 costs about $2 per horsepower. The cost of a Wankel engine probably could be reduced to half that figure, or $1 per horsepower. This rough estimate is supported by an independent cost comparison study. A 90 horsepower, four-cylinder piston engine of modern design, using cast-iron for both block and head, and requiring the lowest-cost tooling and machinery, was compared with a 110 horsepower twin-rotor Wankel engine with an aluminum housing and cast-iron end covers. Both engines were calculated for the same production rate, and the rates were broken down according to accessories, material, and production time. The results indicated that material cost for the reciprocating engine was 7% higher than that for the Wankel rotary engine. At the same time, labor costs were 57% higher for the reciprocating piston engine, and quality was lower.

More recent experiences by the Japanese manufacturer of Wankel engines, Toyo Kogyo Company, and the German firms, NSU and Fichtel & Sachs, show that, for 100 horsepower engines, manufacturing costs of the Wankel are 26% lower than those of the reciprocating engine when considered at the same skill of production, while for small engines

Comparison of a twin-rotor Wankel engine (Curtiss-Wright's RC2-60 U5) with a 283 cubic inch Chevrolet V8 piston engine. All relevant data are given in the text.

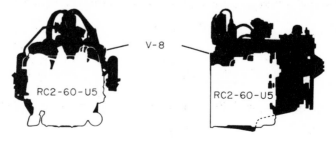

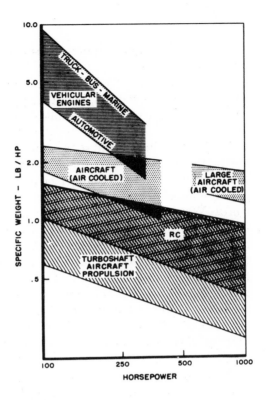

The Wankel engine (RC = rotating combustion) has power-to-weight ratios almost in the same class as shaft-powered gas turbines, and far better than reciprocating piston aircraft engines. Automotive piston engines are far heavier.

of 6 horsepower the total cost is practically the same. The detailed data can be studied in the following tables:

COST COMPARISON (IN U.S. DOLLARS)

	Wankel Engine 115 h.p.	Recip. Engine 100 h.p.	Wankel Engine 6 h.p.	Recip. Engine 6 h.p.
Accessories	$ 80.00	$ 87.00	$ 35.00	$ 29.00
Material	50.00	70.00		
Labor	17.50	25.00	7.50	10.00
Overhead	31.00	44.00	13.25	17.50
Total	$178.50	$226.00	$ 55.75	$ 56.50

COST COMPARISON (%)

	Wankel Engine 115 h.p.	Recip. Engine 100 h.p.	Wankel Engine 6 h.p.	Recip. Engine 6 h.p.
Accessories	100	104	100	61
Material	100	140		
Labor	100	142	100	133
Overhead	100	142	100	133
Total	100	126	100	100

The Wankel engine requires no exotic materials and its components can be manufactured on conventional machine tools (with certain specialized adaptations). Production methods are of a conventional nature, and most of the operations, in assembly as well as in manufacturing, are familiar to the average automobile factory worker. Due to the small number of moving parts, the maintenance required is limited, mechanical reliability is higher and service life extended. These factors, in combination, result in low operational costs.

The reduction in the number of parts also indicates less power loss due to internal friction. Measurements of the various friction losses by Curtiss-Wright show that seal friction is the biggest single friction factor —about 50% higher than friction in the bearings and gears. The second greatest friction in the Wankel engine occurs on the interface between the rotor and the eccentric bearing. This friction is comparable to friction between cylinder walls and pistons, and although it reaches approximately the same magnitude so far as loads are concerned the temperature problems of the piston engine do not exist. The piston engine's valve gear absorbs a lot of power, while it costs nothing to open and close the ports in a Wankel engine because the rotor movement that performs this function is part of the basic power-generating process.

The piston engine is subject to vibrations in several planes. The first- and second-order forces of inertia at play in a piston engine never can be equalized in a single-cylinder engine. Even an inline six, or a V12, which has all primary forces in perfect balance, also has unbalanced secondary forces. Primary balance is achieved by using the motion of one piston to counteract the opposite motion of another. But the secondary forces, due to the swinging motion of the connecting rods on their crankpins, lack symmetry and the result is engine vibration. The Wankel engine has no such problems because it is free of reciprocating motion— all motion is strictly rotary. Although the large angle of each operational phase in the Wankel engine poses a handicap to its low-speed performance, it is advantageous for torque fluctuation. The torque fluctuation rate of the twin-rotor Wankel engine is almost equal to that of a four-stroke, six-cylinder reciprocating piston engine.

By eliminating reciprocating motion in the working parts of the engine, the Wankel avoids all problems of alternating inertia stresses because there are no unbalanced inertia forces. In a piston engine, all reciprocating parts must be designed with adequate material strength to withstand these inertia loads. Reinforcements often have the disadvantage of adding to the weight of the parts in question, thereby further increasing the inertia stresses. Inertia loads also impose a limit on the rotational speed of the engine. Even though rotor eccentricity is responsible for setting up a slight rotational imbalance, this is easily corrected by adding counterweights on the mainshaft and the Wankel engine can develop high rotor speeds.

Over-revving a piston engine can result in bent and burned valves, damaged valve gear, broken pistons, broken connecting rods and crankshafts. Letting a Wankel engine spin at rotational speeds beyond its design limit cannot result in any mechanical breakdown or damage. The only deterioration that will result is a higher rate of wear and increased oil consumption.

The Wankel engine encourages high volumetric efficiency because gas flow into and out of the combustion chamber does not proceed through loops or right angle turns, but in a smooth sweep. Moreover, the Wankel engine aspirates the explosive charge over 270 degrees of mainshaft rotation, compared with an intake stroke duration of 90 degrees of crankshaft rotation in a piston engine. This gives the Wankel

These charts explain why the Wankel engine runs so smoothly. A single-rotor Wankel engine (below) is compared with a single-cylinder, four-stroke piston engine (above). The piston engine has only two power sequences during four crankshaft revolutions, while the Wankel engine receives four power impulses in the same time. Each power phase in the Wankel engine covers a much wider shaft angle than is true of the piston engine. The Wankel engine has shorter dwell periods for intake and exhaust, reducing gas losses and providing smoother operation.

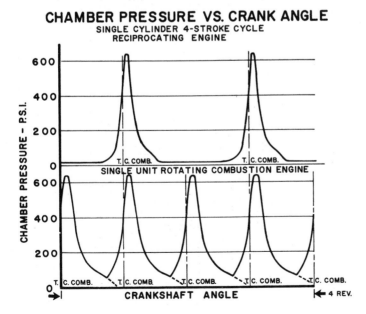

CHAMBER PRESSURE VS. CRANK ANGLE
SINGLE CYLINDER 4-STROKE CYCLE
RECIPROCATING ENGINE

engine a substantial volumetric efficiency advantage. The change from minimum to maximum volume in the working chamber takes place during three-quarters of a mainshaft revolution, compared to one-half crankshaft revolution in a piston engine. Consequently, it is logical to expect at least 50% better breathing in a Wankel engine. It could be even more, because the intake port is fully open for the greater part of the 270 degrees intake phase, while the intake valve in a piston engine opens and closes more gradually. The rotor sweeps burned gases out of the combustion chamber at the end of the power phase, and the Wankel has no quench areas or other corners where inert gases may remain. In addition, centrifugal force assists exhaust gas evacuation, which in turn contributes to higher volumetric efficiency—there is no similar effect in a reciprocating piston engine. Power output in a Wankel engine is smoother than in a piston engine because the engine produces positive torque for about two-thirds of the operating cycle as opposed to one-quarter or less of the cycle in a four-stroke piston engine.

In this chart the torque variations in a Wankel engine are compared with those of a reciprocating piston engine. Peak torque output of a twin-rotor Wankel engine never exceeds 200% of mean torque output, while in a four-cylinder piston engine, peak torque varies between 200 and 300% of mean torque output.

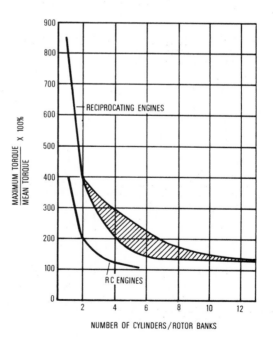

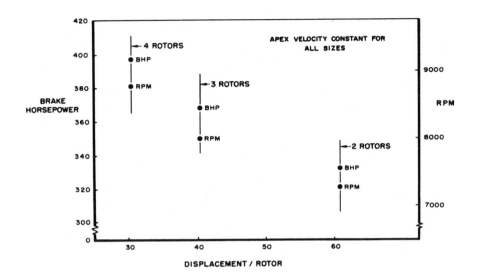

For higher power requirements, additional rotors and housings can be stacked on a basic Wankel engine. Adding rotors adds horsepower and speeds up the engine in an almost linear curve.

The seals in a Wankel engine can be compared to the piston rings in a piston engine. At excessive crankshaft speeds, which produce extreme piston accelerations, piston rings begin to flutter. There is no corresponding phenomenon in the Wankel engine because the seals slide unidirectionally. There are accelerations and decelerations even at steady output shaft speed, but no complete stops and no reversals of sliding direction.

Gas pressure loads on the bearings are higher in a Wankel than in a piston engine, but centrifugal loads are lower; therefore, the risk of bearing failure is far smaller. One reason is that rotor eccentricity is smaller than the crank throw in a piston engine. Negative torque loads, such as are present in a piston engine during the compression stroke, do not exist because the Wankel has no reversal of rotor rotation at any time during its operational cycle. Moreover, the Wankel engine has no parts that must be accelerated, stopped, and returned in linear, reciprocating motion. Also, the output shaft makes three revolutions for each rotor orbit, which means the Wankel can be thought of as having a slow-moving piston, which immediately introduces certain advantages:

1. Lighter inertia loads. . . .
2. Reduced need for lubrication. . . .
3. More freedom in seal design.

In short, the Wankel engine is clearly superior to the reciprocating piston engine in most ways, in spite of the technological problems that have hindered its development. How these problems were solved is covered in the next section.

Section II

DEVELOPMENT

8

Felix Wankel

THE VILLAGE OF LAHR in Germany's "Black Forest," is situated in scenic countryside in southwestern Germany, close to the Swiss border and not far from France's Alsace province. Lahr is near Donaueschingen, the town where the confluents Breg and Brigach form the beginning of the river Danube. The region is called Swabia, and forms the heartland of what is now the German state of Baden-Württemburg. The land is hilly, but not savage, and, although close to the Alps, the altitude of the Black Forest mountains never approaches the tree line. The hilltops are wooded, the valleys dark, the climate extreme. Winters are bitterly cold and summers very hot. The people of Swabia are known to be taciturn, stubborn, unapproachable, industrious, stolid, unimaginative, slow, practical, non-intellectual, shrewd, subtle, settled in their ways, and of an unimpeachable morality. This is where Felix Heinrich Wankel was born on August 13, 1902.

A boy growing up in this area could become a farmer, blacksmith, lumberjack, wine-grower, butcher, baker, or go into related activities. Land, vegetation, botany and life were close to young Felix Wankel. His father, Rudolf Wankel, was a forest commissioner. His mother, Gerty (née Heidlauff) was from a local village family. For a boy with this background to become a major inventor of mechanical devices is as unlikely as the story of a medical doctor in the American midwest building a machine which was to become a principal factor in the formation of the automobile industry. Yet Elwood Haynes, a general practitioner in Kokomo, Indiana, built one of this country's first practical automobiles. And, Felix Wankel *did* become an inventor.

His father was killed in the beginning of World War I, when Felix was only 12. He completed grade school locally and stayed in high school until 1921. At this point in life, it was customary for a young man with a mind of mechanical bent to seek an apprenticeship with an industrial enterprise, usually in a foreign country, but frequently in another province in his own nation. This would lead to a series of subordinate positions, which in combination made up the learning of a trade, and qualified the man for his chosen profession.

Gottlieb Daimler, Carl Benz, August Horch, Emil Hermann Nacke, Heinrich Kleyer, Richard Stoewer, Heinrich Büssing, Nikolaus Dürkopp and other founders of the German automobile industry had launched their careers through engineering experience obtained in the shops of a variety of far-away industries. This, too, would have been the logical procedure for Felix Wankel, when he graduated from high school at the age of 19. But economic conditions did not allow him to follow the customary course. Postwar inflation in Germany hit hard a widow with a son in high school, and Felix had no choice but to go to work and earn a living. He moved to Heidelberg and found employment as a salesman for a publisher of scientific books. In this ancient university town, close to such industrial centers as Karlsruhe, Darmstadt, Frankfurt and Stuttgart, young Wankel was not devoted solely to learning the skills of salesmanship. He was forever experimenting in his mind with new ideas for machinery of various types.

He had always worked with things and insisted on trying out everything in practice—theoretical answers were not good enough for him. It was this frame of mind, his basic mental makeup, that drove him to take one of the most important steps in his career. At the age of 22, Felix Wankel opened his own mechanical workshop in Heidelberg. His first money-making contracts involved such prosaic machine work as grinding cylinders. But the importance of this stage of Wankel's work in terms of his later achievements lay in the fact that in his own shop he learned the key factors of machining and production, the meaning of precision, and thought about improvements to the internal combustion engine. As early as 1924, he began to make sketches for a rotary piston engine. He had, even at this stage, a keen sense of its shortcomings. He evaluated his own ideas without prejudice and concluded that his designs were not suitable for further development.

Throughout this period, Felix continued to study. He went to night school and took correspondence courses, and the idea of the rotary engine would not leave him. Although he wanted desperately to build a rotary engine, he didn't know where to start. He began by experimenting with rotary valves on conventional motorcycle engines. In 1926, when he

Felix Wankel with the inner rotor of the DKM-54.

was working as an independent consultant on high-pressure lubrication apparatus, he started to make suggestions for improvements in the sealing area. This work made him very conscious of the problems of sealing the gaps between two or more surfaces in direct contact. It became obvious to him that the same principles could be applied to obtain better sealing in internal combustion engines. Soon, he devoted all his time to the task of improving cylinder sealing in piston engines, and, in 1928, he set up a test establishment for motorcycle engines.

Felix Wankel took out his first patent (DRP 507 584) in 1929—it covered an engine with a reciprocating piston inside a horizontal cylinder. The piston had a crown at each end, and the cylinder had a combustion chamber at each end. The piston's wrist pin was mounted eccentrically in relation to the mainshaft, so that piston movement produced mainshaft rotation. There was no connecting rod—the wrist pin doubled as crankpin for the mainshaft. This engine called for no new sealing techniques, and enabled him to use standard piston rings. He soon realized, however, that this was sidestepping the issue, and that without ideal sealing his rotary engine could never become a reality. He recognized that many earlier types of rotary engines could not work with high internal gas pressures because the seals were not of sufficiently robust construction. He also realized that sealing elements were dependent on the presence of an oil film to fulfill their intended function.

From 1929 to 1931 Wankel built and developed twenty devices to test his ideas concerning rotary valve seals. He made a disc valve engine and installed it on a motorcycle that ran satisfactorily. He also succeeded in running a test engine with a cylindrical rotary valve. The next two years were spent evaluating the results and applying for patents. In his search for the basic concept of a rotary engine, Wankel defined the heart of the matter. He needed a rotary piston that would also rotate the shaft. But all attempts, all his tests, failed for the same reason—he was unable to provide proper sealing of the working chamber. In 1933 he was still far removed from a successful engine.

Then began a new and bright period. In 1933 he was engaged by Daimler-Benz for research in sealing components, rotary valves and rotary engines. The engagement ended within a year when he fell out with the management of Daimler-Benz, but he soon signed a contract with BMW for similar work. His principal assignment at BMW was the development of a piston engine with rotary valves. The result of his work in the area of sealing finally bore fruit.

In his patent (DRP 637 701), dated November 6, 1936, Wankel said that sealing is not ideal if the seals are formed by the sliding surfaces themselves, no matter how high the precision of machining and assembly because this only results in high friction and rapid wear. Sealing duties should be performed, Wankel concluded, by special sealing elements designed for that purpose only. Wankel applied for this patent on September 24, 1934, covering certain new improvements in gas sealing techniques. He called his sealing element a "packing body."

The packing body is defined as being closely fitted in its supporting body and pressed with its end surface against the machine part which slides over them. Earlier attempts at gas sealing, Wankel pointed out, had failed because gas or liquid pressure could act on the whole underside of the packing body in question. This offered a very large exposed surface, making it necessary to dimension the tip surface on the friction side of the packing body of correspondingly large surface area in order to prevent overloading the working surfaces. Sealing surfaces or packing edges around the openings were therefore much wider than was necessary for sealing purposes alone. Any wide packing surface produces a considerable power loss in high-speed machines because of friction. It is also difficult to cause a wide packing surface to bear uniformly tightly because the small heat warpings of the material, which are unavoidable in practical operation, have a particularly disturbing effect. Certain earlier seal types tried to circumvent these difficulties by not allowing any gas pressure to act on the underside of the packing bodies. The pressure on the sealed interface was obtained solely by mechanical spring pressure.

Wankel saw that a pressure-tight seal permitting heat expansion while working under low friction conditions could not be obtained solely by spring pressure because the gas pressure, which tends to force its way between the sliding surfaces, is capable of exerting a considerable power action in a direction which separates the sliding surfaces. A spring force sufficiently strong to counteract this gas pressure would cause too high a pressure load on the sliding surfaces, resulting in wear and considerable friction when the gas pressure ceased. Wankel's idea was to direct gas pressure so that it would improve seal tightness and act in the direction of any necessary spring loading rather than opposing it. His new packing body obviated the disadvantages of too wide a seal edge and did not require any high spring force. It simply drew upon the operating gas to apply pressure, and spring-loading was necessary only to a suitably adjustable degree for supporting the seal in its position. According to Wankel's invention, the seal was effected by a packing body which surrounded a cavity or opening with a jointless sealing edge. For the purpose of reducing friction, the packing surface was limited to a narrow strip. The packing body was found suitable for both disc-type and barrel-type rotary valves.

In the case of the disc valve, the lower face covers the entire cylinder bore and the packing body takes the form of a ring with an inverted L-shaped cross-section. A spur on the base of the L provides the sealing edge. The disc valve does not rub against the top of the cylinder block, but bears against the packing body which is mounted in the block. The base of the L is spring-loaded against the sliding surface and the mast of the L is guided in a slot which is sealed by two small piston rings. Gas pressure admitted from the cylinder to a chamber below the mast of the L-shaped packing body will act in the direction of the spring-loading and add to sealing tightness. The amount of gas pressure could be regulated by an adjusting screw controlling orifice size. The principle is the same in draw-type rotary valve applications but the packing body takes a different shape.

It does not matter whether the machine is a compressor or a heat energy engine, the sealing elements have to be united so that they form a system that completes a full sealing path, end to end. The various sealing elements must be interlocked to block leakage through corners or joints. In many earlier engines a sealing path could not be completed because of certain design or constructional characteristics, but the Wankel patent covered a complete sealing path. It specified possible applications to both compressors and internal combustion engines. Wankel claimed his seals would withstand gas pressure loads of 30 atü (1.0 atü = 14.7 psi) with a seal surface contact of 5 mm. with no greater leakage loss than exists in a reciprocating piston engine.

The sealing grid effectively covers an entire annular chamber of rectangular cross-section; the top cover revolves while the sides and bottom are stationary. The top cover is sealed by complete seal strip rings, inside and outside the annular chamber. The seals are backed up by spring-loaded seal carriers and the sealing path continues around the bottom in a U-profile—this is the direct basis for the rotor side seal system in the Wankel engine. Joints in the seal strips are made by trunnion pieces and the seals are lightly spring-loaded so as to provide effective sealing despite variations in gap size. Such variations can occur as a result of axial thrust or thermal expansion. The narrow seal contact width of 5 mm. was chosen to give small friction losses. There are no centrifugal loads on the seal strips because they are located in the stationary parts.

Between 1934 and 1936 Wankel succeeded in taking out sealing patents that gave him blanket coverage of all possible rotary valve, compressor and internal combustion engine applications. One type of seal Wankel designed for use in engines with rotary valves consists of two rings, placed one on top of the other. One is solid, the other is split. Both have a conical profile, so that their interface is inclined. It is a face-type seal, sealing a cylindrical cavity against its own recess walls as well as against the lower face of the head or lid. Gas pressure tends to force the rings apart, because the inclined interface produces opposite reactions in the two rings creating a more effective seal. Relative rotational movement in the two rings is prevented by a small dowel pin inserted at one point on the interface.

Wankel invented another type of split-ring seal where the gap is not straight-cut, but stepped. The steps run in opposite directions, so that the gap is T-shaped. The crossbar on the T is filled by an insert which effectively blocks the leakage path, except for the small slot formed by the stem of the T. A seal for the sides of interacting disc valves was also developed where two seal ring segments are combined to form a figure-eight perimeter. The joints are sealed by gas pressure, forcing the open ends of the segments together.

Wankel's first rotary engine, developed during this period, was a far

Wankel developed this sealing grid for a rectangular opening. A = seal strips. B = joint trunnion. This invention was the basis for the side sealing of the rotor in the Wankel engine.

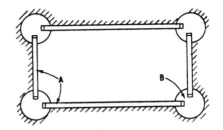

Two views of a split-ring seal patented by Wankel. The insert almost blocks the gap. A = split ring. B = L-shaped sealing insert. C = leakage gap.

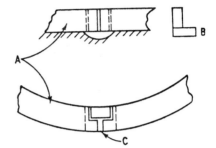

Wankel invented a dual seal for disc valve applications. Both rings have conical profiles, and the upper ring is solid while the lower ring is split. The two rings are interlocked by a dowel pin to prevent their rotating relative to each other. A = solid ring. B = dowel pin. C = split ring. Below, plan view of the split ring.

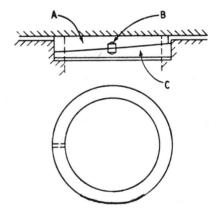

In case of overlapping circular openings Wankel had a sealing system, too. Gas pressure from inside the chamber was used to make the seals more effective, pressing the inside members at A and D against the outside members. B and C represent segments of sealing rings.

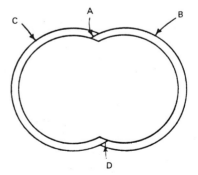

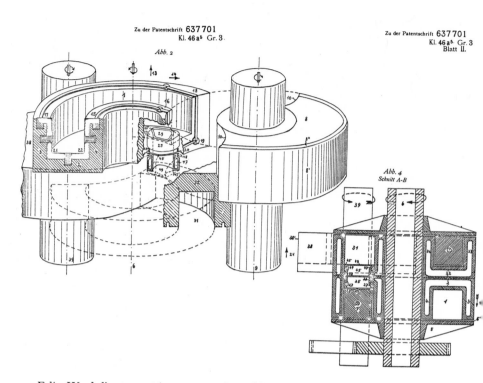

Zu der Patentschrift 637701
Kl. 46 a⁵ Gr. 3.

Abb. 2

Zu der Patentschrift 637701
Kl. 46 a⁵ Gr. 3
Blatt II.

Abb. 4
Schnitt A-B

Felix Wankel's concentric rotary engine with rotary valves at two different levels led to the development of a disc valve engine for BMW in 1934.

cry from the refined machinery now being produced by NSU and Toyo Kogyo. Application for patent was filed on September 24, 1934 for a two-rotor rotary engine, in which all rotation was concentric. Two rotors were necessary to make the engine run on the four-stroke cycle. The two annular working chambers had a rectangular cross-section. Three sides of the working chamber were made up of a cast-iron housing with cooling water jackets. These three sides were the bottom, plus the inner and outer peripheries. The fourth side was a revolving cover—to which the rotor was fastened (or with which it was integral).

The rotor filled less than half the working chamber and had concave faces with a rectangular cross-section nearly 180 degrees apart. The rotors were concentrically mounted on the rotor shaft, one above the other. Engine timing was provided by counter-rotors, spaced 120 degrees apart, revolving in a direction opposite to the unidirectional engine rotors. The counter-rotors overlapped with the engine rotors and formed a combustion chamber. The counter-rotors were kept in phase by gearing (to make sure they did not touch) in a 1:1 ratio. It is the relative movement of the engine rotors and the counter-rotors that provide the necessary pumping action for engine operation.

Abb. 1

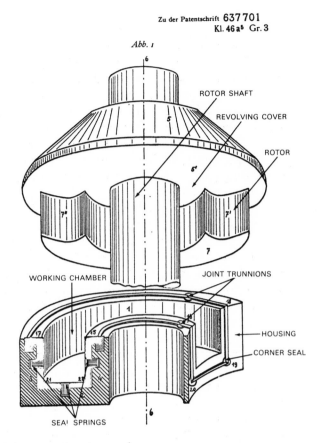

ROTOR SHAFT

REVOLVING COVER

ROTOR

WORKING CHAMBER

JOINT TRUNNIONS

HOUSING

CORNER SEAL

SEAL SPRINGS

Central parts of the Wankel concentric rotary engine.

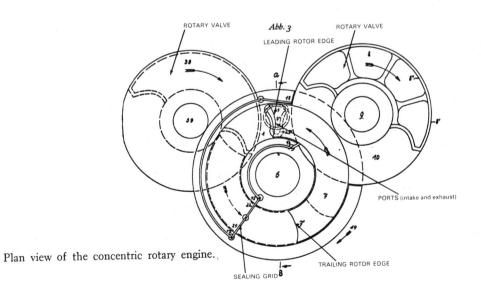

ROTARY VALVE

Abb. 3

LEADING ROTOR EDGE

ROTARY VALVE

PORTS (intake and exhaust)

TRAILING ROTOR EDGE

SEALING GRID

Plan view of the concentric rotary engine.

The lower of the two chambers charged and compressed while expansion and scavenging took place in the upper chamber. The operational principles of this engine involved using the rotor in the lower chamber as the compressor rotor and the rotor in the upper chamber as the power rotor. Both were mounted on the mainshaft and were concentric with it—the end covers were integral with the rotors and revolved with them. The lower chamber had an intake port, while the exhaust port was in the upper chamber. The two chambers were linked by a communicating port, and the opening and closing of all the ports were controlled by rotor movement.

With this design, the counter-rotors provide differences in chamber volume. One counter-rotor intrudes into a part of the compression chamber that has just been vacated by the compressor rotor and sweeps through it in the opposite direction of travel. That movement creates a rapid expansion against the trailing faces of both the compressor rotor and the counter-rotor, which results in the aspiration of a fresh charge from the intake port. As the compressor rotor continues its rotation, its leading face meets the incoming charge, shuts off the intake port, and begins to compress the fresh charge against the leading face of the opposite counter-rotor. At the point of maximum compression, the movement of the compressor rotor opens the communicating port, and the compressed gas is admitted into the power chamber, where it is ignited as soon as the port closes.

Gas pressure from the combustion propels the power rotor. Expansion is allowed to take place due to the relative movement of the power rotor and the counter-rotors. For every revolution, the compressor rotor admits a fresh charge on its trailing face while its leading face is busy compressing the preceding charge. For every revolution the power rotor is driven by gas pressure against its trailing face while its leading face expels the burned gases.

The engine had many drawbacks, such as complicated machinery, excessive bulk and weight, and questionable efficiency. Its significance was mainly in the sealing system developed for it. Here, Wankel developed a brilliant solution to the persistent problem. Complete sealing grids were formed around the working chambers, comprised of a number of curved and straight sealing elements arranged end-to-end and in series, interlocked by cylindrical plungers, with freedom for the entire sealing grid to move axially and exert pressure against a sliding surface from its mounting in the stationary housing. This sealing system was to have a great effect on future developments.

These patents were a triumph for Wankel, but he had moments of despair during the same period because of the political situation. Prior to

Hitler's assumption of power, Wankel had helped uncover an embez-
zlement case that went against the NSDAP (National-Sozialistiche
Deutsche Arbeiter Partei). When he became State Chancellor, Hitler
sought revenge and Wankel was branded a traitor to the party and imprisoned for months.

Free again in 1935, he moved his workshop from Heidelberg south to
his native Lahr. In 1936, Hermann Goering's air ministry heard about
Wankel's work and invited him to conduct his experiments at the DVL
(Deutschen Versuchsanstalt Für Luftfahrt), in Berlin's Adlershof. Wankel declined because he did not want to live in Berlin and persuaded
Goering to set up a separate institute for him: WVW (Wankel Versuchswerkstätten) in Lindau on Lake Constance. The ministry had
high expectations and invested several million marks in this research
institute and its experimental shops. Here, Wankel was to build a
rotary valve engine for the Luftwaffe. Dr. Bensinger, the key man behind the Wankel engine that powers the Mercedes-Benz C-111, was one
of his assistants. Research was directed toward rotary engines during
1938, but 1939 found Wankel occupied with more routine tasks. He
invented a system for improved piston cooling in an aircraft engine and,
in association with Daimler-Benz, he developed a rotary disc valve,
which went into production on the DB601 V12 aircraft engine in 1942.
The reasons for using the disc valve instead of the rotary tubular valve
(which Wankel favored) were not technical—in 1940 it was decided that
the simple design and ease of manufacture of this type valve over the
much more complicated tubular type was to take priority over all technical considerations. The government was confident that Wankel could
solve the problems that had cropped up during the early development
of the disc valve in 1936. The Wankel disc valve was a plate with one of
two open sectors placed between the cylinder head and the block, covering the cylinder bore. As it revolved, the open sectors worked as ports to
admit fresh mixture and allow burned gas to be expelled. Wankel had
been discouraged by problems with overheating—cooling the large-
diameter disc valve was extremely difficult because it had little direct
contact with the cylinder head for heat dissipation, and it had a large
exposed area. He knew that overheating would certainly lead to distortion and leakage and a serious loss in volumetric efficiency.

The Daimler-Benz type DB601 V12 aircraft engine had been originally designed in 1935 as the prototype for a whole new family of powerplants for the aviation industry. It had twelve cylinders in an inverted
V-formation, was water-cooled, and had a light-alloy block. Cylinder
displacement was 2,020 cubic inches (33.4 liters), and output was 1,050
horsepower at 2,400 r.p.m. The initial prototype had carburetors, but

direct fuel injection was developed for subsequent versions. In 1939, a DB601 with turbocharging (rated at 2,770 horsepower developed at 3,100 r.p.m.) was installed in a Messerschmitt Me 209 fighter plane for speed record attempts. The advent of the war prevented any international records from being set, and its performance in test flights became a military secret.

Using this basic engine, Wankel went to work. Special cylinder heads were designed in which a nitralloy steel disc with two ports was centrally mounted on a hollow trunnion in line with the cylinder bore. These discs themselves performed double duty as wheels. Each disc meshed with the adjoining discs to form a train of gears. The drive was taken from the crankshaft via a pinion mounted in place of the camshaft and the discs were geared to rotate at one-quarter crankshaft speed. Fuel injection was retained for the disc valve version, the injection nozzle passing through the center of the trunnion that carried the disc. The decision to place the disc valve model DB601 in production for the Luftwaffe's fighter planes and light bombers was made only after the experimental engine had completed a 200-hour, full-load test and it was proved that the disc valves were still in satisfactory condition.

The sealing of the disc valve was effected by an assembly of circular piston rings for the circular cylinder head ports and the disc carried an expansion sleeve. The two small piston rings were assembled outside the expansion sleeve and pegs on the sleeve fitted into the gaps of the rings to prevent the gaps from moving into line. This assembly was mounted in circular recesses surrounding the cylinder head ports. The flat upper face of the valve disc bore against the flat face of one of the piston rings and the inner expansion rings. End pressure on the disc was caused only by the part covering the valve port, the remainder of the disc being exposed to combustion pressure on both sides, therefore being in balance. A small metering valve, driven from the cylinder head side of the disc, regulated the supply of oil to the disc surfaces.

Daimler-Benz discontinued production and development work on the DB601 in 1944, but another company was still actively engaged in experiments on the Wankel disc valve—Junkers. The engineering business started by Hugo Junkers in 1893 had become a major factor in aviation when Junkers invented the all-metal airplane in 1910. In 1933 the Nazi government reorganized the Junkers industries for a variety of military production needs. Junkers was not interested in the Wankel disc valve for an aircraft engine, but for a torpedo engine.

Between 1937 and 1945 Junkers built and tested a number of experimental engines using Wankel's rotary valves. When World War II came to an end a production order had been issued for the manufacture

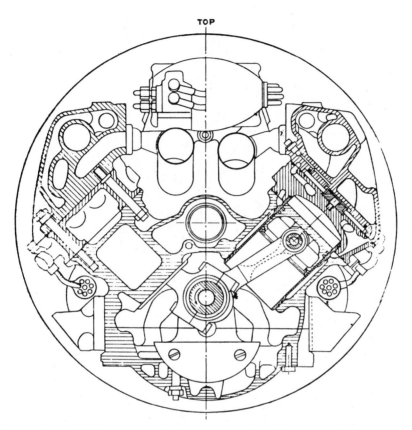

Junkers Jumo KM8 engine installed in its torpedo. This cross-section shows very well why there was no space for conventional valve gear.

of 100 JUMO KM8 engines designed for the special purpose of propelling torpedoes. This engine was considered by experts to represent progress in automotive development. High output, coupled with light weight, in the smallest package space is an important objective in automobile as well as torpedo design. The KM8 engine was required to perform only for a very short time under abnormal operating conditions. There is no doubt that with suitable changes in design, and under less adverse operating conditions, longer life could be expected.

Design objectives for the Junkers KM8 engine called for capability to drive the torpedo at 40 knots, which translated into a power requirement of 275 horsepower at 3,650 r.p.m. Minimum weight was essential, and therefore there was a strong leaning towards Junkers standard aircraft practice in design. The engine had to operate under full load immediately after start-up and accelerate to maximum speed in the shortest possible time. It was started by a compressed-air started motor and had

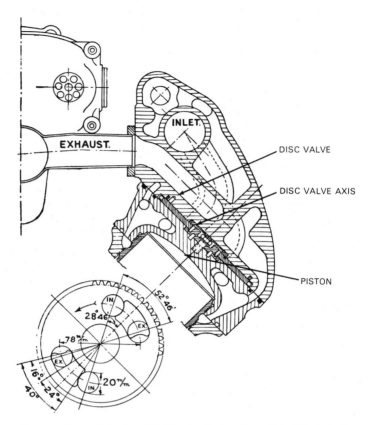

Cross-section of Junkers KM8 torpedo engine with Wankel disc valve. Plan view of the valve disc and its ports (lower left).

to reach maximum speed and power within one to two seconds from cold.

The fuel included a mixture of exhaust gas and oxygen, which meant that the engine had to be capable of withstanding higher temperatures and pressures than exist in gasoline engines. Because the required life of the engine was only a few hours, hard wear of individual parts could be tolerated and, because the torpedo was only 21 inches in internal diameter, the overall cross-section of the engine had to be extremely compact. Wankel's disc valve was chosen for this engine because the planned cylinder displacement of 260 cubic inches (4.3 liters) precluded the use of normal overhead valve gear in the available space.

The resulting KM8 was a liquid-cooled, four-stroke, V8 engine with magneto ignition, 90 mm. bore and 85 mm. stroke. Each cylinder had a displacement of 32.6 cubic inches (0.542 liters), resulting in 261 cubic inches (4.34 liters) displacement for the whole engine. The compression ratio was 6.6:1 and the cylinder heads were silicon light-alloy castings. The same material was used for the integral cylinder block and crankcase.

Each head was divided into an upper and lower section to accommodate the flat disc valves. The heads were attached to the cylinder block by means of studs which passed through both sections. Valve cooling was recognized as a particular difficulty, therefore the lower cylinder heads contained water passages to allow the coolant (sea water) to flow from the cylinder blocks to the upper head section and the upper cylinder heads had inlet and exhaust ports corresponding with those in the lower sections. The cylinder barrels were screwed into the underside of the lower cylinder head section. This type of joint required no further sealing arrangement. The lubrication system was conventional for the most part, although the disc valves were lubricated by a separate multi-piston pump which delivered a fixed quantity of oil to each disc and to each disc bearing.

The disc valves themselves had teeth cut on their peripheries similar to the disc valves on the DB601 engine, and formed two gear trains without any need of gear wheels or idlers. The valve discs were located between the upper and lower cylinder head sections. Cavities were machined in the upper sections to provide suitable clearances and two inlet and two exhaust ports per cylinder were cast into the lower cylinder head. All ports were circular and each had a steel sealing insert, held against the disc by the gas pressure in the cylinder. The outside diameter of the seal insert was made gas tight by two ordinary piston-type com-

Detailed section of the Wankel disc valve showing how critical the clearances were.

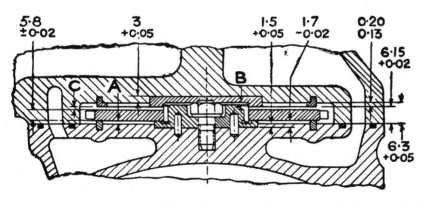

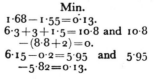

		Max.	Min.
Clearance at A.		$1\cdot7-1\cdot5=0\cdot2$	$1\cdot68-1\cdot55=0\cdot13$.
„	„ B.	$6\cdot35+3\cdot05+1\cdot55=10\cdot95$ and $10\ 95-(8\cdot7+1\cdot9)=0\ 35$	$6\cdot3+3+1\cdot5=10\cdot8$ and $10\cdot8$ $-(8\cdot8+2)=0.$
„	„ C.	$6\cdot17-0\cdot13=6\cdot04$ and $6\cdot04-5\cdot78=0\cdot26$	$6\cdot15-0\cdot2=5\cdot95$ and $5\cdot95$ $-5\cdot82=0\cdot13$.

pression rings. The upper sides of the lower cylinder head sections were machined out above the center of each cylinder to make room for the valve bearings, which were pegged to prevent rotation. A set of needle rollers reduced friction between the inner diameter of the disc and the outer diameter of the bearing. The circular opening through the sealing insert and the narrow section that carried the compression rings minimized the area under pressure and reduced friction. There were two port apertures in each disc, shaped to provide quick opening and closing. The gas pressures acting on the disc valves were counterbalanced by means of thrust bearing rings installed in the upper cylinder head sections.

The disc trains were driven from the crankshaft via pinion gears at the rear of the engine and bevel gears. The disc valve spindles were provided with splined couplings which allowed precise valve timing. Shear pins were fitted to the valve train gears to prevent damage of the discs and gears in case of disc seizure. The target output of 275 horsepower at 3,650 r.p.m. was obtained when running on a mixture of oxygen and exhaust gas with a supercharge pressure of 1.5 atmospheres. Running the engine on gasoline fuel without supercharging, the engine delivered 110 to 120 horsepower, which represented a specific output of only 0.423–0.461 horsepower per cubic inch (25.4–27.8 horsepower per liter).

When running on the oxygen mixture, there were problems with the disc valve bearings and the gas tightness of the sealing inserts. When running the engine on "free air," however, practically no sealing difficulties were encountered—the engine was run as long as 50 hours without any trouble. An experimental Junkers 210-S single-cylinder engine had similar disc valves and was reported to have run trouble-free for more than 200 hours.

During the tests several design changes were introduced. Two bronze thrust rings with provision for pressure lubrication were mounted in the upper cylinder head section to avoid wear on the discs because of excessive pressure loads directed against the lower surface of the upper cylinder head. As a precautionary measure, the discs were made thicker to strengthen the bearings.

About this time, Wankel's sliding seals also got into production. The Mercedes-Benz DB601 disc valve engine was used until 1944 and led to joint contracts between DVL, WVW and such companies as Auto Union, Hanomag and NSU. By 1945, Wankel had firm contracts also with Daimler-Benz and Borsig, Germany's largest manufacturer of railway rolling stock and other railroad equipment. Wankel himself spent the war years 1940–1944 in Berlin-Adlershof, engaged on a research program for DVL.

Wankel built an adjustable rotary compressor in 1944, but the end of the war interrupted his work. From 1945 to 1946 he was imprisoned by the French occupation forces. The Wankel Versuchswerkstätten was dismantled in 1945 and, even after his release, he was forbidden to conduct research and experimental work. This turned out to be something of a blessing in disguise, for it gave him time to review his work, write several papers, and produce initial designs for a rotary piston engine with his own sealing system.

The event that was to enable Wankel to make a fresh start was the release of a man named Wilhelm Keppler from the War Criminal's Prison at Landsberg in 1950. Keppler had been Hitler's business consultant, State Secretary without portfolio in the Nazi heyday. Keppler had known Wankel since they were both young men. When they first met in 1928, Wankel had voiced his ambition to ensure the greatness of the German nation by building a superior type of engine. He was, of course, referring to his rotary engine ideas. It was Keppler who had arranged, at the instigation of Hans Nibel, chief engineer of Daimler-Benz, to rescue Wankel from prison after Hitler had had him jailed. Keppler also protected Wankel in all ways during the years that followed. Thus, again aided by Keppler, Wankel opened his new establishment in 1951—the Technische Entwicklungsstelle in Lindau on Lake Constance, not far from the site of the original WVW.

Borsig, in Berlin, was among the first to give the new company a contract. The contract specifically covered sealing systems for air compressors. Wankel was also to do some work for Goetze, of Burscheid, in connection with improved piston ring sealing. Keppler, still extremely well connected within Germany industry, steered Wankel to NSU. They were then doing well with racing motorcycles, and in 1951 NSU took notice of Wankel's work on rotary valves, with the idea of using them on racing motorcycle engines. This resulted in an initial agreement between Wankel and NSU that was to be followed by a long line of contracts.

The technical aspects of the development of the workable Wankel engine are covered in more detail in the chapter dealing with NSU. It is interesting to note that Felix Wankel never went on the NSU payroll as an employee, but negotiated contracts with NSU for the rights to his engine and future research work. These contracts have since brought him considerable revenue. He gets a substantial share of NSU's income from the selling of licenses to manufacture and market Wankel engines, and he retains the right to engage in work for other clients as he wishes—he heads his own small establishment in Lindau.

Wankel always has been removed from the day-to-day testing of the engine he invented, and not at all concerned with the tooling required for its production. But, he has had the satisfaction of having the engineers from NSU come to him with their problems. He continues to experiment with new types of seals and with new geometrical shapes in his own laboratory. He has also done considerable research work on adapting his engine to diesel fuel (compression-ignition) operation. Under his research contract with Borsig in Berlin, Felix Wankel was still engaged on development of large rotary compressors as late as 1960, and, in December, 1969, he was awarded an honorary doctorate at the Technical Institute of Munich—late recognition of a lifetime devoted to the unwavering pursuit of one basic idea.

9

NSU Develops
the Wankel

As was mentioned previously, Felix Wankel's rotary valve designs created an interest at NSU. This, in turn, led to experiments with the Wankel compressor—the management felt that the device had potential as a supercharger for motorcycle engines. In 1954, several test units of various sizes were built, and proved remarkably effective. One test compressor delivered air to the engine at pressures up to 8 atü (120 psi), with an overall adiabatic efficiency exceeding 70%. A small NSU engine equipped with a Wankel compressor was installed in a motorcycle designed especially for setting speed records. The basic powerplant had quite humble origins—it was the 50 cc. unit that normally propelled the NSU Quickly Moped. The compressor was belt driven and delivered air at 40–45 psi to a pressurized carburetor, thus increasing the engine's power output to 13 horsepower from its stock 1.6—a power gain of 812%.

It was in these experimental compressors that Wankel's complete seal grid, the result of his long-time work on sealing systems, and the two-lobed epitrochoidal working chamber and triangular rotor were united.

The Wankel compressor consisted of two rotors, one inside the other, both located within a common stationary outer casing. The outer rotor had an epitrochoidal inner surface, which provided a working interface with the inner rotor. The inner rotor was shaped like a triangle with curved sides, with a port in each face, halfway between the apices. The central shaft did not rotate, but was hollow and served as an air intake duct. Using a hollow mainshaft was not Felix Wankel's idea; it was invented by one of his closest associates, Dipl. Ing. Ernst

Hoeppner, chief draftsman of Wankel's technical development center. The mainshaft carried a control cylinder which, as it turned through a certain angle, timed the opening and closing of the ports in the inner rotor. The shaft also carried an adjustable control shell piece which separated the atmospheric-pressure intake duct from the high-pressure outlet duct and effectively blocked off the port during compression. Air entered the intake duct from the end cover, was fed into the inner rotor and escaped through the face port into the epitrochoidal working chamber. Continued rotation brought the inner rotor to the point where the port was blocked by the adjustable control sleeve, signaling the end of the intake phase and the beginning of the compression phase. After maximum compression, the port was aligned with the outlet duct and the compressed air was released. To ensure adequate out-

Basic patent drawing covering the Wankel rotary compressor. The circle R lies inside the circle K. The numeral 2 is the radius of circle R and the numeral 3 denotes the radius of circle K. Wankel proposed to let circle K remain stationary and letting circle R revolve inside it. Allowing points D_1 and D_2, on extended radii, to roll with the circle R, produces the three-lobe hypotrochoid T. If the circle K is allowed to rotate around circle R, different curves result: a two-part hypotrochoid consisting of an outer lobe H_a and an inner lobe H_1.

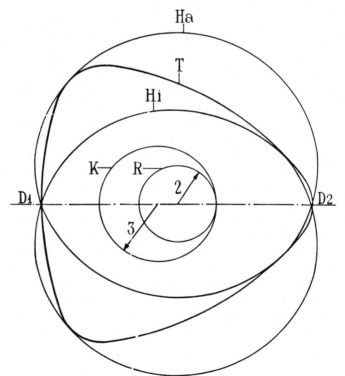

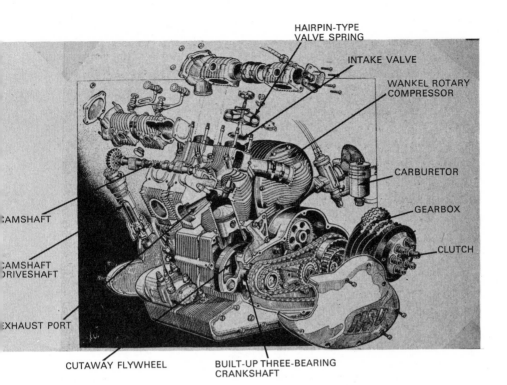

HAIRPIN-TYPE
VALVE SPRING

INTAKE VALVE

WANKEL ROTARY
COMPRESSOR

CARBURETOR

GEARBOX

CLUTCH

CAMSHAFT

CAMSHAFT
DRIVESHAFT

EXHAUST PORT

CUTAWAY FLYWHEEL

BUILT-UP THREE-BEARING
CRANKSHAFT

The 500 cc. two-cylinder engine used in the record-breaking motorcycle had a rotary
Wankel compressor. Bore was 63 mm. and stroke 80 mm. giving 499 cc. displacement. Power output was 110 net horsepower at 8,000 r.p.m.

let phase duration, the central control cylinder would swing along with
the inner rotor for about 10 degrees, then return to its normal position
in anticipation of the next intake phase.

With this design, one phase was in progress at all times on each of
the inner rotor's three faces. When one face completed an exhaust
phase, the next was about to complete its intake phase, and the third
was about two-thirds into its compression phase. The inner and outer
rotors revolved in the same direction. Both rotors had simple rotation
around the mainshaft center; however, their rotational speed differed
to produce the volumetric changes that make compression possible. The
speed ratio was 3:2, the inner rotor being controlled by gears in the
outer end cover. Every time the outer rotor made three revolutions, the
inner rotor revolved twice.

Of course, Wankel realized almost from the outset that the epitrochoidal chamber and the triangular rotor invented in March of 1954
was adaptable to operation on the four-stroke principle. The Wankel
compressor was, after all, only an air pump and, like all air pumps, the
addition of fuel and spark converted it into an engine. The develop-

The DKM-54 completely assembled, but partially cut away to show both rotors.

ment of such an engine was the next step that Felix Wankel and NSU were to take.

Experiments with port locations showed that the four-stroke cycle was, indeed, the most suitable mode of operation for this engine. A decision to concentrate engine development based on the four-stroke cycle was made by April 13, 1954. Wankel spent the following $3\frac{1}{2}$ years on intensive research and development of his engine. At this time he was assisted by a staff of NSU engineers, led by Dr. Walter Froede.

Drehkolbenmotor (usually shortened to DKM) is a German word which simply means "rotary piston engine." It was the name given to the motorized version of the Wankel rotary compressor, which was first tested in the NSU factory on February 1, 1957. At first, it could not be made to run under its own power. After some carburetor adjustments, the engine ran for several minutes—long enough to get a torque output reading. The Wankel engine was a reality.

In the course of his geometric studies, Felix Wankel came to recognize that epitrochoids with inner envelopes were the only type that promised a combination of realistic displacement for automotive engines and useful compression ratios. He arrived at the shape of the working chamber empirically and did not identify the curve described

Cross-sectional aspect of the DKM-54.

by the rotor apices during rotation. It was Professor Othmar Baier of the Technical College in Stuttgart who demonstrated that the shape was, in fact, an epitrochoid. This was of more than academic interest, because it was to facilitate mathematical analysis of the engine and simplify the working out of practical methods for machining the working surface.

Inner rotor of the DKM-54 carried spark plugs in its faces and had ports in its sides.

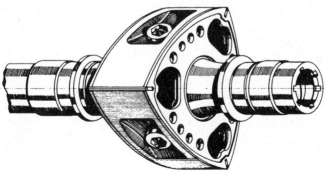

Geometrically, the DKM Wankel engine was identical to the rotary compressor, but there were notable differences in their mechanical make-up. For comparison with current Wankel engines, it is important to note that there was no eccentricity in the rotor motion of the DKM. The volume variations were due only to the relatively flat faces of the inner rotor. The output shaft was connected to the outer rotor. The outer rotor transferred torque—the inner rotor was incapable of such duty because it did not run eccentrically on its shaft.

The first development engine, DKM-54, was a very complicated design, built to test the trochoidal concept and a variety of proposed sealing systems. The number 54 was simply an indication of its chamber volume, measured in cubic centimeters. At this stage, no thought was given to production methods and costs—the object was to establish feasibility and obtain confirmation of the basic theory.

The fresh mixture entered through a carburetor at the side of the engine and was admitted to the working chamber via the hollow shaft of the inner rotor, intake ports in the inner rotor, and ports in the inner rotor faces. The spark plugs were mounted centrally in the faces of the inner rotor and revolved with it. The spark plug location ne-

Elevation of the first Wankel engine, the DKM-54.

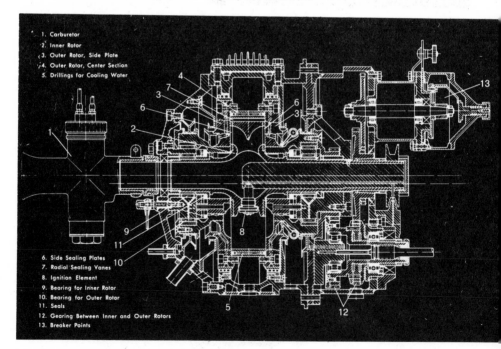

1. Carburetor
2. Inner Rotor
3. Outer Rotor, Side Plate
4. Outer Rotor, Center Section
5. Drillings for Cooling Water

6. Side Sealing Plates
7. Radial Sealing Vanes
8. Ignition Element
9. Bearing for Inner Rotor
10. Bearing for Outer Rotor
11. Seals
12. Gearing Between Inner and Outer Rotors
13. Breaker Points

cessitated slip-ring electrical connections, and made cleaning or replacing the plugs a major operation.

One major difference between the compressor and the engine was that
the rotor faces had no ports, the intake charge was led through side
ports in the inner rotor into channels in the outer rotor. Admission was
controlled by the relative movement of the two rotors and the exhaust
port was located in the periphery of the outer rotor. Sliding speeds of
the apex seals in the DKM-54 varied between 3,700 and 9,600 feet per
minute at 15,000 r.p.m. Sliding speeds in such an engine are highest
around the major axis and lowest around the minor axis—the *average*
sliding speed in the DKM-54 was 6,700 feet per minute.

Tests of the DKM-54 proved feasibility beyond a shadow of doubt—
the validity of the fundamental principles had been established. While
Wankel continued to study all possible geometrical configurations in an
endless search for the ideal shape for both rotor and working chamber,
the NSU engineers under Walter Froede concentrated on developing
the existing concept.

Among the early Wankel engine configurations examined was a design with a three-lobe working chamber and a four-lobe rotor. This ne-

NSU Wankel KKM-125

cessitated a six-phase operational cycle, and the idea was rejected. Development concentrated on the two-lobe chamber with the triangular
rotor because the two-lobe design provided shorter overall sealing length

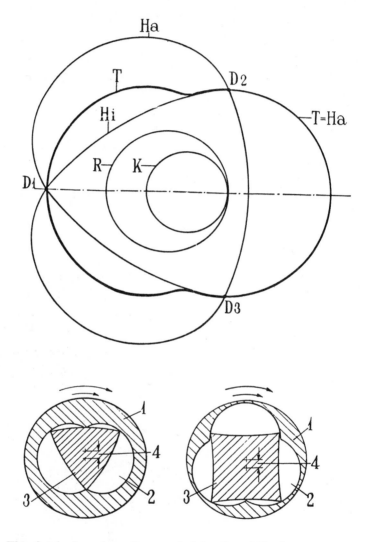

This sketch shows how the use of circles R and K, alternating in
the roles of base circle and generating circle, can create both
two-lobe and three-lobe epitrochoids.
Below: Proposals for inner rotors revolving in the same direction
as the outer rotor with the inner rotor doing two-thirds the speed
of the outer rotor at left and three-quarters the speed of the outer
rotor at right.

for a given chamber displacement than other configurations having more lobes. This proved to be a great advantage in reducing the size and number of leakage paths.

In mid-1957, a new and larger Wankel engine was designed and built—the DKM-125. Its chamber volume was 125 cc. (7.6 cubic inches). Ultimately, its power output was raised to 28.4 horsepower at 17,000 r.p.m. (outer rotor and output shaft), which equalled 11,300 r.p.m. at the inner rotor. The inner rotor had a 65 mm. radius and a width of 38 mm. The R/e ratio was 6.85:1, which allowed compression ratios up to 17.6:1.

Spring-loading was not required for the apex seals, because the apex seal swing angle was limited to 26 degrees. The DKM-125 was fully balanced and did not have any mechanical vibration. Bearing loads resulted from gas pressure only, because both the inner and outer rotors revolved around their own centers of gravity. Both rotors revolved in the same direction and were connected by phasing gears that positioned the two rotors but had nothing to do with transmission of the engine torque. The outer rotor was made of nitrided steel but, despite the use of high-grade metal, it tended to distort from true epitrochoidal shape under centrifugal force. The rotating inertia in the outer rotor was so high that it had a detrimental effect on the engine's ability to accelerate—it wanted to run at a constant speed. The inner rotor, instead of being one complete casting, was built up from a number of components, including separate side plates and a variety of seals and seal parts. Initially, the inner rotor had no provision for cooling (the outer rotor was water-cooled). Later, water cooling was added to the inner rotor—with considerable complication.

Port sealing was a big problem because of the high rubbing velocity of the outer rotor. It was found to be almost impossible to make a satisfactory gas seal for the exhaust port, where high gas temperature aggravated the already unfavorable conditions. The problem also was aggravated by the fact that friction losses in transmissions increase with a rise in gear rotational speed, and the output shaft revolved at outer rotor speed. The stationary nature of the mainshaft added to the difficulty of providing a suitable transmission. Efforts were made to limit the operational speed without engendering unacceptable sacrifices in power output.

Volumetric efficiency was found to be 98% at 7,000 r.p.m. and 70% at 16,000 r.p.m. The DKM was tested at speeds up to 25,000 r.p.m. in order to prove the mechanical safety of the engine. Peak torque was reached at 8,000 r.p.m. and peak mean effective pressure was 120 psi. The engine ran most economically at 12,000 to 14,000 r.p.m.

The DKM-125 had been designed without regard to installation problems and transmission requirements. The dual rotation made a stationary
outer casing necessary if the engine was to be installed in a vehicle.
This would have added substantially to the weight and cost, and the
obvious course was to redesign the power unit with vehicle installation
and power transmission in mind. This work was carried out by Dr.
Froede and the research staff at NSU, the result of which was a concept
that made the Wankel engine truly practical. Dr. Froede discarded the
idea of adding an extra casing and concentrated on redesigning the

Walter Froede shows off the KKM-125.

DKM with a stationary outer housing. In other words, he decided to prevent the outer rotor from rotating, and to confine rotation to the inner rotor and the shaft.

This line of thought led him to the invention of what he called "kinematic inversion" and to the introduction of eccentric rotor movement. These principles were first combined in a new engine called the KKM. KKM stands for *Kreiskolbenmotor*, which means "circuitous piston engine."

One of the versions of Froede's KKM, shown in his basic patent covering the kinematic inversion.

Aug. 2, 1960   W. G. FROEDE   **2,947,290**

HEAT GENERATING ROTARY INTERNAL COMBUSTION ENGINE

Filed Nov. 18, 1958   12 Sheets—Sheet 12

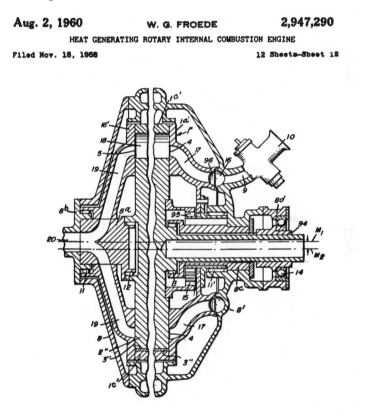

INVENTOR
WALTER G. FROEDE

BY

ATTORNEYS

The kinematic inversion consisted of transforming the outer rotor of the DKM into a stationary housing and mounting the inner (and only) rotor on an eccentric. The mainshaft also became the output shaft, and, due to the use of a stationary block, the installation problems associated with the DKM were eliminated. To keep the rotor in phase, a reaction gear was shrunk into the end cover. Its function was merely to keep the rotor in orbit. Reaction loads were low, since the gears carried no torque. Theoretically, no forces acted on the phasing gears except

One of the configurations covered by Dr. Froede's patent is this DKM with a two-lobe inner rotor and an almost circular outer rotor, eccentrically mounted in relation to each other.

Aug. 2, 1960   W. G. FROEDE   **2,947,290**

HEAT GENERATING ROTARY INTERNAL COMBUSTION ENGINE

Filed Nov. 18, 1958   12 Sheets—Sheet 10

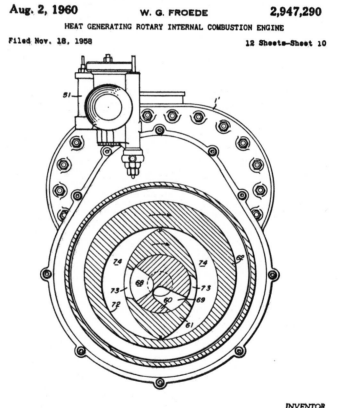

INVENTOR
WALTER G. FROEDE

ATTORNEYS

very slight friction loads and effects of random variations in engine r.p.m. The annular gear on the rotor's inner surface had 45 teeth. It was meshed with a stationary 30 tooth reaction gear locked into the end cover to give a 2:3 ratio with the rotor.

Dr. Froede's first KKM was a 125 cc. single-rotor unit, built and tested in 1957. The KKM-125 weighed 37.4 pounds with a cast-iron housing,

Froede's patent covered multi-lobe rotors and epitrochoids with up to six lobes.

Aug. 2, 1960   W. G. FROEDE   **2,947,290**

HEAT GENERATING ROTARY INTERNAL COMBUSTION ENGINE

Filed Nov. 18, 1968   12 Sheets—Sheet 11

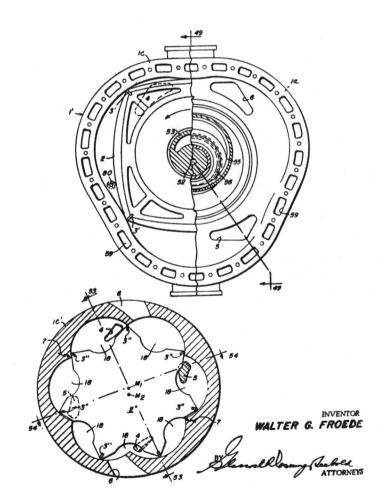

INVENTOR
WALTER G. FROEDE

BY
ATTORNEYS

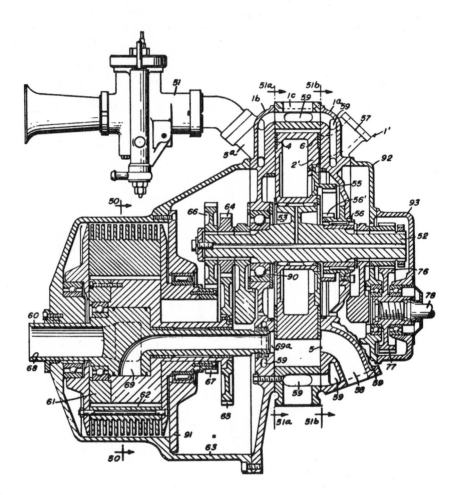

INVENTOR
WALTER G. FROEDE

BY
ATTORNEYS

An elaborate version of a KKM with side ports, patented by Dr. Froede in 1960.

Closeup of the KKM-125, with one end cover cut away to show the rotor.

and only 23.2 pounds with an aluminum housing. This engine represented the first combination of Felix Wankel's sealing grid, the epitrochoidal working chamber, and *eccentric* rotor movement, and was the basic Wankel engine as we now know it.

The invention of the KKM led to simplified breathing and ignition systems. The mixture was fed into the working chamber through a manifold and ports in the housing, instead of through the shaft as in the DKM. The spark plugs were removed from the rotor faces and installed in the housing (made stationary) which had great operational advantages for the ignition system.

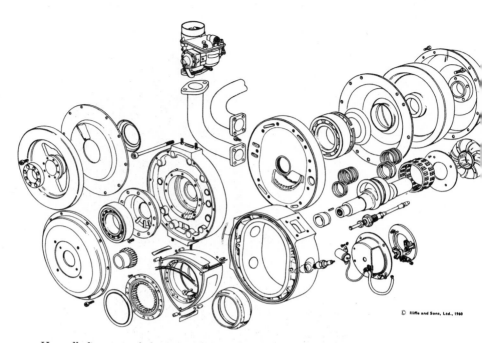

Here all the parts of the KKM-125 are placed in proper relationship to each other, spaced out to show each individual part more clearly. (© *Iliffe Technical Publications*)

At first, Felix Wankel strongly opposed the KKM because he felt that the introduction of eccentricity to the movement of the rotor placed his entire invention in jeopardy. He insisted that his sealing system, as applied to the DKM-125, was the basis of the success of the early prototypes, and he did not approve of the sealing modifications necessary for KKM operation. The eccentricity of rotor movement gave rise to fluctuations in apex seal loadings, including negative centrifugal force loading every time an apex approached the minor axis, while the apex seals in the DKM engine operated under stable conditions with unidirectional centrifugal loads. When it turned out that there were no fundamental problems with the sealing system of the KKM-125, Wankel was mollified and began joint experimental work with NSU on new KKM designs.

The sealing grid which Dr. Froede developed for the KKM-125 was both simple and effective. The corner seals brought the apex seals in direct contact with the side seals. The trunnion which carried the apex seal had two small segments cut from its end to allow access to the apex seal for the side seal ends. The trunnions provided a longer leakage path for gases trying to escape and thereby slowed down the leakage gas flow. Apex seal angular swing was restricted to between 20 and 30 degrees per side. This cured 90% of the oil problem. When the oil seals on the eccentric were relieved of pressure, serious leakage stopped. Improving the oil seals themselves was not regarded as a high-priority matter. One of the biggest parasites to power output was found to be the turbulence of the oil inside the rotor. NSU instituted a new line of research in order to achieve better control of the oil flow. This led to the development of a phased oil extractor inside the rotor, extending to each apex.

Another problem, rotor bearing seizure, was cured by adopting needle roller bearings in place of the plain bearings. This led to a complex assembly, which was in turn replaced by two tracks of double rollers,

The housing for the KKM-125 had water passages contoured for heat distribution differences. Both ports were peripheral. The spark plug was positioned almost exactly at the minor axis. (© *Iliffe Technical Publications*)

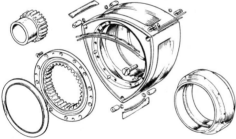

This exploded view gives clear details of the rotor sealing system used on the KKM-125. The rotor's inner gear was a separate piece which bolted in, meshing with a stationary reaction gear. (© *Iliffe Technical Publications*)

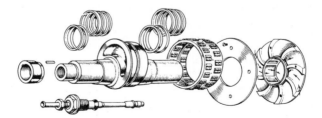

The rotor runs on a roller bearing carried on the sculptured eccentric, which was integral with the main shaft. (© *Iliffe Technical Publications*)

staggered in relation to each other. Later, NSU reverted to plain bearings, after it had been established that the type of rotor bearing used had not been the principal cause of the failures. The bearings froze not because of defective bearing operation but because of centrifugal distortions in the rotor. The rotors that suffered from this trouble were made of light alloy. At that time, light alloy was considered preferable to cast-iron because of its light weight and high conductivity. It had been adopted because it was thought that rotor temperature distribution would be very even and no distortion problems would crop up. The rotor was cooled not only by the oil inside, but also by the incoming air-fuel charge, and to a small extent by the apex seal tips through their contact with the water-cooled working surface. The spark plug hole in the housing presented an interruption in the surface swept by the apex seals. This was no problem in terms of seal durability and sealing effectiveness, but it did mean a high risk of local overheating of the housing metal around the spark plug hole. That in turn led to increased risk of cracking in that part of the housing. The problem ultimately was solved by developing a cooling system with higher effectiveness in high-temperature areas and lower effectiveness where little cooling was needed.

When Dr. Froede was satisfied with the state of development of the KKM-125, he designed a new engine with twice the displacement, the

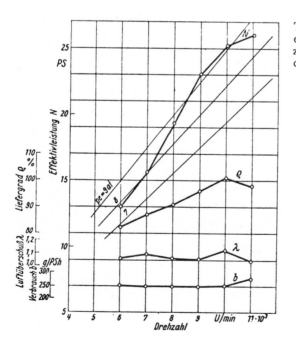

Test data on the KKM-125 engine. N = horsepower. Drehzahl = r.p.m. Verbrauch = fuel consumption.

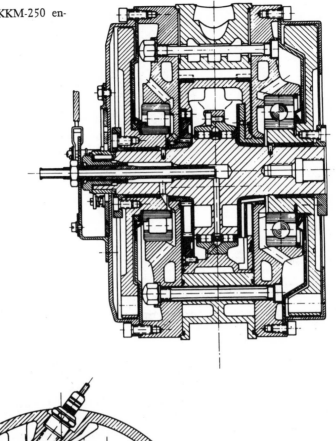

Elevation of the KKM-250 engine.

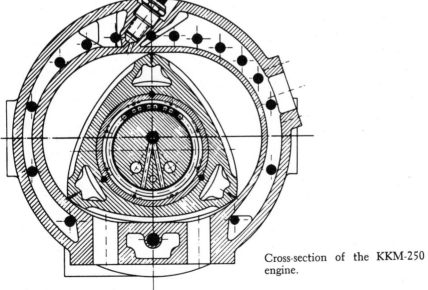

Cross-section of the KKM-250 engine.

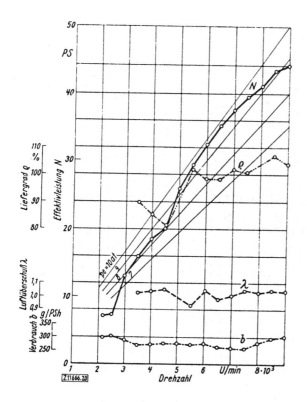

Test data for the KKM-250 engine. N = horsepower. P_e = mean effective pressure. Verbrauch = fuel consumption. Drehzahl = r.p.m.

KKM-250, drawing fully on all the lessons learned from experimental work with the earlier model. It was designed as a test engine to establish the best spark plug position and to explore new combustion chamber configurations. The mainshaft ran in one plain bearing on each side of the rotor, while the rotor ran on plain bearing shells on its eccentric. (The initial design carried the rotor on a needle roller bearing, but was later modified to use a four-race roller bearing.) Peripheral ports were used, and lubricating oil was brought to the bearings through the partly hollow mainshaft. Radial channels in the eccentric then led the oil into the rotor for cooling.

The KKM-250 was about 9.5 inches in outside diameter, with a width of 7 inches. The two peripheral ports were positioned fairly close together near the minor axis, which produced high phasing overlap. An aluminum rotor was used, and the compression ratio was 8.0:1. The spark plug position was 20 mm. behind the minor axis. The engine was equipped with a 32 mm. Solex carburetor and the flywheels were fixed to the output shaft entirely by the friction of the cone clusters set up by end thrust loads. An R/e ratio of 7.77:1 was used, which offered the possibility of running with compression ratios up to 19.5:1. The maximum apex seal swing angle was 23 degrees, rotor radius was 85.5 mm.

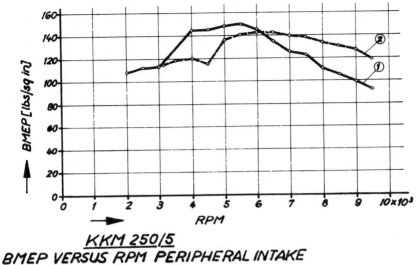

KKM 250/5
BMEP VERSUS RPM PERIPHERAL INTAKE
PORTS OF VARIOUS SIZE
1 = 22mm DIAM   2 = 18x27mm RECTANGULAR

Mean effective pressure graph for the KKM-250 with two different size intake ports, throughout a 2,000 to 9,500 r.p.m. speed range. Port 1 was circular with a diameter of 22 mm. Port 2 was rectangular, 18 x 27 mm.

Fuel consumption plotted against mean effective pressure in the KKM-250 at a constant speed of 5,500 r.p.m., with a peripheral intake port of 22 mm. diameter.

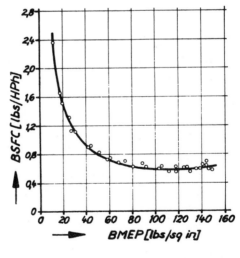

KKM 250/5

BSFC AT PART LOAD
CONSTANT SPEED n=5500RPM
PERIPHERAL INTAKE
22mm DIAM

and eccentricity was 11 mm. As in all NSU engines, rotor width was about four times the eccentricity.

The KKM-250 underwent endurance testing in July, 1959. These tests confirmed the basic soundness of both concept and design, and successful 1,000-hour tests were completed by 1960. Another series of tests were made to find ways to reduce material and machining costs, with particular reference to the trochoidal bore and the finish of the end-covers.

This graph shows the effect of variations in air-fuel ratio on specific fuel consumption and mean effective pressure in the KKM-250 engine.

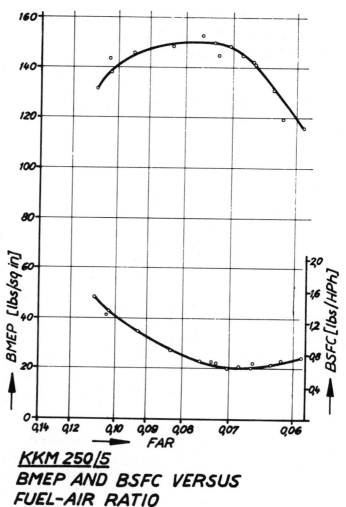

KKM 250/5
BMEP AND BSFC VERSUS
FUEL-AIR RATIO

The KKM-250 had an installation weight of 48.4 pounds. Power output rose in an almost straight line—10 horsepower at 2,700 r.p.m., 20 horsepower at 4,300 r.p.m., 30 horsepower at 5,500 r.p.m., 40 horsepower at 7,600 r.p.m. and a peak of 44 horsepower at 9,000 r.p.m. Peak torque (28 foot pounds) was developed at 6,000 r.p.m. Mean effective pressure rose in waves from 80 psi at 2,500 r.p.m. Above 6,000 r.p.m. pressures remained high, falling off slightly to 125 psi at 9,000 r.p.m. Volumetric efficiency was 100% or better at all speeds above 5,500

This graph shows the effect of changes in ignition advance on specific fuel consumption and mean effective pressure in the KKM-250 engine.

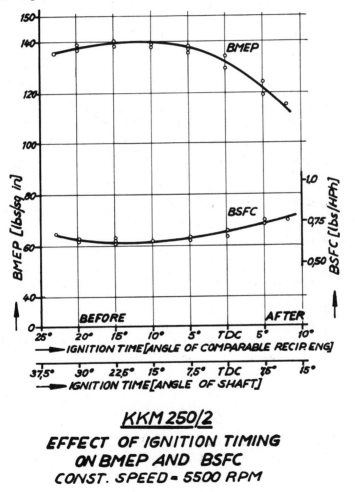

KKM 250/2

EFFECT OF IGNITION TIMING
ON BMEP AND BSFC
CONST. SPEED = 5500 RPM

r.p.m., the low point was 82% at 4,500 r.p.m. At 3,500 r.p.m., the engine breathed at 90% of capacity. Specific fuel consumption was almost constant at .675 pounds per horsepower hour from 3,500 r.p.m. to 7,500 r.p.m. Above 7,500 r.p.m., however, fuel consumption rose sharply to 0.78 pounds per horsepower-hour.

NSU used the KKM-250 to test hundreds of different shapes and materials for both single-piece and complex apex seal assemblies. With some systems, complete failure occurred after only 15–20 minutes running; other systems that were identical geometrically, but had different material combinations, survived endurance tests of 1,000 to 2,000

This graph shows how exhaust gas temperature in the KKM-250 rose with an increase in r.p.m.

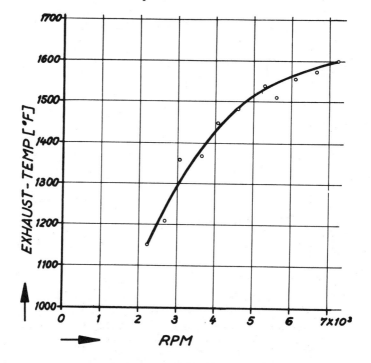

KKM 250-7

EXHAUST TEMP AT FULL THROTTLE

hours duration. The apex seal wear rate, originally 0.0018 inches in 400 hours, was much reduced by better combinations of materials for the seal and cylinder. Seals made from materials with very high wear resistance caused considerable wear on the working surface, while a working surface of great hardness wore the seals rapidly. NSU decided in favor of softer seals and harder working surfaces, reasoning that it was better to fit new apex seals than to replace the much more expensive and complicated housing which contained the epitrochoidal surface.

This graph shows how heat transfer to the cooling media (water and oil) varied according to r.p.m., relative to the heat value of the KKM-250's output.

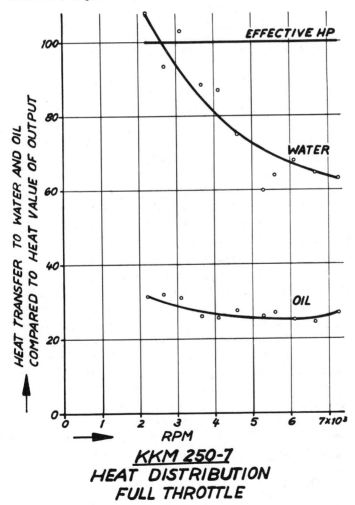

KKM 250-7
HEAT DISTRIBUTION
FULL THROTTLE

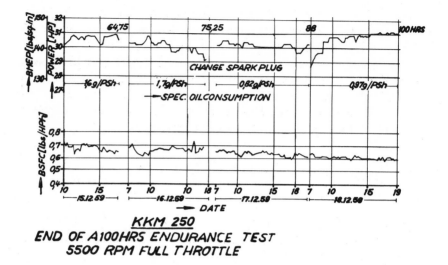

KKM 250
END OF A 100HRS ENDURANCE TEST
5500 RPM FULL THROTTLE

This graph shows how mean effective pressure and specific fuel consumption patterns evolved during a 100 hour endurance test of the KKM-250 engine.

The magnitude of the forces that worked to circulate the oil inside the rotor was first established on a KKM-250. In one instance, a small foreign body was left within the rotor. It wore away so much of the rotor wall near an apex seal that a failure occurred and a considerable amount of oil was lost. NSU then began to investigate how this natural circulation could be best utilized to cause oil to flow to and from the rotor without the assistance of pressure-feed or scavenge pumps.

NSU also developed a way to study flame propagation. The method was based upon conveniently spaced ionization detection plugs. A special KKM-250, with five detection plug positions, was prepared for this investigation. The detection plugs were interchangeable with the spark plugs. Position 1 was 56 mm. before the minor axis, position 2 was 24.5 mm. before the minor axis, and position 3 was 45.5 mm. after the minor axis. The fourth position was 77 mm., and the fifth position 154 mm., after the minor axis. Separate tests confirmed that the beginning of ionization coincided with the arrival of the flame front. Quartz windows with photocells behind them were installed to observe the combustion process, then tests were made at various speeds and with various compression ratios with the spark plug mounted in one of the five optional positions. With rising rotor speed, gas velocity increased, as did flame front velocity. Consequently, the combustion process was by no means as slow as the elongated combustion chamber shape indicated. Dr.

Froede concluded that flame speed was superimposed on the tangential gas velocity as the mixture left the trailing part of the working chamber past the minor axis and entered the expanding, leading portion. Gas velocity was almost invariably greater than the speed at which the flame could spread. The travel time for the flame front varied considerably depending on the air-fuel ratio.

In this endurance test, the KKM lost power as the apex seals wore out. When new apex seals were installed, full power was restored.

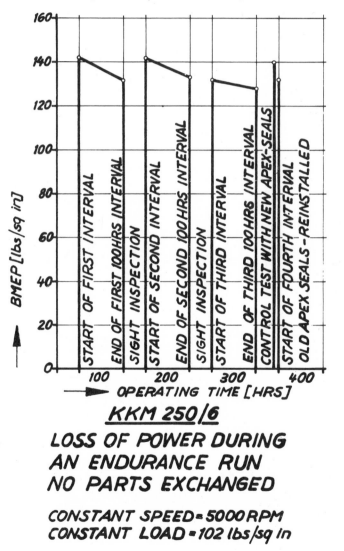

Many tests were performed to find the best spark plug position and the highest practical compression ratio. Compression ratios varying between 6.8:1 and 12.6:1 were used with a symmetrical combustion chamber in each rotor face. The effects of placing the spark plug in two alternative positions in front of the minor axis, and similarly, in two

Wear of apex seals on the KKM-250 (thousandths of an inch) in a 300-hour endurance test. Only the seal tip in contact with the epitrochoidal surface has significant wear. The sides of the apex seals wear as well as the rotor side seals, on completion of an initial 200-hour period of breaking-in.

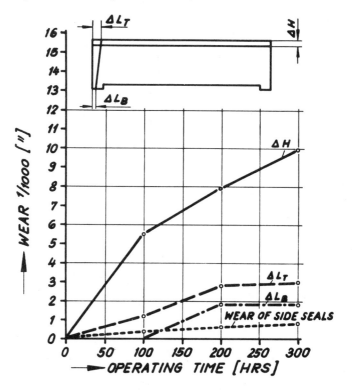

KKM 250/6

WEAR OF SEAL-COMPONENTS
DURING ENDURANCE RUN
CONST. SPEED = 5000 RPM
CONST. LOAD = 102 lbs/sq in

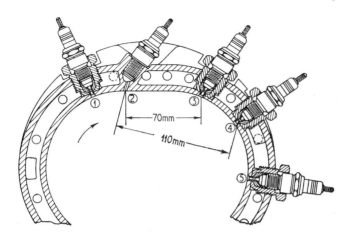

NSU devised a special experimental casing for the KKM-250, with a number of holes for ionization detection plugs to be used for a study of flame propagation. The detection plugs are threaded in just like spark plugs, and are interchangeable with the spark plug. Here the spark plug is in position 2 (just ahead of the minor axis).

positions after the minor axis, were tested. Until the onset of knock both spark plugs were fired simultaneously, and power output, efficiency and fuel consumption curves were plotted for the whole speed range and for lean mixtures. Maximum mean effective pressure was obtained with compression ratios between 9:1 and 11:1, while minimum fuel consumption was obtained with a compression ratio of 9.5:1. The highest pressure at the lower compression ratio was obtained with two spark plugs firing simultaneously, while at higher compression ratios the best performance was achieved with a single spark plug in front of the minor axis. Dr. Froede concluded that a single spark plug placed after the minor axis allowed lower pressure and ignited air-fuel mixtures of varying ratios with greater certainty.

Tests with multiple spark plugs failed to show conclusive evidence that the combustion process could be accelerated by multiple ignition, but it seemed that power output could be raised by about 5% by installing two spark plugs spaced at certain intervals. Extensive experimental work was necessary before satisfactory low-speed operation and idling were attained. Finally, it was possible to maintain 600 r.p.m. output shaft speed, at a corresponding rotor speed of 200 r.p.m., for many hours without irregular ignition and combustion.

The experimental department of Goetzewerke, the leading piston ring manufacturer in Germany, played an important part in the development of the various oil sealing principles. One type of oil seal was so effective that even after 200 hours running the amount of oil lost was too small to measure. At the end of 500 hours, the combined losses of the lubrication and rotor cooling systems amounted to less than 0.00441 pounds per horsepower-hour.

Cooling oil circulating through the rotor also wets the end-cover surfaces, and thus lubricates the side sealing strips and joint trunnions. Until 1963, 1–2% lubricating oil was added to the fuel to ensure lubrication of the apex seals. Then NSU realized the advantage of metering the oil separately into the cylinder and arranging for the metering control to respond to variations in load and speed. One KKM-250 engine with a cast-iron housing and nitrided working surface underwent a 500-hour endurance run at a constant 5,500 r.p.m. with the fairly high mean effective pressure of 130.8 psi. For this test, 2% lubricant was added to the fuel. A test of 2,000 hours was successfully completed by a similar engine. Shaft speed was maintained at 5,500 r.p.m., with a mean effective pressure of 105.8 psi. Although chatter marks appeared on the working surface, to a depth of 0.0031 to 0.0035 inch, regrinding the surface restored the original output characteristics. The longest full-load test before 1963, at a mean effective pressure as high as 147 psi, lasted for 700 hours. By reducing the mean effective pressure to 105.8 psi, an endurance test of 1,400 hours was successfully completed. By then, the cast-iron seals were worn out, but the side-sealing elements and joint trunnions remained good for twice this period.

As a passenger car manufacturer, NSU naturally pursued Wankel engine research and development with automobile applications in mind, but the company remained sensitive to other potential areas of application, and designed and tested many units far below practical motor vehicle size. For example, there was the air-cooled KKM-60 unit, de-

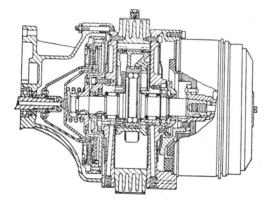
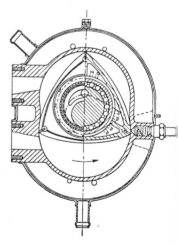

Elevation and cross-section of the KKM-150.

Test data on the KKM-150 engine. PS = horsepower. g/PS-hr = grams per horsepower-hour. Spec. fuel cons. = specific fuel consumption. lb./in.^{2} = pounds per square inch (psi).

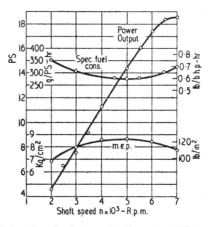

signed to run with a lubricant added to the fuel in a 1:40 ratio. This lubricant proportion proved adequate for all bearings and sealing elements. The engine developed about 5 horsepower at 6,000 r.p.m. and its axial fan supplied sufficient air for adequate cooling of the end covers and central casing. Another, the KKM-150W engine, was water-cooled and capable of developing between 12 and 20 horsepower according to the cross-sectional area and the timing of the ports. The rotor of this engine was, of course, oil-cooled. Another version of this engine was developed specifically for marine applications.

The shape and proportions of the epitrochoidal surface of the KKM-150L engine were identical to those of the KKM-150W engine. This engine, however, was air-cooled. After preliminary studies, preference was given to circumferential cooling fins and a radial blower. The engine proved capable of a continuous power output of about 12 horsepower with housing temperature limited to 200°C. The weight allowance of the KKM-150 was as follows:

Housing light alloy	2.38 lbs.
Rotor, complete with seals	2.27 lbs.
End cover—pump take-off end—complete with oil pump, race and connectors	2.27 lbs.
Eccentric and bearings	1.87 lbs.
End cover—dynastart end—with pinion and pipe connectors	4.19 lbs.
Flywheel—output end	4.96 lbs.
Miscellaneous small parts	1.32 lbs.
Total	19.26 lbs.
Dynastart, blower and balance weight	14.73 lbs.

It was found that friction losses varied almost linearly between 2,000 r.p.m. and 6,000 r.p.m.; the corresponding mean effective pressures for these speeds were 22.05 psi and 36.75 psi. The change in the rate of oil

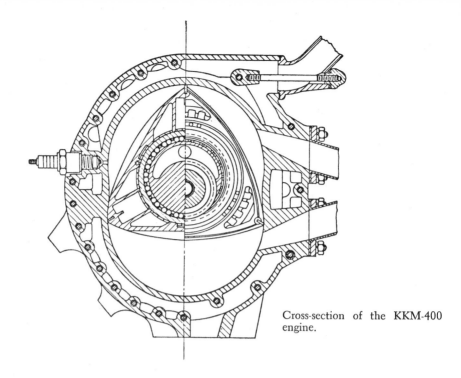

Cross-section of the KKM-400 engine.

Elevation of the KKM-400 engine.

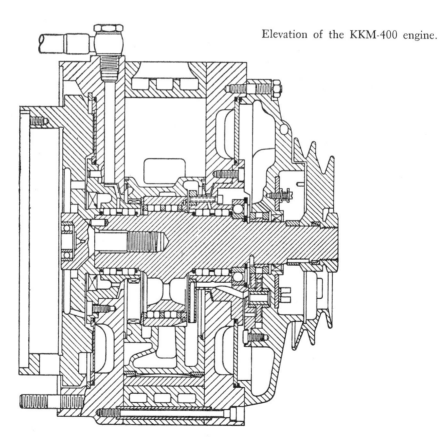

circulation was virtually linear—between 2.64 and 6.16 pounds per minute. Power output exceeded 24 horsepower at 8,000 r.p.m. However, this potential power output never was fully exploited because the port sizes and timing were altered, which, in conjunction with the installation of an air filter and an exhaust muffler, considerably reduced the effective power output. A version of this unit for marine propulsion developed over 18 horsepower at 7,000 r.p.m. Between 3,000 and 6,500 shaft r.p.m. the mean effective pressure exceeded 117 psi and the fuel consumption was 0.691 pounds per horsepower-hour.

Test results obtained with the KKM-400. lbs./HPH = pounds per horsepower-hour. BSFC = specific fuel consumption.

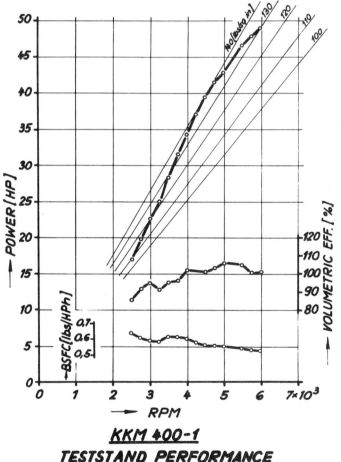

KKM 400-1

TESTSTAND PERFORMANCE
WITHOUT MUFFLER

Normal high-temperature spark plugs were used extensively for the development of the engine. Special plugs, with long central electrodes that reached close to the working surface, were also very successful. These were superior for starting, idling, and running under no-load or negative-load conditions. A variant of this type plug was developed in collaboration with Robert Bosch G.m.b.H. and it gave good starting and idling characteristics, in addition to having a life of more than 200 hours under full load. This experimental plug was chosen for development into a production design. Ignition by a standard contact breaker assembly was used, but experiments also were conducted with transistorized electronic and piezoelectric systems.

But it was the KKM-400 that was to lead NSU directly to the first Wankel-powered automobile in the world. The KKM-400 engine had a single chamber with a displacement of 400 cc. (24.4 cubic inches). Its general configuration was identical to that of the KKM-150 unit. The KKM-400 was shown to have a more favorable surface-to-volume ratio than its predecessor, the KKM-250. It had a ± 28 degree angle of obliquity, as opposed to the 23 degree angle of obliquity of the KKM-250. Power output varied between 40 and 50 horsepower. Differences in power output and fuel consumption could be obtained by simple changes in the carburetor settings to give either maximum power output or minimum fuel consumption. The lowest fuel consumption, 0.487 pounds per horse-power-hour, was recorded at 4,000 r.p.m—this increased to 0.536 pounds per horsepower-hour at maximum output.

By 1962, the engine had reached a high level of efficiency. If fuel heat input is defined as 100%, it was found that up to 28% of this energy was converted to useful work. Heat lost to the cooling oil remained practi-

These heat balance charts compare the KKM-250 and the KKM-400. The larger unit had higher heat losses to the exhaust, but lower heat losses to the housing cooling system, yet it had more of the fuel's heat energy available for useful power.

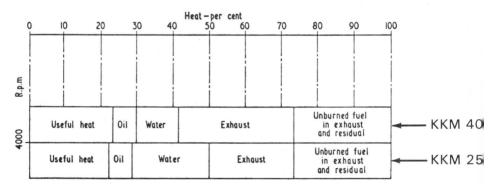

cally constant at about 7%. Within the normal performance range of the engine, 13% of the heat was lost to the cooling water and 36% expelled with the exhaust gases. The amount of heat lost by convection and radiation was not determined readily; however, it was thought to be about 2 to 5%.

Tests with the KKM-250 and KKM-400 engines showed that there were notable differences in the relationship between cooling water temperature and exhaust gas temperature. The temperatures were very close in engines with high R/e ratios, such as the KKM-250, but the coolant temperature remained about two-thirds the exhaust gas temperature in engines with low R/e ratios, such as the KKM-400. Both engines were tested under full load at 4,000 r.p.m. shaft speed, with similar air-fuel ratios. Dr. Froede concluded that the KKM-400 required a smaller radiator, when installed in a car, than did the KKM-250. Other differences were that the KKM-400 ran with slightly higher oil temperature but derived more useful energy from the heat value in the fuel.

Some experimental engines suffered rotor and gear failures. Several research teams measured the destructive forces present and attributed them mainly to shock loads created when the pressure rose during combustion and to shock waves set up when the exhaust port was uncovered. Other possible causes were gas pressure acting on the rotor, and rotational forces. None of the early NSU Wankel engines had any of the cold-starting problems that the experts had predicted. Dr. Froede reported in 1963: "Contrary to popular prediction when the engine was first introduced, there has never been any starting problem either at normal ambient temperatures or during very cold weather. Slightly higher rotational speeds may have been required for starting, but, on the other hand, the breakaway torque is decidedly lower than that of the conventional reciprocating piston engine. Consequently about the same power is required for starting the two types of engines. A KKM-400 engine starts at 130 to 150 shaft r.p.m. and a 0.9 h.p. starter motor fitted to an experimental installation in a car has proved entirely satisfactory under all conditions encountered over a very long period."

There was only one problem to which NSU had no possible solution in sight by 1962—seal tip wear. Tip wear on the apex seals was so critical that in July of 1962 NSU could not run an engine for more than 100 hours at 5,000 r.p.m. without replacing the apex seals. Between that date and January, 1963, a series of new materials was tested—and rejected. In May and June new carbon compounds were developed, and some ran over 250 hours at 5,000 r.p.m. Wear was within 10–15% of the previous year's engines. Further development work led to seals that wore 2–3% of the 1962 wear rate, before production of the KKM-502 engine began.

This was confirmed in repeated tests exceeding 200–220 hours; after January, 1965, tests up to 700 hours showed even more wear resistance, and seal tip wear was finally judged "satisfactory."

Up to 1961, all NSU experimental engines had been built on the single-rotor principle. When plans were formulated for a 1962 production car with a new Wankel engine, it was felt that a chamber volume of 500 cc. would be sufficient to meet the performance goals of a special sports version of the Prinz. But there was considerable discussion as to whether the 500 cc. should be all in one chamber or split between two working chambers in a twin-rotor arrangement. NSU built its first twin-rotor Wankel engine in 1961. It was based on the KKM-400, having the same geometry, but had narrower chambers of 300 cc. displacement each. This KKM-2 x 300 was designed to develop about 60 horsepower at 5,000 r.p.m. With a displacement equivalent to 1.2 liters (73.5 cubic inches), it actually developed 66 horsepower. The engine had been originally designed for automobiles and marine propulsion, but was soon superseded by more advanced projects.

With this experience to go on, the engineering office began a thorough study of both types. Basic geometry was identical, with an R/e ratio of 7.15:1, a maximum swing angle of 25.5 degrees, and rotor widths fixed at 4.8 times eccentricity. The single-rotor design was preferred, and ultimately selected, for a number of reasons. It offered a considerably shorter sealing path, had a more favorable surface-to-volume ratio, and cost less to produce.

Road test mileage shows fuel economy obtained with three different NSU cars using the KKM-250 and KKM-400 engines.

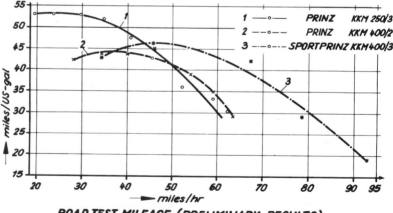

ROAD TEST MILEAGE (PRELIMINARY RESULTS)
AVERAGE RESULTS ON LEVEL-ROAD BOTH DIRECTIONS

The prototype single-rotor 600 cc. engine had an R/e ratio of 7.68:1, which permitted compression ratios up to 19.7:1 and gave a maximum apex seal swing angle of 23 degrees. Actual compression ratio was 8.5:1, and seal tip sliding velocity varied between 44.7 and 108 ft./sec. at 6,000 r.p.m. A slightly smaller design had a chamber volume of 447 cc. This design's 5.66:1 R/e ratio gave a more pronounced waistline and a maximum apex seal swing angle of 32 degrees. Maximum possible compression ratio was 14.6:1, and seal tip sliding velocity varied between 28 and 90.5 ft./sec. at 6,000 r.p.m.

The engine destined for the NSU Wankel Spider was the KKM-502, an enlarged and improved version of the KKM-400. NSU Sport Prinz cars equipped with the KKM-400 engine were test driven for more than 500,000 miles while the KKM-502 was on the drawing board. The initial version, the KKM-500, was the very first unit designed and developed as an automobile engine. Its power output was limited to about 44 horsepower and particular attention was paid to high torque at low speed—obtained by the appropriate choice of port opening periods, spark plug position, combustion chamber shape and compression ratio—without straying from the principle of the peripheral intake port.

The KKM-502 engine was designed in different versions, with both peripheral and dual-side intake ports. Power output was 30% greater with peripheral ports at 6,000 r.p.m. Because the KKM-502 was intended for use in a sports car, where maximum power was required, NSU committed itself to the peripheral intake port. The resultant lack of smooth-

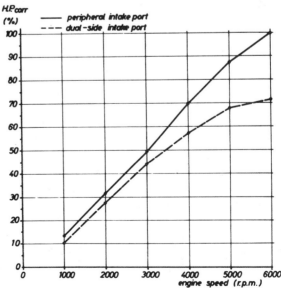

Comparison of power output curves for the KKM-502 engine with peripheral and dual side intake ports. The solid line represents the peripheral intake port, the broken line represents dual side ports.

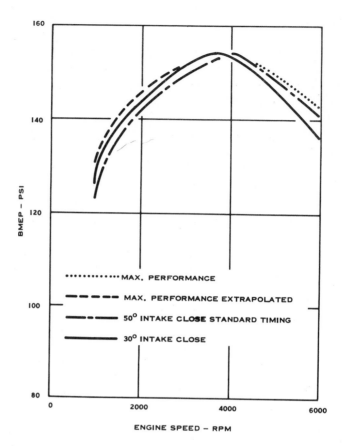

This graph shows the effect of intake port timing on mean effective pressure, throughout the r.p.m. range of the KKM-502. Earlier intake closing improves low-speed performance, while later intake closing raises high-speed power output.

ness at idle and under part-throttle conditions was not considered objectionable in a sports car which was intended for operation mainly at higher r.p.m. The ports were cast-in, and their positions were chosen to provide a symmetrical port-opening diagram. Port sizes, opening periods and carburetor throat size were chosen to obtain adequate torque at relatively low shaft speeds. The maximum torque was 57 foot pounds at 3,500 r.p.m., and the corresponding mean effective pressure was 147 psi. At higher speeds, the torque dropped to 47 foot pounds at 6,000 r.p.m., which indicated a maximum of 53.3 horsepower. Fuel consumption at speeds between 2,000 and 5,000 r.p.m. was below 0.58 pounds per horsepower-hour and the lowest figure, measured at 2,500 r.p.m., was 0.503 pounds per horsepower-hour with the carburetor set for maximum power. Highest mean effective pressure was obtained at 4,000 r.p.m. with the intake port so positioned that the rotor closed it off and began the compression phase 50 degrees after the major axis—earlier or later closing caused a loss of power at lower shaft speeds.

Both cast-iron and aluminum housings were tried. The aluminum housing was far less susceptible to temperature variations between various areas, keeping between 100°C. and 150°C. over about 120 degrees of trochoidal surface, from the middle part of the compression phase to the beginning of the exhaust phase. Temperature never exceeded 150°C., even during combustion. The cast-iron housing temperature exceeded 300°C. during the combustion phase, with fairly steep build-up and fall-off. From 50 degrees before the minor axis on the port side up to 50 degrees before the minor axis on the ignition side, temperatures in the cast-iron housing closely paralleled those recorded with the aluminum housing. It was found that housing temperatures in excess of 200°C. led to a breakdown of the oil film on the trochoidal surface. The aluminum housing was preferred, despite its higher cost, on the basis of its superior heat dissipative qualities. In the production engine, both the end covers and the housing were light alloy parts designed for low-pressure die-casting, and the coolant passages were formed by welding thin sheet metal plates to the end flanges and port stubs. The rubbing surface of each end cover was sprayed with a molybdenum and steel coating by the Ferral

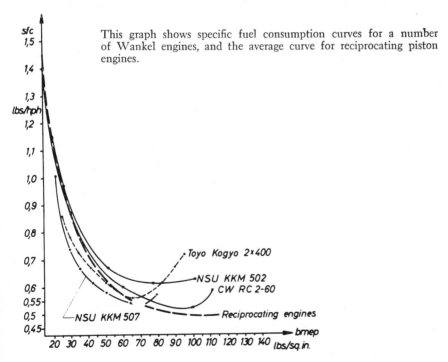

This graph shows specific fuel consumption curves for a number of Wankel engines, and the average curve for reciprocating piston engines.

process. Later, it was established that no molybdenum coating was required. Instead, the inner surfaces of the end covers went through an induction hardening process which made them even more wear-resistant than did the molybdenum spray. The exact housing configuration was dictated by production considerations, although circumferential ribs were added to stiffen the assembly.

The apex seals on the KKM-502 were made of carbon; the side seals were made of cast-iron. The apex seals were radially disposed, and the side sealing strips were arranged to overlap their leading joint trunnion tangentially and abut normally against their trailing link block. This arrangement prevented damage to the sealing surfaces even when wear developed. However, it did open up a new leakage path, and this configuration was discontinued when better methods were developed. The failure of many durability tests during 1963 stemmed from burned or broken apex seals (carbon type). Careful selection of impregnating materials, close attention to precision machining and finish and stricter inspection and quality control resulted in an improvement in seal life from the beginning of 1965. By 1968, engines ran routine tests of over 1,000 hours duration without any further apex seal failure.

The KKM-502 rotor was shaped to transfer a minimum of heat from its faces to the hub bearing. The hub bearing ring part of the rotor was connected to the inner walls of the rotor faces by a series of bridges spanning the flow of the cooling oil. A roller bearing with hardened steel outer races was pressed into each end cover and carried the eccentric, which was located axially by a separate ball bearing. Spur teeth were machined on the external periphery of one of these races, to mesh with teeth machined in the annulus of the rotor. An external-internal spur gear oil pump was accommodated in one end cover. Because of fractures experienced with earlier designs, this annular gear was made an integral part of the rotor, a malleable cast-iron part with more than the required fatigue strength.

Oil consumption had been the second largest problem with the KKM-500. The main cause of oil burning was leakage of cooling oil from the rotor. This oil could be sealed off from the combustion chamber at two places—either between the rotor and the eccentric, or between the rotor and the sidewall. When the Spider went into production, the rotor had an oil seal between the rotor and the end cover, carried in the rotor flank in a machined groove inside the gas-sealing side seals. The ring face had a 6 degree taper, as do some oil control rings used in modern piston engines. The metallic ring had a rubber O-ring inserted into a channel on its inside circumference, which blocked the oil from passing under the ring and onto the end cover wall. The gas pressure that leaked beyond the side seal got into the rotor segment next to the oil seal and threatened to cause further oil consumption problems. The solution, worked out theoretically before a prototype was built, was to mount the metallic ring and its attendant rubber O-ring in a seal carrier designed to equalize gas pressure on its leading and trailing sides, without reducing the spring pressure on the oil seal. This meant providing a gas leakage path behind the seal carrier, opening when pressure built up ahead of it and closing when equal pressure had been established. This oil seal worked effectively when new but deteriorated rapidly to cause high oil consumption at an unusually low mileage in the life of the car. A new system was developed and soon replaced the original design in production—the solution was to seal off the rotor from its eccentric bearing. The rotor was modified to carry a slip ring, matching up with a conical neck on the adjacent bearing carrier. Two seal rings, both in radial tension, were mounted in the slip ring and kept the cooling oil from leaking out.

The NSU engineers were still concerned about gas leakage past the side seals and its effect on oil sealing. The new rotor had an oil seal inserted in a circular slot inside of the side seal grooves, tapered to block radial movement of any oil that might escape that far. This two-piece oil seal was spring-loaded against the end cover to relieve gas pressure build-up ahead of it by providing a leakage path around the oil sealing interface. In addition, there was a gas pressure relief valve, carried in the end cover, providing an escape route to the exterior for gases trapped between the rotor-bearing oil seals and the rotor-end cover oil seal.

This was not the final version of the KKM-502 oil sealing system. One remaining problem was oil leakage through the dual radial seals when the engine was at standstill. The outer oil seal helped keep oil from entering the combustion chamber. However, this was an expensive sealing system that had to be simplified. In the last version, used on the KKM-502 engine, the piston ring-type oil seal in the bearing carrier neck was replaced by a carbon-type unit similar to those used in water pumps. This seal was spring-loaded and fully independent of gas pressure. Its weakness was its dependence on a back-up seal in the form of a rubber diaphragm. The rubber diaphragm was vulnerable to distortion due to local overheating, but this design effectively cured oil leakage even at 8,000 r.p.m.

The cooling system used a circumferential-path coolant flow pattern. Cool water entered the top of the housing near the major axis, on the ignition side, flowed down through channels in the direction of rotor movement, then flowed up again on the port side to an outlet immediately opposite the intake pipe. Two side-flow patterns diverted cool water to the end covers via a T-junction in the passages, at the bottom of the

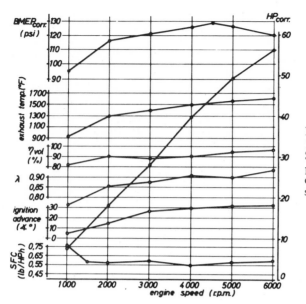

Performance data for the KKM-502 as installed in the Wankel Spider. BMEP = mean effective pressure. λ = volumetric efficiency. SFC = specific fuel consumption. HP = horsepower.

housing near the major axis. This portion of the coolant ran upwards in the end covers and was routed back to the outlet pipe through a T-junction in the top of the housing.

A two-stage, two-barrel Solex carburetor was developed especially for this Spider engine. It was built as a unit with the intake manifold. The primary throat was 18 mm. in diameter and contained the idle speed fuel feed system as well as the choke mechanism. The primary throat had its throttle valve about an inch behind the choke. The secondary throat was larger, 32 mm. in diameter, with a throttle valve very close to the port. It was positioned to prevent exhaust gas from lodging in the port area when the secondary throat was closed. Under part-load conditions only the primary throat was active; at a certain point the secondary throat opened.

Another item that was designed especially for the Wankel engine was the ignition system. The spark plug itself received most of the attention. In a Wankel engine, the spark plug never gets the cooling effect of freshly aspirated, cold air-fuel mixture. This makes it necessary to use a cool plug—a plug that dissipates heat faster than others. This need for a cool plug in the Wankel engine caused some ignition irregularities (difficult starting and uneven running at part throttle and on the overrun). The KKM-502 used a single-surface discharge 12 mm. spark plug with an attendant costly Bosch high-tension condenser-type discharge system. NSU found this system essential with a single spark plug and peripheral intake port because a high voltage rise rate was required to prevent high-tension leakage as a result of plug fouling. The plug was positioned before the minor axis and its points were contained in a small antechamber connected to the combustion chamber by a small-diameter "shoot hole."

By 1968, NSU felt that the engine could be adapted to normal cool ignition. Advancing the spark plug position from 20 to 28 mm. before the minor axis gave a 10% power increase throughout the useful speed range (2,500 to 6,000 r.p.m.). Fuel economy, however, had been 20–25% better with the spark plug positioned 20 mm. before the minor axis. To offset this drop in economy, NSU tried dual ignition on the KKM-502, with one plug 56 mm. *before*, and the second plug 20 mm. *after*, the minor axis. This gave higher power output and lower fuel consumption than had any single-plug system.

The KKM-502 combustion chamber had an elongated shape and it was feared the engine would be prone to abnormal combustion phenomena. The flame front had a long way to travel, and temperature rise rates and pressure wave formations were expected to cause spontaneous com-

In this chart, friction pressure in the KKM-502 engine is shown, broken down to the various sources, for the full speed range. MEFP = mean effective friction pressure.

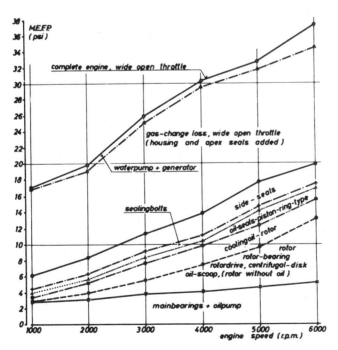

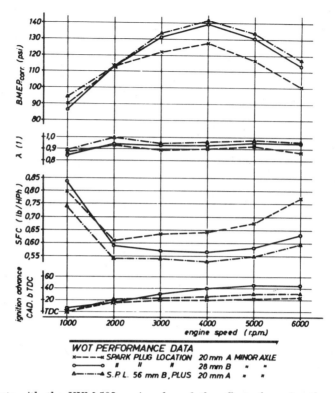

Tests with the KKM-502 engine showed the effect of moving the spark plug forward from 20 mm. after the minor axis to 28 mm. before the minor axis, and using dual plugs, one at 56 mm. before the minor axis, the other 20 mm. after it. One plug, 20 mm. after the minor axis, gave lowest power and highest specific fuel consumption. Dual plugs gave best power throughout the speed range, and lowest specific fuel consumption. A single plug at 28 mm. before the minor axis shows a parallel curve at slightly less favorable levels.

bustion under high-load conditions (such as during acceleration or climbing a gradient). But, early in the program, NSU engineers were pleasantly surprised to find the knock resistance of the Wankel engine quite high and its octane requirements very reasonable.

After the Wankel Spider went into production, research into sealing, production methods and tooling for twin-rotor engines was intensified. In 1964, NSU built a twin-rotor engine with the rotors from the production KKM-502. This experimental engine was called the KKM-509/506. It included an integral water and oil heat exchanger instead of the external unit used with the single-rotor production engine. All the accessory drives were placed at the front, and a cutaway section on the flywheel, plus a semi-circular weight on the mainshaft nose, provided generous counterbalancing outside the engine casing.

The mostly aluminum KKM-509/506 weighed only 154 pounds; including the transmission, its weight was only 210 pounds. Displacement was equivalent to 2.0 liters, or 122 cubic inches, while output was 110 horsepower at 6,000 r.p.m. with an 8.7:1 compression ratio. The apex seals in these new engines were allowed less maximum angularity to the working surface. Experimental engines had used 28.2 degrees— 24.8 degrees was the limit in the KKM-502 and KKM-509/506. One twin-rotor Wankel engine was exhibited at the International Auto Show in Frankfurt in September, 1965. It was redesigned during the following winter and became the KKM-612. By April, 1966, units were running on the test bed and in a DKW Munga light military vehicle.

The KKM-612 was the basis for the Ro-80 automobile. This was unusual in that the engine was not installed in any existing vehicle—the car was designed around the engine. Utilizing the extra freedom provided by the engine's small bulk and low weight, the designers achieved exceptional vehicle architecture with great success. The car was the end

Cutaway view of the KKM-622 (612) from the spark plug side. The rotors are 180 degrees out of phase.

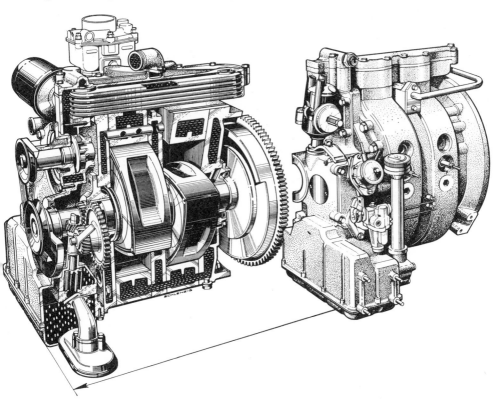

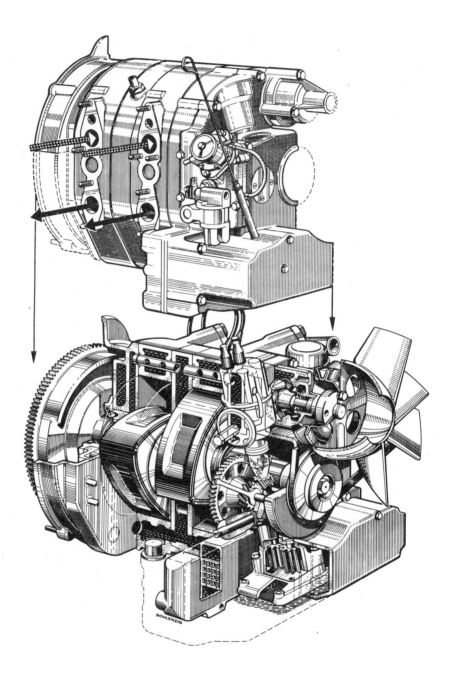

Cutaway drawing of the KKM-612. Dotted arrows represent fresh mixture; black arrows represent exhaust gas.

product of many years of concerted engineering effort by an outstanding team. Ewald Praxl was the overall coordinator of the total vehicle concept, while the power unit was designed and developed by Walter Froede and Georg Jungbluth. The key chassis engineers were Rudolf Strobel and Herbert Brockhaus. Development was the responsibility of Hans Georg Wenderoth.

Chassis testing with twin-rotor engines for the Ro-80 began early in 1965. The KKM-612 put out 136 horsepower at 5,500 r.p.m. and produced a peak torque of 117 foot pounds at 4,500 r.p.m. It was essentially two KKM-502 units placed end-to-end on a common eccentric shaft. The KKM-612 had the same chamber dimensions as the Spider: radius 100 mm., eccentricity 14 mm., and width 67 mm. Its R/e ratio was 7.14:1. This engine was the first NSU Wankel engine with cast-iron apex seals, replacing the KKM-502's carbon compound seals which had a tendency to break up under detonation. The carbon seals worked out fine in laboratory tests, but on the road, they did not hold up because at extremely high rotational speeds the centrifugal advance mechanism would advance timing to the point of detonation, and this form of abnormal combustion immediately cracked the brittle carbon seals. The improved KKM-612 metallic apex seal was a link-type seal. Unlike the KKM-502 seal, the split between the center and corner pieces was on the tip rather than on the axial face of the seal strip. It was partially self-adjusting to wear on both axial faces, and to conform to any convex or concave de-

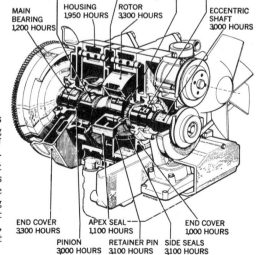

ROTOR BEARING CENTER SEPARATION ROTOR GEAR MAIN BEARING
1,500 HOURS   2,000 HOURS   1,550 HOURS 1,700 HOURS

MAIN   HOUSING   ROTOR   ECCENTRIC
BEARING   1,950 HOURS 3,300 HOURS   SHAFT
1,200 HOURS   3,000 HOURS

END COVER   APEX SEAL   END COVER
3,300 HOURS   1,100 HOURS   1,000 HOURS

PINION   RETAINER PIN SIDE SEALS
3,000 HOURS   3,100 HOURS   3,100 HOURS

Reliability and durability tests with the KKM-612 were going on when the author visited NSU in the summer of 1967. A number of engines were tested at 85% of full load at various speeds up to 5,000 r.p.m. The figures indicated on the drawing are averages for the many test engines. Under such conditions, 3,000 hours is roughly equivalent to 200,000 road miles.

formation of the seal strip groove. The seal was installed oversize and had a break-in period of 30 hours. Small slots in the leading face of the apex seal ensured rapid pressure build-up beneath the seal to prevent "spitback" (firing across the apex seal from the leading combustion chamber into the following one, i.e., the chamber under compression).

It was found that the cast-iron apex seal was incompatible with the chromed working surface used with the previous carbon compound seals; therefore NSU developed an entirely new coating material for the working surface called EINSIL. This material, a nickel plate containing fine particles of silicon-carbide suspended in an electrolyte, is integrated with the surface during the plating process. The plating layer is 200 microns thick when the working surface is machined. This coating is superior to chrome in terms of cost, quality, consistency, and reliability. With this combination, NSU achieved 3,000 to 9,000 hours seal tip life. Under deliberate test bed knocking, the old-type carbon seals stayed together for only 10 seconds; under identical conditions the iron seals lived for 10 hours.

Largely as a result of these changes, the KKM-612 had 25% less internal friction than the KKM-502. The new apex seals reduced friction by 17% under wide-open throttle conditions, and new oil seals cut friction loss at the side surfaces by 59%. Parasitic losses in main bearings and oil circulation were reduced by 46.7%. Still, the NSU engineers were sur-

On the KKM-612 prototype, the triangular corner seals for the apex seal did not reach the working surface. They allowed a small passage, parallel to the rotor sides, at both apex corners. In the course of normal wear, only gradual deterioration in seal effectiveness could be expected.

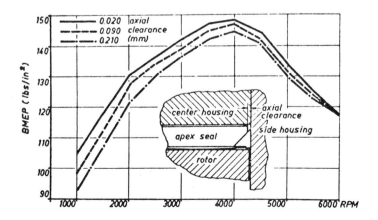

In later versions of the KKM-612, a new type of apex seal was used. In the course of approximately 20 hours of normal wear, this seal would be worn down to the level of the corner seals, and thereby close off the gas passage in the two corners.

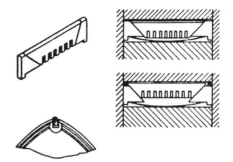

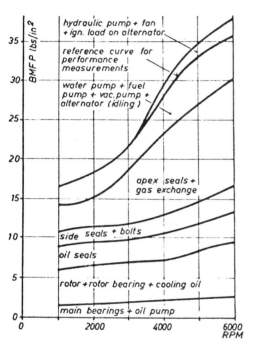

In the KKM-612, mean friction pressures on all parts rise sharply when r.p.m. exceed 3,500, approximately the same point at which torque becomes truly significant for acceleration.

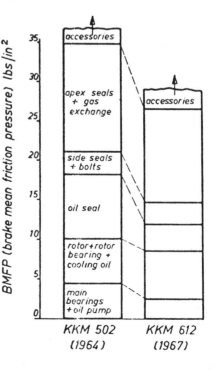

This graph compares the mean effective friction pressure of the KKM-502 engine with the twin-rotor KKM-612, at a steady speed of 5,000 r.p.m., on wide open throttle. Highest gains were in the oil seal, main bearing, and oil pump areas.

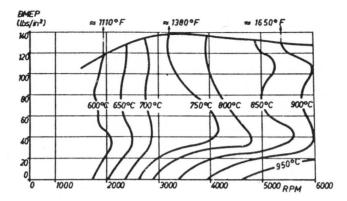

This graph of the exhaust gas temperatures in the KKM-612 shows a high general level, which is favorable for complete combustion and low emissions.

Distribution of housing skin temperature in the KKM-612, measured across the leading spark plug at 6,000 r.p.m. on full throttle, shows that temperatures in the immediate corner of the shooting hole and the epitrochoidal surface may become so high that the stresses may cause cracks in the casing. The copper alloy bushing around the plug takes away the temperature peak from the aluminum.

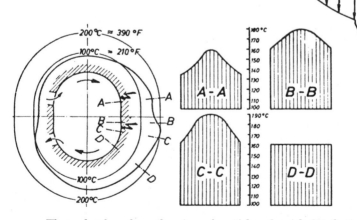

These sketches show the circumferential and axial distribution of rotor housing skin temperature in the KKM-612 engine, at 6,000 r.p.m., on wide open throttle. The cross-section on the left shows the intake and exhaust ports, in addition to the location of the two spark plugs.

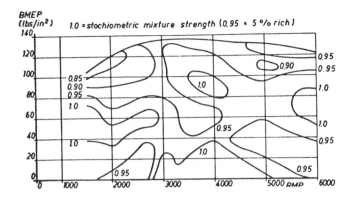

This graph shows the differences that result in mean effective pressure with variations in air/fuel ratios and r.p.m. You can read the graph as a topographical map showing altitude differences.

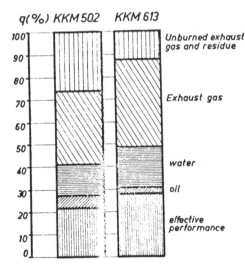

Comparison of heat balance between the KKM-502 and the KKM-612 engines at 4,000 r.p.m., show higher effective performance in the newer unit and far less unburned exhaust gas. Less heat is dissipated by the oil, and more by the water. NSU says that the 4,000 r.p.m. curves are representative for the full speed range.

These graphs show how horsepower and specific fuel consumption are affected by a step-by-step leaning out of the air-fuel mixture. Curves are shown for a constant 2,000 r.p.m. under variable load.

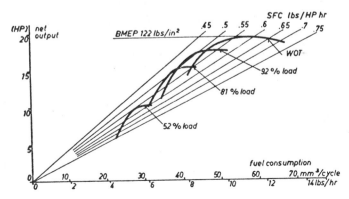

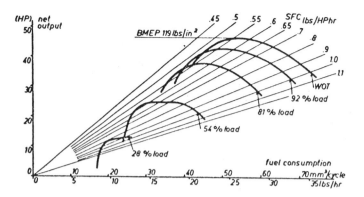

The wide open throttle curves show a drop in power with rich mixtures. Puzzling? Misfiring occurred sooner when running rich. Curves are shown for a constant 5,000 r.p.m. WOT = wide open throttle.

This graph compares temperature of the spark plug center electrode and ignition advance of the KKM-502, with that of the KKM-612. All tests were made at wide open throttle. The KKM-502 required a fairly cold plug, but with dual ignition, a plug of normal heat rating could be used. The 14 mm. plug of the earlier engine was dropped in favor of two 18 mm. plugs, which gave far superior fouling resistance.

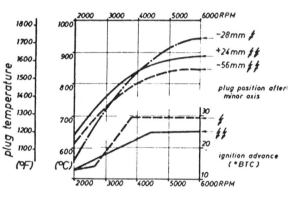

This graph shows the fuel consumption of the KKM-612 (complete with accessories) in a load vs. speed diagram. NSU points out that the engine requires a fairly rich mixture to assure driveability, even with an accelerator pump in the carburetor.

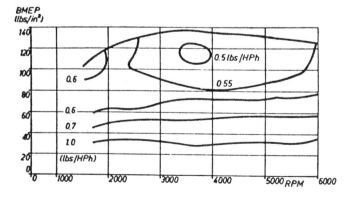

prised at the amount of wear found in the engines of cars that had been running on a stop-start test cycle. To cure this problem, they twice changed the kind of cast-iron used for the rotor tip seals, each time gaining a significant improvement in cold-start wear, and they are at present satisfied that the trouble has been cured.

There were also ignition difficulties at first, due to servicing troubles and poor electrical design, but these were cured by changing from two spark plugs *per rotor* to one, and from conventional contact-breakers to those of the transistor-assisted type. The dual-plug ignition system of the KKM-612 had two spark plugs, one located 56 mm. before, and the other 24 mm. after, the minor axis. They were fired simultaneously. The use of dual ignition gave a number of advantages in view of the long and flat combustion chamber:

1. Steeper pressure rise.
2. Faster combustion.
3. Higher thermal efficiency.
4. Lower exhaust gas temperature.
5. Better cold starting ability.

The change to single-plug ignition was made mainly because twin spark plugs needed separate ignition systems. Two plugs could be wired in series only if one was completely insulated from the engine; if they were wired in parallel, the one with the smallest gap would fire first and short out the second. Also, two coils were needed, and the special distributor had two sets of breaker points and a double-wiper rotor with two opposed electrode arms. The single-ignition system substantially reduced production costs, and the transistors provided a strong spark despite plug fouling, in addition to extending the speed range of the

Comparison of spark plug and seal configuration between KKM-502 and KKM-612. The tapered seat is required to assure identical and reasonably small breathing volumes. It was moved from bottom to top, because the earlier type brought a danger of deformation of the housing surface as a result of overtightening. The new plug has a longer thread, which aids heat transfer. On the KKM-502 the plug was screwed into the aluminum housing; on the KKM-612 it is lodged in a copper alloy bushing. The new plug is of the surface gap type, the old one conventional.

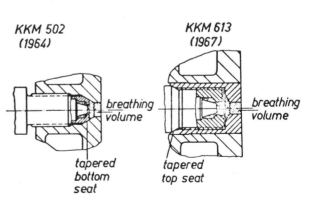

KKM 502
(1964)

KKM 613
(1967)

breathing volume

breathing volume

tapered bottom seat

tapered top seat

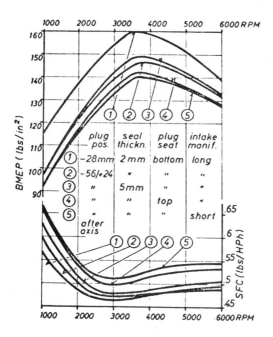

	plug pos.	seal thickn.	plug seat	intake manif.
(1)	-28mm	2 mm	bottom	long
(2)	-56/+24	"	"	"
(3)	"	5mm	"	"
(4)	"	"	top	"
(5)	"	"	"	short

after axis

Performance curves for the KKM-612 prototype with five stages of engine modifications during its development. The upper curves show effective pressure in psi (pounds per square inch). The lower curves show specific fuel consumption in pounds per horsepower-hour. (1) = Spark plug positioned 28 mm. before the minor axis, apex seal thickness 2 mm., plug seat at the bottom, with long intake manifold. (2) = Dual spark plugs, 56 mm. before and 24 mm. after the minor axis; seals, plug seat, and manifold unchanged. (3) = Plugs as (2), but 5 mm. thick apex seal; plug seat and manifold unchanged. (4) = Plug seat changed to top, otherwise as (3). (5) = Short intake manifold, otherwise as (4).

This sketch shows how gas pressure is utilized to improve apex seal tightness in the KKM-612. Had the corner seal edge (1) been solid, it would have blocked the high-pressure combustion gases, because of the tilt forced upon it by the sudden buildup of pressure in the leading chamber. This phenomenon was cured by machining a number of small slots along the inner leading edge of the seal. This allowed combustion gases free entry into the volume behind the seal, where they would force the seal strip radially outward, as well as against the trailing wall in the seal slot.

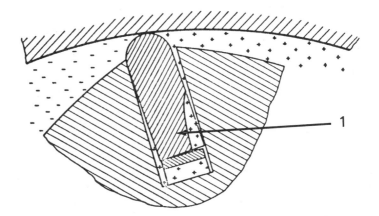

engine by several hundred revolutions. The present Ro-80 KKM-612 plugs have tapered seats, which control plug position more accurately than do compressible gaskets. The tapered seat is located outside of the threaded section because overtightening could produce distortion of the trochoidal

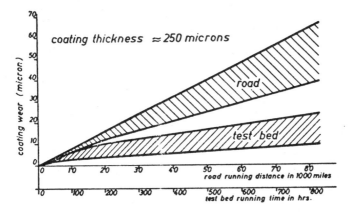

The Einisil coating on the working surface of the KKM-612 yielded average wearing qualities in the range of 50 microns per 80,000 miles, in actual road use. The test bed field in the chart includes many different conditions of operation, including wide open throttle at 6,000 r.p.m. These results were obtained with the latest form of apex seal (slotted on the leading face).

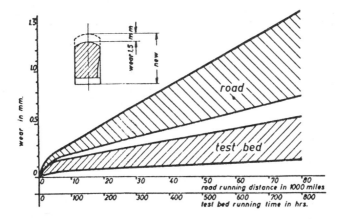

Wear tests on the bench and with KKM-612 engines installed in cars, showed such low wear on the new apex seals against the Einisil surface of the epitrochoidal surface, that the functioning of the apex seal is never in danger. Even 3 mm. wear cannot influence performance characteristics.

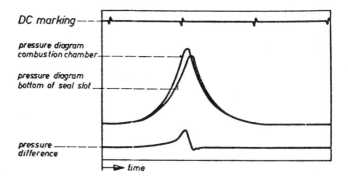

DC marking

pressure diagram
combustion chamber

pressure diagram
bottom of seal slot

pressure
difference

time

The apex seal is dependent upon the presence of gas pressure behind it to effect a proper sealing face. These curves are an example
of retarded pressure buildup, such as that which existed with the
non-slotted seal.

The old-type apex seal suffered
from the spitback phenomena.
The results are shown in the
lower graph here. The upper
graph shows the results obtained
with the new sealing formation,
which eliminated spitback.

time   pressure pickup location:

TDC

—intake manifold

—intake BDC (cold lobe)
—TDC

no spit-back
(ign. adv. 35° BTDC)

time

TDC

—intake manifold

—intake BDC (cold lobe)
—TDC

spit-back
(ign. adv. 40° BTDC)

track if the seat were positioned too close to the working surface. The
spark plugs are 18 mm. units because of their improved resistance to
fouling. They are not threaded directly into the housing but into a copper alloy bushing, which reduces the temperature peak and the risk of
local cracks in the housing.

In the production Ro-80, about which more will be said later, the
eccentric shaft is supported at each end by a plain tri-metal bearing. In
the interest of simple design and assembly, there is no center bearing
between the rotors. The two malleable cast-iron rotors are identical, and

the combustion chambers, recessed in the rotor faces, are symmetrical. Malleable cast-iron is one of the best materials available for controlling expansion and distortion at high temperature; therefore the rotor runs with adequate clearance for the difference in expansion rates between the rotor and housing parts.

The engine housing is water-cooled with circumferential flow. The working chamber housings are made of aluminum alloy for good heat dissipation, while the center separating wall and the end covers are made of cast-iron. Certain small areas on the inner surfaces of the end covers have a thin coating of molybdenum sprayed on to give added wear resistance. All five housing parts are bolted together with long tension bolts parallel to the engine axis. The water pump delivers 29 gallons per minute at 6,000 r.p.m., and is positioned at the upper front end of the engine. Coolant flows circumferentially in parallel streams through all the housing parts, and through the oil and water heat exchanger embedded in the sump. Coolant returns through a passage at the top of the engine—coolant capacity is 2.3 gallons. The cooling system was modified

The marine version of the Ro-80 engine forms a very compact power unit for an inboard/outboard installation.

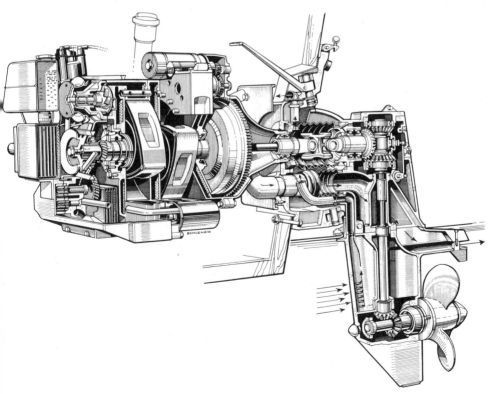

In 1968 NSU installed an Ro-80 engine in this small motorboat, for test and demonstration purposes.

This racing boat has distinguished itself in many events. It is powered by a twin-rotor NSU Ro-135 developed from the KKM-612, which drives the Ro-80 passenger car. In seven starts in the two-liter class of international boat races, the Swedish pilot Zetterström took three first places, one second and three thirds. In the 1969 Rouen (France) 24-hour race, it finished third overall.

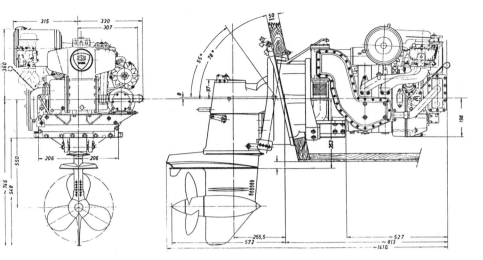

The Ro-135 is rated at 135 horsepower at 6,000 r.p.m. It has a heavy-duty cooling system and a waterproof electrical system. The drive train is made by ZF and the propeller is geared down in a 1.51:1 ratio. The drive can be turned through 26 degrees each way, which offers excellent maneuverability.

in production, and now incorporates a redesigned thermostat, an improved heater valve, and a different fan.

The oil pump in the KKM-612 engine delivers 5.2 gallons per minute at 6,000 r.p.m. The rotor bearings, which have no center groove to increase the oil holding capacity of the bearing shells, are fed through the partly hollow mainshaft. Oil from the eccentrics is also squirted directly into the hollow rotors to provide cooling. An additional small oil pump in line with the engine's main lubrication pump supplies pressure through an external pipe to the torque converter, which also uses engine oil. The return flow combines with the return engine oil in a passage at the bottom of the engine, then is drained to the sump. Total oil capacity is 1.7 gallons. Lubrication of the Ro-80 engine is, therefore, just a matter of putting in a little oil now and then to replace the amount of lubricant metered into the carburetor to lubricate the rotor and the housing. Because the sump oil in a Wankel engine is not subjected to blowby or contamination it was planned at the outset to abandon periodic oil changes, but the car had to pass high-mileage tests before NSU could be sure that there were no ill effects. The decision to actually abandon oil changes was made after more than 3 million miles of test driving over a four-year period, preceded and accompanied by bench testing to evaluate the road test results. NSU has been trying to make the production Wankel engine a maintenance-free powerplant.

In contrast to the piston engine, in which the piston ring performs closely related functions as a gas seal and oil control element and where the piston movement involves blowing combustion gases past the rings, the Wankel engine provides a sharp distinction between the functions and timing of lubrication and gas flow. NSU has designed what they call a secondary compression chamber which retards passage of combustion gases by alleviating marginal overpressure through a relief channel controlled by a ball-operated pressure valve. This provides a neutral zone which impedes the mixing of end gas (volatile combustion residue) with the oil.

Two dual-stage sidedraft carburetors are tuned to the breathing pattern of the KKM-612. The two intake ports combine in a single circular port on the trochoidal surface of the central wall between the two working chambers to give earlier intake port closing and thereby raise low-speed torque. There is a complex of mufflers to silence the exhaust, which is not easy on a Wankel engine that, just like a two-stroke has ports rather than poppet valves and is sensitive to back pressure. The two chambers have independent systems to prevent interference between the trains of power impulses. Separate pipes from each housing lead to a pair of small mufflers that handle the high-frequency components. These are followed by a series of two partitioned units of golfbag proportions that eliminate the lows.

·   ·   ·

From the day the first experimental Wankel engine ran in the NSU laboratory until the day when the first KKM-502 production engine was assembled, over seven years had passed. You have read an account of what some of the technological problems were and how they were solved. Insuperable as they must have seemed to many, they were overcome. It is a small miracle that so much was accomplished in those seven years, and it would seem a fantastic achievement if it had all come about in an environment free from financial worries and stress. The fact is that throughout this initial research and development period, NSU was in a state of perpetual financial turmoil.

The destiny of the company seemed to depend solely on the reactions of the financial world and public opinion as to the worth of the Wankel engine. NSU stock fluctuated in spectacular leaps and bounds on the stock market, depending on the latest news from the laboratory.

When Dr. Froede was given the task of collaborating with Felix Wankel on the conversion of his rotary compressor to an internal combustion engine, NSU still called itself the "Number One Motorcycle Maker in the World," and new racing successes and international class record attempts were maintaining its reputation at top level. But fewer people bought motorcycles, and the Dresdner Bank, which controlled NSU, was not interested in world records and racing successes, but in the company's profit-and-loss statements. While Dr. Froede and Felix Wankel had made great progress on rotary engine development, NSU's income had dwindled as a result of the motorcycle boom's coming to an end. It was doubtful whether NSU had enough capital strength to bring both the new engine and the Prinz automobile to a state of readiness for production. To conduct large-scale experimentation with an unconventional powerplant concurrently with the design, testing and development of a completely new car required investments beyond the scale on which von Heydekampf and his directors could operate. However, the NSU Prinz did go into production, and Wankel engine development was never interrupted. How did NSU manage both? Felix Wankel simply went out in search of new sources of financing.

Wankel associated himself with a post-war millionaire, Ernst Hutzenlaub. Trained as an architect, turned inventor, Hutzenlaub had created a small conglomerate by uniting small companies in the plastics and rubber fields, which were exploiting his inventions. He sold licenses worth $500,000 outside Germany, and saw a legal opportunity to reduce his taxable income through such operations. This did not fall into the category of tax evasion because, at that time, the German Government wanted to encourage exports, and revenue from invoices earning foreign currency were only taxed at the special low rate of 25%. The experienced Hutzenlaub staked $250,000 of his own money on the Wankel engine, then he and Wankel formed a company known as Wankel G.m.b.H., which was to act as a sort of cash collecting office for future licensees. Wankel and Hutzenlaub each held a 50% interest in the company, which guaranteed 50% of its income to each. As the next step, Wankel and Hutzenlaub tried to persuade dozens of industrial firms to buy a license for the "engine of the future." Their activity impressed the management of NSU. Von Heydekampf of NSU felt encouraged; he saw hope for fresh financing through the proposed licensing agreements.

The sale of licenses was held up by many factors. One was lack of faith in the engine. That was natural—investors are traditionally skeptical about inventions. But NSU's involvement with the Wankel engine tended to be more of a hindrance than a help. NSU was known as a small firm with good products but a dwindling market. Its management was untested in that kind of situation. At that time, no leader of industry could imagine that NSU, with its 7,000 employees and an annual gross turnover of less than $50,000,000, was capable of producing a technical revolution. It was felt that such a revolution could only come about with the financial aid of, and the prestigious association with, a world famous company. One of the key men in the Dresdner Bank, Hermann Richter, outspokenly advised von Heydekampf to do something to improve NSU's reputation, something to make the Wankel engine more palatable to the cautious industrialists approached by Wankel and Hutzenlaub.

The NSU Prinz was von Heydekampf's answer, but the question of whether that would be enough to entice Wankel engine backing from outside remained unanswered.

Wankel tried to negotiate a deal with Krupp in 1957 and 1958, but Krupp did not want any part of it. The first bite came from Curtiss-Wright whose president, Roy T. Hurley, thought they could take the experimental engine, refine it, and mass-produce it. Curtiss-Wright signed an agreement on October 21, 1958 that they would openly share all additional patents and design changes with NSU. The agreement also gave Curtiss-Wright all rights for Wankel engine manufacture, in all sizes and for all applications, in the United States. NSU is reported to have received $2.1 million from Curtiss-Wright for the license, which also included a 5% commission (on the selling price) to NSU on all Wankel engines that Curtiss-Wright might build and sell.

As soon as Hutzenlaub heard of the agreement between NSU and Curtiss-Wright, he went to Neckarsulm (West Germany) with strong demands. He wanted 40% of all NSU's income from Curtiss-Wright, and a contract stipulating the same percentage on revenue from all future license takers. The NSU directors put up a tough fight. They said Wankel had been on the payroll since 1961, with a monthly salary of DM 3,000, and his accomplishments at NSU were the property of the NSU company. Hutzenlaub explained that Wankel had brought them his life's work, unique experience and ability, the essence of lifelong experimentation—in fact, assets that could not be overestimated at 40%. NSU gave in, mainly to avoid the unfavorable publicity of a court process against the inventor, and Wankel and Hutzenlaub received their 40% of the initial payment from Curtiss-Wright. Roy T. Hurley paid little attention to this last deal. To him, the money he paid was a small stake on the fantastic fortune to be reaped through Wankel engine applications for aircraft, boats and automobiles, as well as for stationary purposes.

Because of its unique position relative to Wankel G.m.b.H., NSU was not regarded as a regular licensee but was allowed to manufacture and sell Wankel engines against a fee specified in a separate agreement. NSU's planning was cautious. At first, NSU was to manufacture stationary Wankel engines, to use as pumps for instance, then they would produce the 400 cc. Wankel engine destined for the NSU Prinz car. All other sizes and versions of the Wankel engine were to be manufactured by other license takers.

What profit potential was there? A friend and advisor to Felix Wankel figured that if in a few years only 10% of all engines worked on the Wankel principle, it would mean a production of between 7 and 8 million Wankel engines per year in the Western world. If $18.50 per engine could be counted on as the combined license and royalty fee, NSU and Wankel would collect about $150 million annually. Of this amount, NSU would retain $90 million, and after taxes, would still have a surplus of $50 million—a figure almost as high as its entire capital stock.

Up to this point, not much was known about the Wankel engine and the agreement between NSU Motorenwerke AG and Felix Wankel, but when the contract with Curtiss-Wright was announced, an avalanche of publicity followed. The press reaction was overwhelmingly favorable, and both speculators and old established brokers began to buy up NSU stock. Even conservative bankers and critical investors fell for the "engine of tomorrow" promise.

The Dresdner Bank held 51% of the NSU stock and controlled the company, the other 49% was held by 5,000 individual shareholders. Nominal value of the capital stock was equivalent to only $4,500,000—NSU stock prices had remained sadly low even in the period when NSU motorcycles dominated the market. A new low was reached in 1957, when NSU stock was quoted at 124.5% of par. (European firms often issue stock in lots with different par values. The stock exchange listings refer to percent of par value, whether the share has a face value of $25 or $1,000.) Suddenly, anonymous advertisements began to appear in German newspapers: "NSU *shares bought at top prices*." Observers guessed that they were inserted by someone who was anxious to unload a package of NSU stock and wanted to first drive up the price. Von Heydekampf got worried and tried to find out who was behind the advertisements. They were traced as far as a lawyer in Düsseldorf who would not reveal the identity of his client. Answers to the advertisements were refused or unanswered. NSU stock continued its upward trend and, at the end of 1959, it stood at an even 1,000.

This rise was completely against the expectations and calculations of the Dresdner Bank. They were unwilling to face the possibility of spending year after unproductive year with this revolutionary new powerplant on speculation that the Wankel engine might one day replace the existing piston engine. While NSU was getting great publicity thanks to the Wankel engine, and morale among the workers as well as management was enormously strengthened, the Dresdner Bank officials decided it was time to end their long-established mesalliance with NSU. Because of the high stock market quotations they saw a chance to sell their NSU stock at a high profit and free themselves of the small, but embarrassingly busy, industrial operation. The bank first offered its NSU holdings to Friedrich Flick, chairman of Daimler-Benz AG, who refused. He was at the time busy with the Auto Union acquisition and was making arrangements to sever the DKW motorcycle division to concentrate on passenger cars and light commercial vehicles. Then, feelers were put to Roy T. Hurley of Curtiss-Wright.

Hurley was interested in taking over the entire 51% at the current value, but this was not to be. Holding 51% would mean that Curtiss-Wright would, in effect, acquire control of NSU Motorenwerke AG. Germany has no restrictions regarding foreign ownership of its industry, but there were more subtle complications. With control of NSU, Curtiss-Wright also would be in a position to dictate to whom future licenses would be granted. It was the shrewd Hutzenlaub who put a stop to the deal. The Dresdner Bank gave him the cold shoulder, but went along with his protest, pretending that they were acting in order to prevent an American capital invasion. There were no further attempts on the part of the bank to find a new parent company for NSU.

Instead, the Dresdner Bank began feeding its NSU stock to the exchange during the week of January 8 to 15, 1960. The average German industrials then lay around 370—the NSU stock sold at 1,025! Director Fritz Andre of the Dresdner Bank defended the action by saying that the bank wanted to act against the mad price pattern of NSU stock. This consideration was felt to be merely an alibi against possible accusations of stock manipulation during 1959. The bank admitted that it sold about half of its NSU holdings in January, 1960, which brought its interest in NSU down to 26%. The bank received some $12,300,000 in the transaction, for stock with a par value of a mere $1,150,000. The following month NSU Motorenwerke AG was recapitalized. New stock was issued to raise its capital from $4,500,000 to $6,700,000. Just how the new stock was to be distributed was a matter of discussion at the general assembly on July 22, 1960—a meeting at which the Wankel engine played an important role.

Hermann Richter and Erich Vierhub of the Dresdner Bank faced 350 stockholders. As a rule, only about a dozen stockholders showed up for the meetings, and business was usually concluded in a quarter of an hour. But not this time. From the way their questions were answered, the stockholders reached the conclusion that they were not going to be invited to share in the new stock issue. They noted that holders of 27% of NSU stock were not represented at the meeting, and they guessed that the bank might have sold blocks of some size to Curtiss-Wright or to Mr. Flick, who could have changed his mind about the Wankel engine since 1958. That started a riot, and Richter calmed down the audience only by showing them the experimental Wankel engine and explaining that more work was needed before it could be exploited.

By mid-year, 1960, the stock hit the 3,000 mark. The record was reached on June 30, 1960, at the Munich stock exchange, with quotations of 3,200! The end of the NSU stock market romance came when the Dresdner Bank sold part of its remaining 26%. The result was a sheer drop in the price of NSU stock, which had the effect of destroying the high hopes of the speculators. They in turn began to sell their own NSU holdings, bringing prices further down. By the end of 1960, the stock held at about 1,300 and during 1961, it moved only within the 1,300–1,600 range, continually up and down.

The principal reason for the inherently speculative nature of all dealings in NSU stock was the uncertainty of the Wankel engine's potential. The public had few facts, if any, upon which to base its opinion and judgment. This was partly remedied when the technical press was invited to a seminar on the invention, design and development of the Wankel engine on January 19, 1960.

The meeting was held in the Deutsches Museum in Munich and organized by the V. D. I. (Verein Deutscher Ingenieure) to discuss the invention, design and development of the NSU Wankel Rotary Combustion Engine. The meeting was attended by 1,350 scientists, engineers, teachers and students. Professor Dr. Ing. Ernst Schmidt, of the Technical Institute of Munich, gave a lecture on the historical development of internal combustion engines in Germany, from N. S. Otto and Rudolf Diesel to Felix Wankel. Dr. Ing. Eugen Wilhelm Huber, head of the Huber Power Plant Research Institute, discussed his own work on the Wankel engine and showed a sound film of the engine running on a dynamometer. Professor Dr. Ing. F. Baier, of the Technical Institute of Stuttgart, dealt with the geometry of the epitrochoid. He demonstrated that the working chamber and rotor can have up to at least seven different shapes in working chambers of varying curvature. Felix Wankel briefed the audience on the historical development of rotary engines from the beginning of the century, and presented a lengthy discussion on sealing for gas tightness at the sliding surfaces.

Then Dr. Walter Froede outlined his KKM design and development work at NSU. He showed that with the three-lobe rotor design the compression ratio of the engine was limited to 15.6:1. A four-lobe rotor would allow a 100:1 compression ratio and a five-lobe rotor would allow a compression ratio up to infinity. Dr. Froede went on to say that, in his opinion, the engine could be used for a multitude of power applications. He predicted that the engine would first be used as a small constant speed stationary unit of under 50 horsepower, for such duties as pumping water or generating electricity. Froede gave credit to Goetze Werke for its notable contributions to the development of the sealing system, and pointed out that five of the principal oil companies in Germany had run test engines in their laboratories to determine fuel octane requirements and select the most suitable lubricants.

Soon after, NSU announced that, after spending $8 million on its development, the Wankel engine was still not ready for production. This did not affect the stock price, but stockholders and other interested parties were anxious for some word from NSU as to when production was likely to begin. In desperation, one small stockholder wrote a letter to NSU. In the guise of a potential customer, he asked innocently when he could hope to buy a Wankel-type marine engine. The answer from NSU was dated August 22, 1960, and was signed by one Dr. Hirsch. It said that NSU planned to start production by mid-1961. The engine was to be a 400 cc. unit, developing 45 to 55 horsepower at 5,000 r.p.m. Not primarily a marine engine, a marine version would no doubt follow. Would the customer please write to NSU again in the middle of the following year? The stockholder gave the letter maximum publicity and NSU stock went up again, almost 200 points.

Dr. Hirsch was only an assistant attached to the board of directors, but it did management no good to explain that Dr. Hirsch had vastly overstepped his authority. Management maintained that the timetable for the start of production indicated 1962 or 1963 rather than mid-year 1961. But speculators as well as stockholders preferred to believe the more optimistic Dr. Hirsch rather than the conservative management. The net result was the muzzling of all of NSU's management men, including Dr. Gerd Stieler von Heydekampf.

Von Heydekampf debated with himself whether or not to resign, but he stayed on. There was a complete news blackout on Wankel engine development—no interviews were given, nobody visited the factory, no information was released. Stock prices fell to the 1,800 mark as a result of the lack of press coverage of NSU activities. Then, in November of 1960, the Dresdner Bank began to advise its customers *against* buying NSU stock. On December 12, 1960, the announcement that Fichtel & Sachs AG of Schweinfurt had signed a license agreement with NSU for production of Wankel engines in the 0.5 to 12 horsepower range produced no rise in the price of NSU stock, which on December 7 was listed at 1,755.

The Dresdner Bank gained definite tax advantages from the fall of

NSU stock prices towards the end of 1960. All money institutions in Germany are taxed for the market value of their stocks and bond holdings. This enabled the bank to list a very low value for its NSU stock in its annual tax declaration. A lawyer in Düsseldorf, named Engler, who represented a group of small NSU stockholders, accused the Dresdner Bank of willfully forcing NSU stock prices down, after the bank itself had sold its NSU holdings. The most important of these minority stockholders was a medical doctor from Stuttgart, Dr. Dietrich Albers. His holdings had a par value of $20,500, which gave him just under ½% control of NSU. Dr. Albers was instrumental in the formation of an NSU stockholder association, which published advertisements with the following text in various daily papers: "NSU stockholders, keep your shares. The stock price fall is directed by high finance, so that they can buy your stock as cheap as possible. Be patient. You own one of the most valuable German stocks.—Schutzgemeinschaft der NSU Aktionäre e.V."

Through this association, Dr. Albers formed a united front of 600 minority stockholders. He valiantly pursued all clues to what was going on with the Wankel engine. During the first part of 1961, he discovered that Daimler-Benz had a batch of NSU Wankel engines on test. Daimler-Benz was not yet ready to buy a license though. First, in typical fashion, they wanted to evaluate the Wankel engine as a practical automotive engine. NSU released the test engines to them, having full confidence that they would act honorably, no matter what their conclusions in the technological field. Dr. Albers sounded the alarm and demanded an explanation from NSU, whose management made no comment. Then he started to spread rumors about a payment of $5 million from Daimler-Benz to NSU in the hope that confirmation or denial would be forthcoming. Actually, the deal did not involve any money at all; it was simply a verbal agreement between the directors of the two companies that all Daimler-Benz's contributions to the development of the Wankel engine, including any patents they might take out, were to be made available to NSU at no cost. Dr. Wolf-Dieter Bensinger of Daimler-Benz's engineering department translated the NSU power unit into proper Mercedes-like dimensions and began building test engines of his own. But the acquisition of a license was postponed indefinitely.

On February 25, 1961, NSU issued a Wankel engine license to Yanmar Diesel Company, Ltd. of Osaka, Japan, covering gasoline engines from 1 to 100 horsepower and diesel engines from 1 to 300 horsepower for all applications other than motorcycles, passenger cars and aircraft. Two days later, NSU signed a contract with another Japanese firm. Toyo Kogyo bought a license to manufacture Wankel engines for gasoline fuel with power output from 1 to 200 horsepower, for application within the framework of Toyo Kogyo's product line as of 1961. From these arrangements, NSU was said to have collected cash payments of $1,250,000.

About this time, NSU decided to send a letter to its stockholders, warning them against being overly optimistic about the Wankel engine. This sent the stock down to 1,200, but it was soon to go up again. NSU invited 700 of its dealers to Neckarsulm to show them some Prinz prototypes powered by 400 cc. Wankel engines. These cars were strictly experimental, hurriedly cobbled up with Volkswagen transmissions and Ford radiators, but they would reach 100 m.p.h. on 50–55 horsepower. There was considerable doubt about their longevity, but the net result of the experience was to send NSU stock up about 100 points.

In 1961, Hutzenlaub sent drafts for license agreements to Citroën and Renault, Krupp, M.A.N. and Klöckner-Humboldt-Deutz. Both Rolls-Royce and General Motors sent representatives to NSU to examine and evaluate the engine. Renault had sent a team of engineers to NSU in 1960, where they were permitted hundreds of man-hours to study the Wankel engine. The Frenchmen pretended that Renault had earlier developed a rotary engine of its own—their study of the Wankel engine was purportedly a strictly academic investigation.

Their attention, however, centered on the Wankel engine's sealing system. As soon as the Renault team returned to Billancourt, they filed for French patents for the same sealing system! Renault then brought forth the old Cooley engine and pretended they wanted to try to apply Wankel's sealing system to the American design.

Of the other interested parties, Citroën was not to enter any formal agreement for some years, and the story of its joint ventures with NSU (Comobil and Comotor) is told in the chapter about Citroën's work with Wankel-powered cars. Klöckner-Humboldt-Deutz AG of Cologne acquired a license covering diesel engines of all sizes, for all applications, on October 4, 1961. Daimler-Benz AG formally took out a license for Wankel engines from 50 horsepower up on October 26, 1961. Four days later, M.A.N. Maschinenfabrik Augsburg-Nurnberg AG signed an agreement with the exact same stipulations contained in the Deutz contract. On November 2, 1961, Friedrich Krupp of Essen signed an identical agreement. These agreements with German firms were estimated to have brought NSU a gross license fee income of about $1,500,000. F. Perkins, Ltd. of Peterborough, England, a subsidiary of Massey-Ferguson of Canada, acquired a Wankel license for gasoline and diesel engines up to 250 horsepower on August 8, 1961. This deal brought NSU a sum of about $375,000.

All these license agreements antedate the actual production of Wankel-powered cars by NSU. It was in mid-1963 that von Heydekampf said NSU had plans to make 5,000 Wankel-Spiders in 1964. The car was, in fact, shown at the Frankfurt Auto Show during the week of September 12–22, 1963. It was a white hardtop coupe designed by Bertone, resembling the Sport-Prinz.

With an actual vehicle being readied for production, a few other companies became interested.

Rheinstahl-Hanomag signed a license contract for Wankel engines of 40 to 200 horsepower on December 19, 1963. On April 15, 1964, Alfa Romeo, which had been experimenting with Wankel engines of their own design for several years, signed an agreement covering gasoline-fuel Wankel engines from 50 to 300 horsepower. Alfa's interest dated back to May, 1959, when there were rumors that a merger between Alfa Romeo and NSU was planned.

Rolls-Royce Ltd. of Derby bought an extensive license covering diesel and hybrid Wankel engines from 100 to 850 horsepower on February 17, 1965. The following morning, NSU sold a Wankel engine license to a firm in Communist East Germany (DDR)—Vereinigte Volkseigener Betriebe Automobilbau. The agreement specified gasoline-fuel engines in two ranges, from 0.5 to 25 horsepower and from 50 to 150 horsepower.

Dr. Ing. h.c. F. Porsche K.G. of Stuttgart-Zuffenhausen acquired the rights to manufacture gasoline-fuel Wankel engines from 50 to 1,000 horsepower by its contract of March 2, 1965.

When the Wankel engine was still in the experimental stage at NSU, an Israeli group acquired 5% of the NSU stock. As progress continued, their holdings increased to 12% by the time of the Volkswagenwerk takeover in 1969. This group formed a company incorporated as Savkel, under Israeli laws with headquarters in Tel Aviv, and was able to acquire a license for manufacturing Wankel engines in Israel. NSU/Wankel granted to Savkel the right to manufacture gasoline-fuel engines of 0.5 to 30 horsepower for all applications, except for land vehicles, aircraft and marine installations.

General Motors invited the technical press to its "Progress of Power" presentation at the GM Technical Center in Warren, Michigan, on May 7, 1969. A number of unconventional power units were demonstrated, including steam cars, electric cars, a gas turbine bus, and various hybrid power systems. The Wankel engine, however, was missing, but rumors of GM's interest in the Wankel engine remained strong. On June 1, 1970, General Motors confirmed that it was holding discussions with Wankel G.m.b.H. According to a report from Frankfurt, GM was interested in obtaining 40% of the shares of Wankel G.m.b.H. Industry sources in Europe said GM was conducting the talks through its West German subsidiary, Adam Opel AG. These sources indicated that GM was offering the equivalent of about $27.5 million for the 40% interest. This would let GM in on the ground floor of all present and future licensing arrangements. A GM spokesman said only that "discussions are being held with Wankel G.m.b.H. as part of our stated policy of investigating all possible sources automotive." The outcome of the GM negotiations will be fully discussed in a later chapter.

Before the end of 1970, two other Japanese companies signed agreements for the rights to Wankel engines in certain power ranges. The first was Nissan Motor Company of Tokyo, manufacturer of Datsun cars and Nissan cars and trucks and parent company of an industrial group including Fuji Heavy Industries and Aichi Machinery Company. Nissan is placed fourth among Japanese industrial establishments. The company began to develop into a large-scale enterprise after World War II when Nissan resumed automobile production in cooperation with Austin Motor Co. Ltd., Longbridge, Birmingham, England. Mergers with Prince Ltd., and Isuzu Motors Ltd., launched a steep rise in economic growth about five years ago. Strengthened by an ever-growing production capacity, the company now produces a wide range of models, starting with the small Datsun Cherry, Sunny, and 510, including the Datsun 240Z sports car, the medium-sized Nissan Cedric and the Nissan President with its 4-liter engine. Today, Nissan Motor Co. Ltd. has about 48,000 employees in different factories on the coast and in the adjacent hinterland of Tokyo Bay. A total of 1,230,000 vehicles came off the production lines in 1969. With the extension of the Tochigi works completed, an annual capacity of 2 million cars is planned for 1972. The company had a turnover of $1,850 million for the 1969 fiscal year.

Nissan's license from Wankel G.m.b.H. and Audi-NSU Auto-Union AG entitles the Japanese company to produce Wankel engines from 80 to 120 horsepower for passenger cars. That does not mean there will be Wankel-powered Datsun and Nissan cars in 1972. Mr. Katsuji Kawamata, President of Nissan Company, signed the agreement at the NSU head office in Neckarsulm on the afternoon of October 1, 1970. Nissan has no immediate plan to produce the engine on a commercial basis, but will instead push ahead with research and development in the interest of exhaust emission control. President Kawamata, however, commented that the rotary engine would play the part of a reserve unit rather than the mainstay in the fight against air pollution.

Faced with the prospect of tighter emission standards and the need to make engines run efficiently on non-lead gasoline, Nissan also is engaged in intensive experiments with electric cars, steam engines and gas turbine engines, in parallel with improvements to the conventional internal combustion engine. As part of its program, Nissan had kept tabs on the rotary-piston engine and had been negotiating with the German companies for quite some time.

The other Japanese contract was signed in Japan on November 26, 1970 by representatives of Wankel G.m.b.H. and Audi-NSU Auto-Union AG and the chief executive officers of Suzuki Motor Company Ltd., of Hamamatsu. The licensing agreement covers manufacture and distribution of Wankel engines of 20 to 60 horsepower for motorcycles. Suzuki produces some 500,000 motorcycles a year plus two lines of small passenger cars: the rear-engined Fronte 360 and the front-wheel-drive Fronte 800. Suzuki has a labor force 10,000 strong, and annual turnover for the 1970 fiscal year will approach $300,000,000.

How many of these licensed companies have developed the Wankel engine concept for their own applications is covered in the following chapters.

10

Curtiss-Wright

KNOWN MAINLY FOR ITS contributions in the world of aviation rather than land transportation, the Curtiss-Wright Corporation is a relatively young organization, having been established on August 9, 1929 by a merger between the Aeroplane & Motor Co., the Curtiss-Wright Aeronautical Corp. and their related and subsidiary companies engaged in various branches of the aviation industry. The latter group included Curtiss-Wright Airplane Co., Curtiss-Wright Airports Corp., Curtiss-Wright Export Corp., Curtiss-Wright Caproni Corp., Devon Corp., Keystone Aircraft Corp., Moth Aircraft Corp., N.Y. Air Terminals Inc., and N.Y. & Southern Airlines, Inc.

Both the Wright and the Curtiss interests had traditions starting from the very beginning of controlled flight. It is proper to say that Curtiss-Wright's history began when the Wright brothers made the first powered flight at Kittyhawk, North Carolina, on December 17, 1903. In 1909 the Wright brothers formed the Wright Company of Dayton, Ohio, to manufacture airplanes. In 1911, another intrepid aviation pioneer, Glenn H. Curtiss, sold the Army its second airplane and the Navy its first. In 1916, he formed the Curtiss Aeroplane and Motor Corporation.

In 1907, Glenn Curtiss, together with Dr. Alexander Graham Bell, J. A. D. McCurdy, F. W. Baldwin and Lieut. Thomas Selfridge, formed the Aerial Experiment Association which built a number of airplanes and performed various aerial experiments.

The outbreak of the war in Europe in 1914 gave a decided impetus to the demand for airplanes, and the Curtiss and Wright plants grew rapidly to meet the demand, expanding still further after the United States entered the struggle. Curtiss concentrated on airframes while Wright devoted most of its capacity to engines. A subsidiary, Wright-Martin Corporation, built the famous Hispano-Suiza V8 aircraft engine under license. After the war, Curtiss planes and Wright engines continued to set records. It was a Wright "Whirlwind" engine that powered Lindbergh's plane, "The Spirit of St. Louis," on the first non-stop flight from New York to Paris in 1927. After the 1929 merger, the Curtiss-Wright Corporation became an integrated aviation manufacturing concern, bringing about a new era in air transport.

In 1936 Curtiss-Wright president Guy Vaughn dissolved all existing subsidiaries except Wright Aeronautical Corp., and operations were taken over by the parent company. Wright Aeronautical Corp., 98% owned, was merged into the parent firm on Oct. 31, 1951, the business being continued as Wright Aeronautical Division. During World War II, Curtiss-Wright was a prime manufacturer of pursuit planes, dive bombers, scout observation planes, transports and of aircraft engines and propellers. For a number of years, the company existed mainly on defense contracts from the U.S. government. It let itself be bypassed when the turbo-jet gas turbine engines came in, and its turnover sank in one year from $599,000,000 to $389,000,000—a 35% drop.

Basic components of the IRC-6.

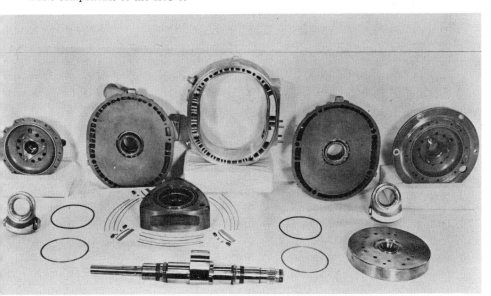

Bucking the trend, Curtiss-Wright continued to manufacture piston-powered aircraft engines long after the turbo-prop gas turbine had proved its clear superiority for applications where turbo-jets were not considered suitable. Radial piston-powered aircraft engines still in production in 1965 were the 3,700 horsepower Turbo-Compound, the 2,800 horsepower Cyclone 18, the 1,625 horsepower Cyclone 9 and the 800

Elevation and cross section of the IRC-6. a = axial width of chamber. R = rotor radius. e = eccentricity. b = maximum clearance along major axis.

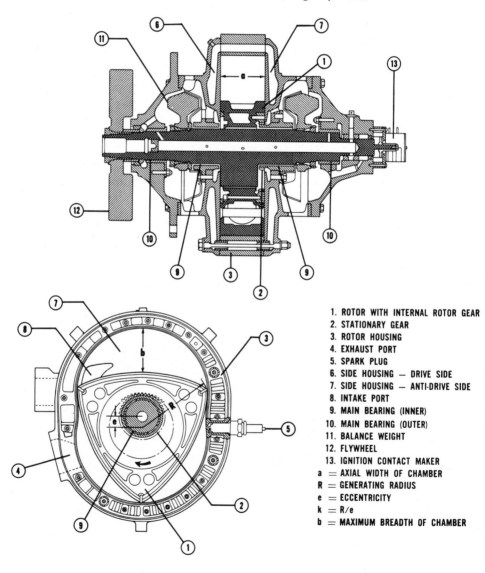

1. ROTOR WITH INTERNAL ROTOR GEAR
2. STATIONARY GEAR
3. ROTOR HOUSING
4. EXHAUST PORT
5. SPARK PLUG
6. SIDE HOUSING — DRIVE SIDE
7. SIDE HOUSING — ANTI-DRIVE SIDE
8. INTAKE PORT
9. MAIN BEARING (INNER)
10. MAIN BEARING (OUTER)
11. BALANCE WEIGHT
12. FLYWHEEL
13. IGNITION CONTACT MAKER
a = AXIAL WIDTH OF CHAMBER
R = GENERATING RADIUS
e = ECCENTRICITY
k = R/e
b = MAXIMUM BREADTH OF CHAMBER

horsepower Cyclone 7. It never really got into competition with Pratt & Whitney and General Electric for the lucrative contracts that were being offered in the late Fifties and early Sixties to supply gas turbine engines for the new-generation civil aviation airplanes. Instead, the corporation started to look at diversification moves as a possible way out of its troubles.

Curtiss-Wright president Roy T. Hurley, a former Ford Motor Company executive, tried hard to generate some enthusiasm for the company's activities by backing technical innovations. The first time was 1956, when Curtiss-Wright built an experimental air-cushion vehicle. It was an automobile without wheels, supported on a layer of air forced underneath the car by fans and ducts and kept from escaping by skirts around the entire vehicle perimeter. It remained purely experimental, but it received favorable press reaction, and quotations for Curtiss-Wright stock went up.

Full throttle performance of the IRC-6 with side intake port. The engine developed 100 horsepower at 5,400 r.p.m. The torque curve stayed within a 20 foot-pound range from 2,000 to 6,000 r.p.m.

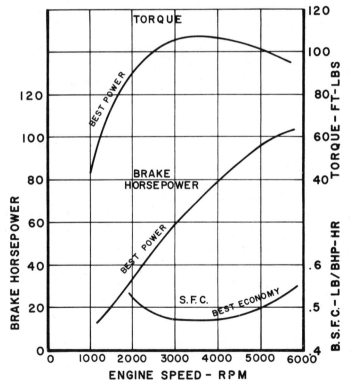

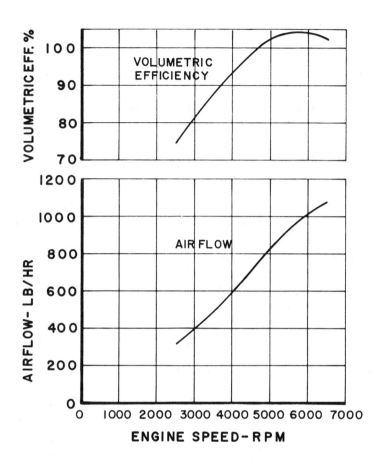

Volumetric efficiency easily exceeded 100% with the peripheral intake port. However, the peripheral intake port proved slightly inferior with regard to air utilization (compared with the side port version), despite air flow characteristics allowing an intake of almost 1,100 pounds per hour at 6,500 r.p.m.

The second time was late in 1958, when Curtiss-Wright purchased the U.S. rights to the Wankel engine from NSU Motorenwerke AG. Again, the press reacted favorably, and the stock market responded in very positive fashion. The $1 shares had been selling for about $30 for a long time. After the Wankel engine deal, the price rose to $40.75. All buyers hoped that through manufacture of the new type power-plant, Curtiss-Wright would soon be able to pay fabulous dividends.

Then the company announced that its quarterly dividend was being cut in half because of other changes in the profit picture, and the stock price fell again. This brought trouble from the government: Hurley was called up to answer to the Securities and Exchange Commission, who suspected stock manipulation. It looked like a classic case of announcing a new technical miracle to boost the stock price so as to sell

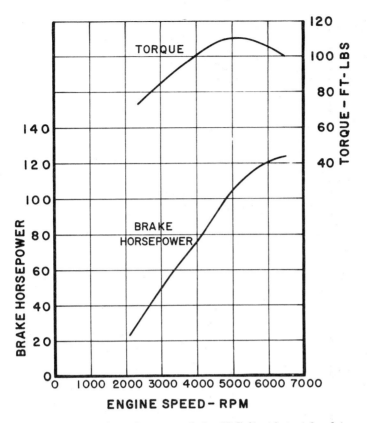

Wide open throttle performance of the IRC-6, with peripheral intake port, shows lower intake losses with correspondingly higher air flow and output.

personal stock at a huge profit—before announcing the cut in dividends. No action was taken against Curtiss-Wright, because Hurley not only had acted in good faith but had also kept his Curtiss-Wright holdings. As if to add further proof of their belief in the Wankel engine, Curtiss-Wright began an intensive research program under the direction of Max Bentele. Bentele was the expert who had sold Hurley on the idea of the Wankel license in the first place.

This German-born engineer was 49 years old when Curtiss-Wright signed the contract with NSU. Bentele had studied at the Technical University of Stuttgart and got his engineering diploma in 1932. Post-graduate studies earned him the title of Doctor of Engineering in 1938. Bentele left Germany in 1956 to accept a position as staff engineer with Curtiss-Wright. Working in the Design Department, he embraced the Wankel engine with great enthusiasm in 1958. He was in charge of all rotating combustion engine projects from 1958 to 1960 when he was promoted to Chief Scientist. This position enabled him to concentrate his efforts on the more intricate problems of the engine and its appli-

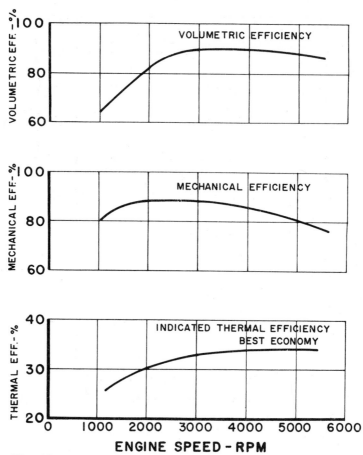

The side port gave high volumetric efficiency over a wide speed range. The peak efficiency was 90%, while the peripheral-port engine exceeded 100%. Mechanical efficiency was low at low speeds, indicating low mechanical friction. Friction losses increased moderately with rising speeds. The thermal efficiency curve for the side intake port version of the IRC-6 shows 34% to be optimum for maximum economy over a wide r.p.m. range.

cation to wider fields. In 1964, his responsibility was extended to all design activities which covered gas turbine, reciprocating and rotating combustion engines as well as space power plants. Bentele left Curtiss-Wright in 1967 when the basic problems of the Wankel engine were resolved and it was ready for production. His place in the rotating combustion engine design group was taken by his assistant, Charles Jones.

Jones was chief project engineer of Curtiss-Wright with responsibility for reciprocating engine design and all rotating combustion engines. He joined Curtiss-Wright in 1950 as a test engineer and in 1955 he was made section head for charge stress and applied mechanics, then he

IRC6 TEST PERFORMANCE
FULL THROTTLE
SMALL VS. LARGE SIDE INTAKE PORT

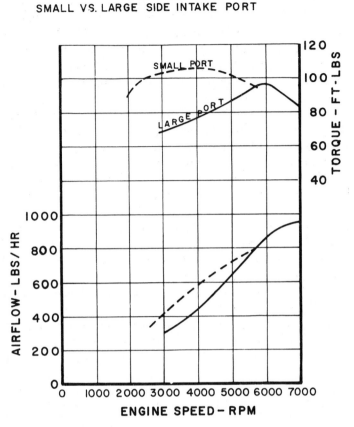

Port size proved to have a great influence on engine power and speed characteristics. The small port showed higher torque at lower r.p.m., air flow being considerably higher in the critical 3,000 to 5,500 r.p.m. range.

worked as a project engineer on the J-65 engine in 1957. The following year, Bentele selected Jones for the design activity of a group of engineers assigned to develop the Wankel engine. Jones was given responsibility for all rotating combustion engine design, including basic engine development applications and advance design. Hurley, who had made the decision to back the Wankel engine, left Curtiss-Wright only about a year after signing the agreement with NSU and Wankel. His place was taken not by an engineer, but by a lawyer, T. Roland Berner.

Apparently it was never Curtiss-Wright's idea to start producing Wankel engines of any kind in its own factories. They planned to study applications, develop basic units, and sell licenses to other companies.

This has not, as yet, resulted in any contracts of major importance. To date, Curtiss-Wright's revenue from the Wankel engine rights is pathetically small. In 1969, the company earned $1,543,000 on the Wankel engine. That means about 12¢ per share was earned from Wankel royalties, against total Curtiss-Wright earnings of $1.36 per share. The overall financial picture at Curtiss-Wright is so complex that it is difficult to sum this up as "good" or "bad."

When you read about the Wankel-powered cars built by NSU, Toyo Kogyo, and Daimler-Benz, the question of what Curtiss-Wright has accomplished in the Wankel engine department since taking out its license more than 12 years ago is inevitable. The engineering developments will be covered later in some depth. On the commercial side, the achievements have been more visible to the public. Three things have happened:

1. Curtiss-Wright succeeded in re-negotiating its license agreement with NSU in 1964.
2. Curtiss-Wright contracted with Westinghouse to develop Wankel-powered generator sets in 1965.
3. Curtiss-Wright sold a sub-license to Outboard Marine Corporation in 1966.

The new contract with NSU called for revenue distribution on license fees and royalties according to the following table:

a) Outside North America—NSU 54%, Wankel 36%, Curtiss-Wright 10%
b) However, if engines are imported to U.S.—Curtiss-Wright 75%, NSU 15%, Wankel 10%

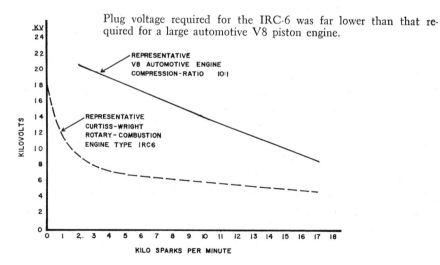

Plug voltage required for the IRC-6 was far lower than that required for a large automotive V8 piston engine.

c) In North America on manufacture by Curtiss-Wright—NSU Wankel 1½% for automotive, 3% for other

d) For sub-licensing entrance fees and royalties received by Curtiss-Wright from sub-licensing—Curtiss-Wright, 60%, NSU 24%, Wankel 16%. However, if engines are exported to Germany—Curtiss-Wright 25%, NSU 45%, Wankel 30%

Experiments with dual side intake ports on the RC-60 (the renamed later versions of the IRC-6) show almost equal airflow characteristics to the peripheral-port engine up to 3,500 r.p.m. The peripheral-port engine has better volumetric efficiency above that speed.

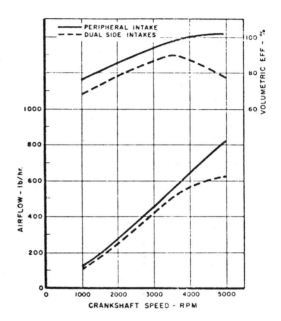

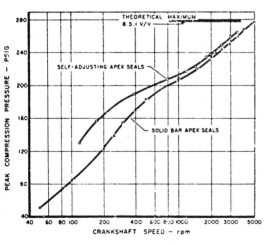

The peripheral-port engine was fitted with an experimental type of self-adjusting apex seals. It is compared with a side-port engine running with solid apex seals, in terms of peak compression pressure. A side-port engine will develop its theoretical pressure rise at lower r.p.m., because the effective low-speed compression ratio is higher, due to earlier intake closing with side ports. But tests showed that with the self-adjusting apex seals, the peripheral-port engine had considerably higher pressures from cranking speed to idling speed. The difference at operational speeds is less than one point (in relation to an 8.5:1 compression ratio).

The deal with Westinghouse was announced on December 3, 1965, and concerned the development of a Wankel engine to provide basic power for a lightweight electric generator set developed by the Aerospace Electrical Division of Westinghouse Electric Corporation.

Being offered initially to the military services, the Westinghouse generator set is seen as meeting the need for a transportable unit to supply precision power to such forward combat area systems as radar, communications networks and mobile missile launchers. Military equipment of this type requires a close-tolerance power supply for electronic components, but the power generator equipment must be capable of being moved on short notice along with the tactical gear. When mobility and transportability become paramount, weight of the equipment and fuel consumption, because re-supply can be difficult, are major considerations in design. The Curtiss-Wright RC engine was

Attempts to develop a high-performance version of the RC1-60 were rewarded by a peak power output of 155 horsepower at 7,000 r.p.m. The torque curve stayed at 130 foot-pounds from 5,000 to 5,500 r.p.m., and volumetric efficiency exceeded 100% between 4,250 and 7,000 r.p.m.

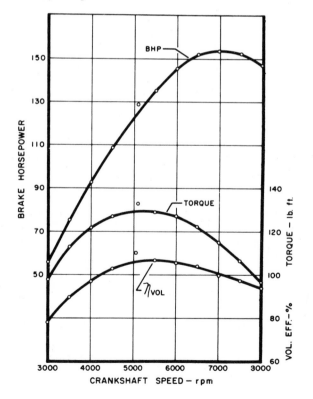

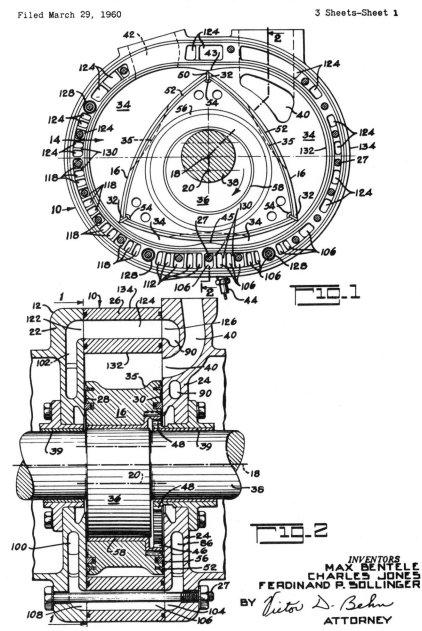

FIG_1

FIG_2

INVENTORS
MAX BENTELE
CHARLES JONES
FERDINAND P. SOLLINGER

BY _Victor D. Behn_

ATTORNEY

Basic patent for the water cooling system developed at Curtiss-Wright, shows passages of varying size according to heat distribution.

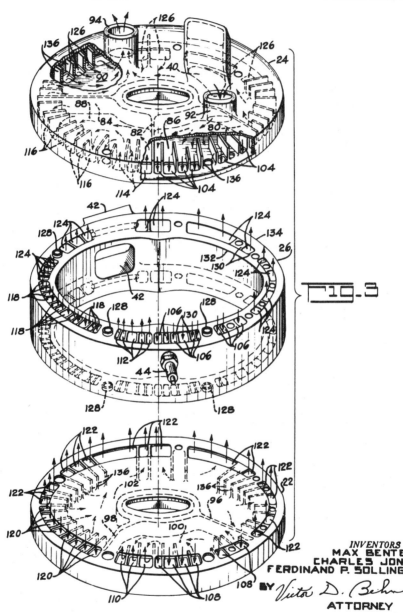

FIG. 3

INVENTORS
MAX BENTELE
CHARLES JONES
FERDINAND P. SOLLINGER
BY Victor D. Behn
ATTORNEY

Proposed coolant guide-vane configuration for the end covers and housing.

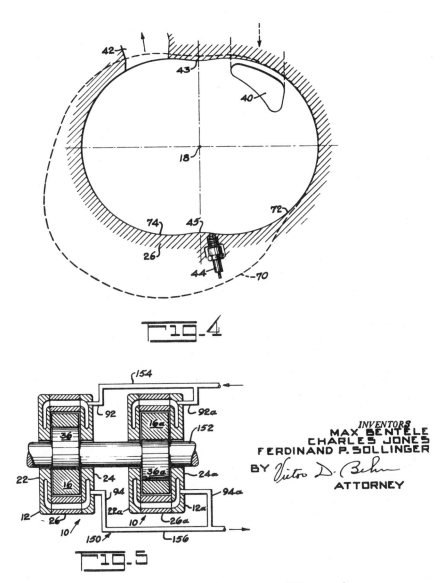

FIG. 4

FIG. 5

INVENTORS
MAX BENTELE
CHARLES JONES
FERDINAND P. SOLLINGER
BY *Victor D. Behn*
ATTORNEY

The upper sketch shows the heat distribution curve (70) as established by experiments at Curtiss-Wright. The lower sketch shows how Max Bentele proposed to cool a twin-rotor housing, using distribution to both units from a common pipe, individual circulation through four end covers, and return flow through separate outlets into a common pipe.

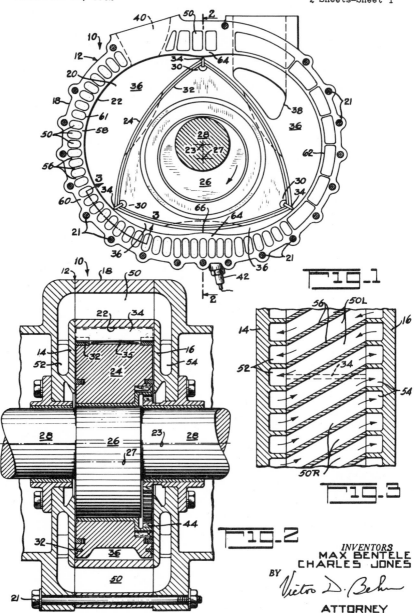

Radial separating ribs between the coolant passages were claimed to reduce the tendency to produce chatter marks on the working surface. The angle of the ribs was intended to add strength to the housing.

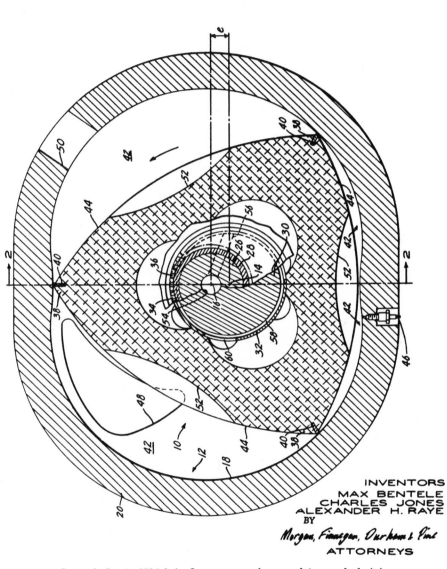

INVENTORS
MAX BENTELE
CHARLES JONES
ALEXANDER H. RAYE
BY
Morgan, Finnegan, Durham & Pine
ATTORNEYS

One of Curtiss-Wright's first patents for supplying and draining rotor cooling oil via the eccentric shaft. e = eccentricity. 2 = minor axis. 10 = rotor. 14 = eccentric axis. 16 = mainshaft axis. 18 = working surface. 34 = ring gear. 36 = rotor gear. 54 = central passage. 56 = radial passage. 58 = annulus inlet.

FIG. 2.

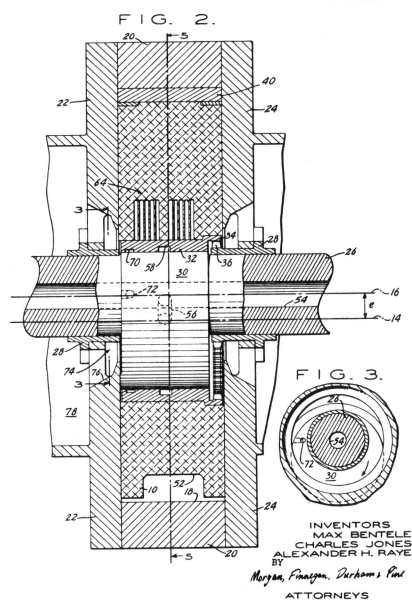

FIG. 3.

INVENTORS
MAX BENTELE
CHARLES JONES
ALEXANDER H. RAYE
BY

Morgan, Finnegan, Durham & Pine

ATTORNEYS

Elevation of the rotor cooling system. Inside the rotor are three sets of finned passages (64), located adjacent to, but radially inward from the apex portions. All passages are connected to an inlet manifold and an outlet manifold.

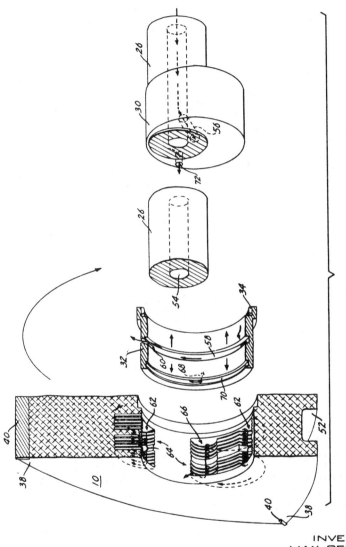

INVENTORS
MAX BENTELE
CHARLES JONES
ALEXANDER H. RAYE
BY
Morgan, Finnegan, Durham & Pine
ATTORNEYS

Exploded view of the rotor cooling system.

F I G. 5.

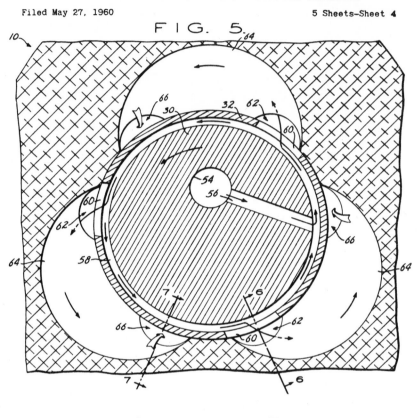

F I G. 7.   F I G. 6.

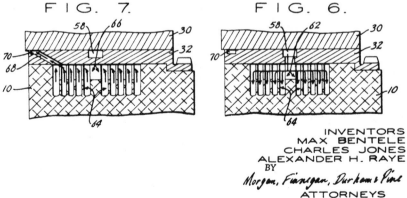

INVENTORS
MAX BENTELE
CHARLES JONES
ALEXANDER H. RAYE
BY

Morgan, Finnegan, Durham & Pine
ATTORNEYS

Detail of the rotor cooling system in cross-section (above) and elevation (below).

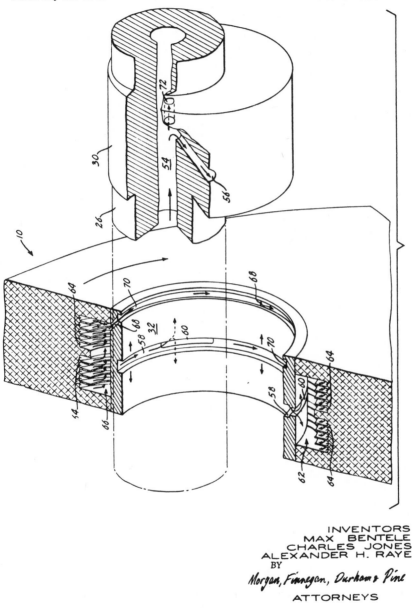

INVENTORS
MAX   BENTELE
CHARLES JONES
ALEXANDER H. RAYE
BY
Morgan, Finnegan, Durham & Pine
ATTORNEYS

Arrows mark the direction of cooling oil flow to and from the rotor.

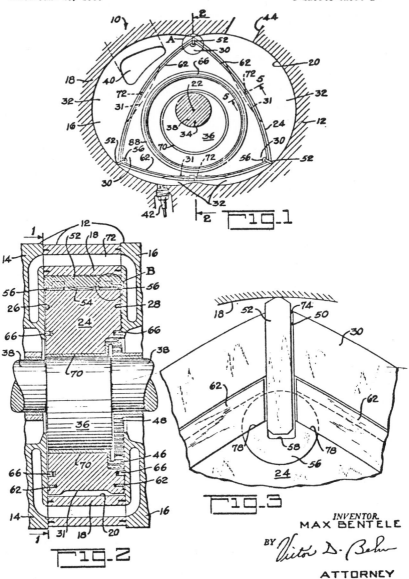

INVENTOR.
MAX BENTELE

BY

ATTORNEY

Max Bentele applied for a patent for this sealing configuration in January 1960. The solid, full-width apex seals had an intricate internal spring-loading. 52 = apex seal. 56 = joint trunnion. 54 = spring. 62 = side seals.

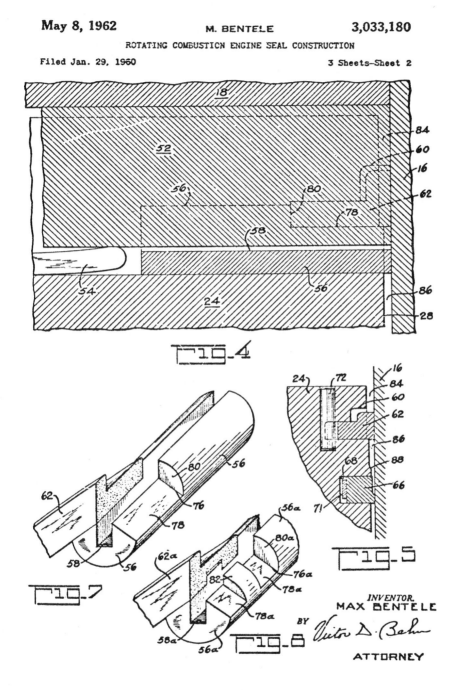

INVENTOR.
MAX BENTELE

BY *Victor D. Behm*

ATTORNEY

Enlarged detail of the apex seal corner in Bentele's sealing system.
52 = apex seal. 54 = spring. 56 = joint trunnion. 62 = side seal.

May 8, 1962 M. BENTELE 3,033,180

ROTATING COMBUSTION ENGINE SEAL CONSTRUCTION

Filed Jan. 29, 1960 3 Sheets-Sheet 3

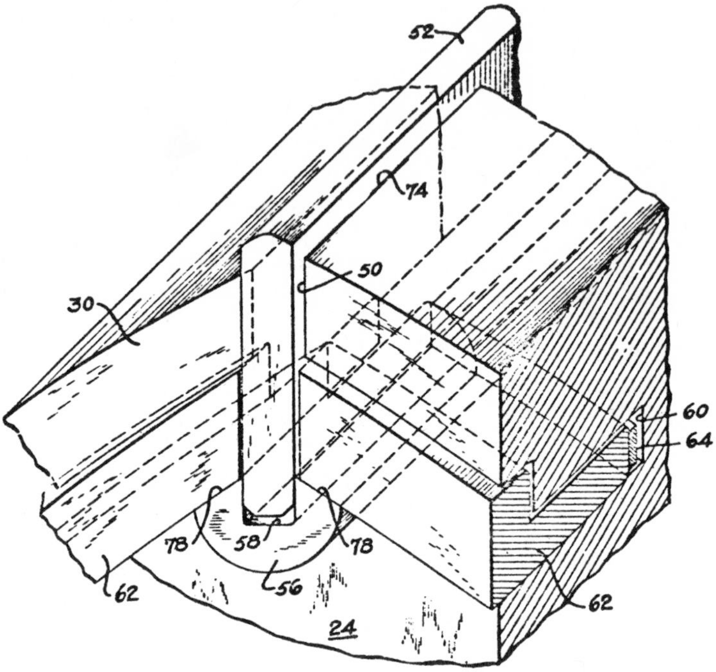

FIG. 6

INVENTOR.
MAX BENTELE
BY Victor D. Behn
ATTORNEY

Apex and side seal assembly according to Bentele's patent.

FIG. 1.

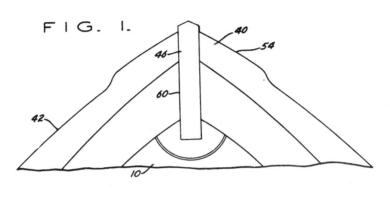

FIG. 2.

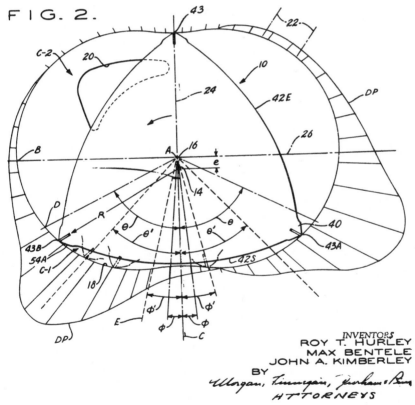

INVENTORS
ROY T. HURLEY
MAX BENTELE
JOHN A. KIMBERLEY
BY
Morgan, Finnegan, Durham & Bu
ATTORNEYS

Hurley's patent covered a raised land near each apex, to aid sealing
and gas flow characteristics.

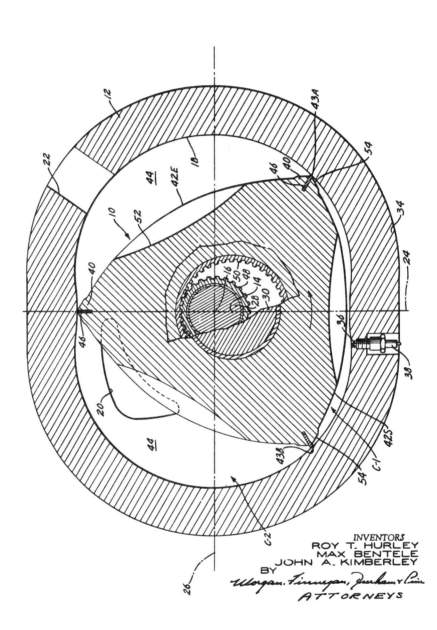

INVENTORS
ROY T. HURLEY
MAX BENTELE
JOHN A. KIMBERLEY
BY
Morgan, Finnegan, Durham & Pine
ATTORNEYS

Cross-section of the Hurley rotor with raised apex lands.

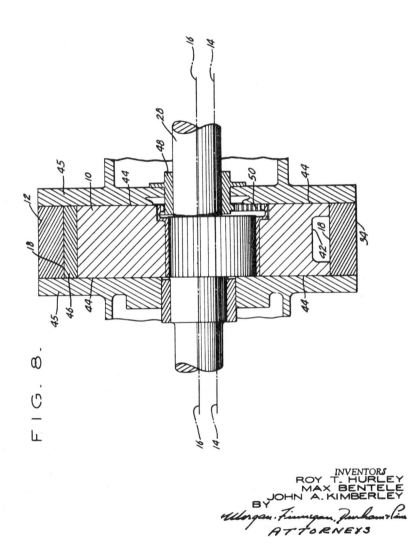

FIG. 8.

INVENTORS
ROY T. HURLEY
MAX BENTELE
JOHN A. KIMBERLEY
BY
Morgan. Finnegan, Durham + Pine
ATTORNEYS

Elevation of the Hurley rotor with raised apex lands.

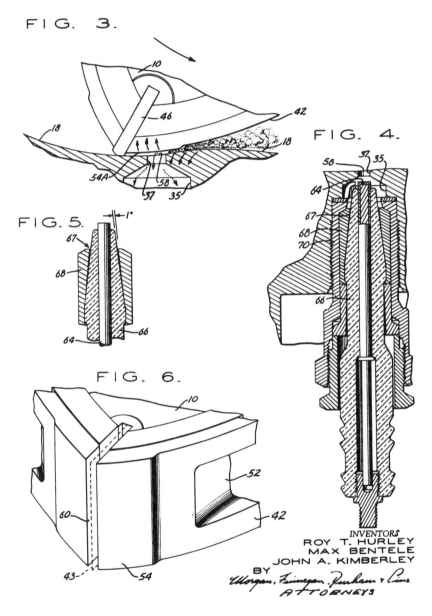

FIG. 3.

FIG. 4.

FIG. 5.

FIG. 6.

INVENTORS
ROY T. HURLEY
MAX BENTELE
JOHN A. KIMBERLEY
BY
Morgan, Finnegan, Durham & Pine
ATTORNEYS

Details of the raised apex land and seal configuration. The top drawing shows how exhaust gases are to be prevented from flowing beyond the exhaust port.

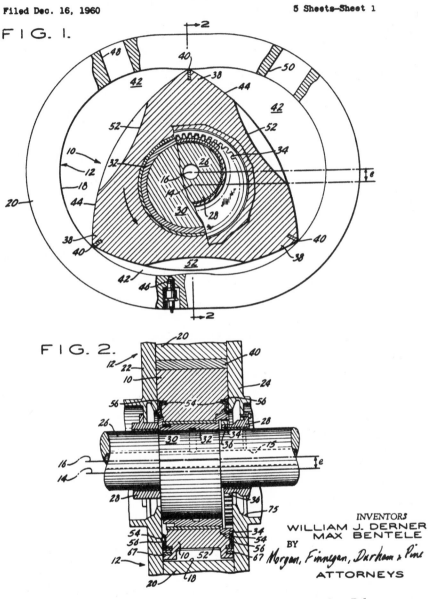

FIG. I.

FIG. 2.

INVENTORS
WILLIAM J. DERNER
MAX BENTELE
BY Morgan, Finnigan, Durham & Pine
ATTORNEYS

Bentele's patent for improved oil sealing. Top: cross-section. Below: elevation. The principle is two-stage sealing. The inner seal is marked 56 and the outer seal 54. Between them, overlapping with the phasing gears, is a crescent shaped vent opening that is present only at the point of inward radial acceleration. Any oil trapped between the two seal rings is therefore evacuated through the vent hole, in a volume and force dependent on rotor acceleration loads.

FIG. 3.

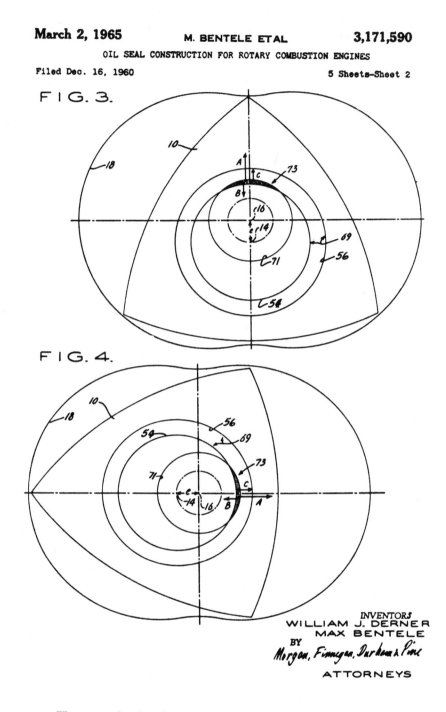

FIG. 4.

INVENTORS
WILLIAM J. DERNER
MAX BENTELE
BY
Morgan, Finnigan, Durham & Pine
ATTORNEYS

These two sketches show the movement of the oil seal vent hole during rotor rotation.

FIG. 5.

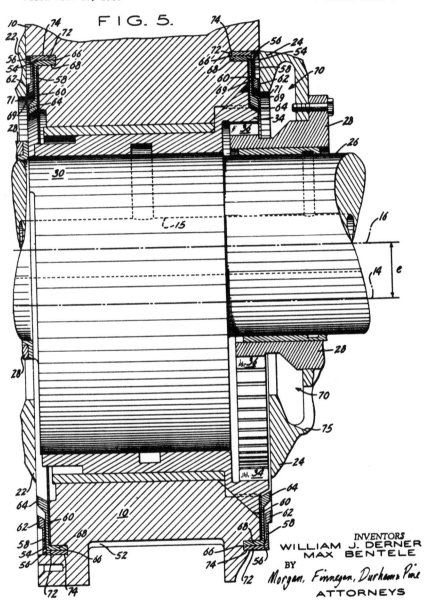

INVENTORS
WILLIAM J. DERNER
MAX BENTELE
BY
Morgan, Finnegan, Durham & Pine
ATTORNEYS

Closeup of the basic dual oil seal configuration patented by Max
Bentele. The annular legs extend inward, and are axially flexible to
insure contact with the end cover walls. The legs of the inner seals
are stiffer than the legs of the outer seals and do not adapt to changes
in tolerance, since they operate with oil on both sides. The outer
seals will conform to thermal distortions between the rotor and the
end covers.

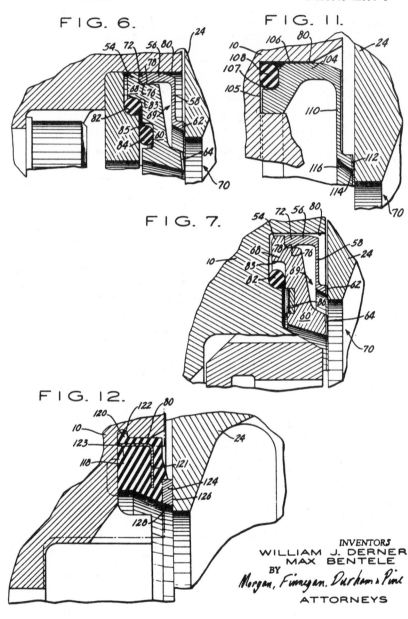

FIG. 6.

FIG. 11.

FIG. 7.

FIG. 12.

INVENTORS
WILLIAM J. DERNER
MAX BENTELE
BY
Morgan, Finnegan, Durham & Pine
ATTORNEYS

Variations on the dual oil seal.

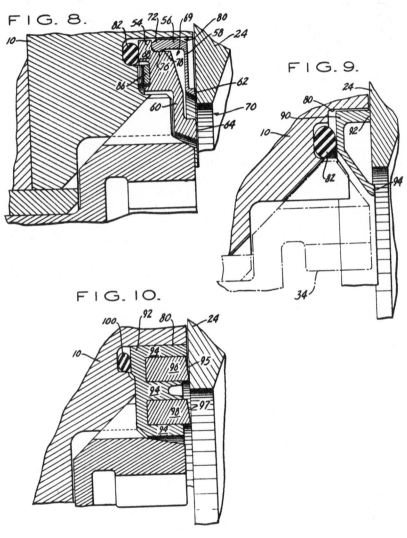

FIG. 8.

FIG. 9.

FIG. 10.

INVENTORS
WILLIAM J. DERNER
MAX BENTELE
BY
Morgan, Finnegan, Durham & Pine
ATTORNEYS

Variations on the dual oil seal.

Curtiss-Wright's 4 RC-180-2 air-cooled engine.

chosen because of its relative simplicity (only two moving parts in the engine itself) and because of its weight, size and fuel consumption advantages.

The model chosen for the Westinghouse unit is designated the Curtiss-Wright RC2-60-N8. Curtiss-Wright engineers say the N8 engine can be operated at a constant speed of 4,800 r.p.m. and is adaptable to generating equipment in the 45 to 60 kilowatt range. The Westinghouse generator set, including the RC engine, weighs under 1,100 pounds, compared to about 4,700 pounds for a unit with a diesel engine. Fuel consumption with the Curtiss-Wright RC engine is expected to be about 9.5 gallons per hour, compared to a gas turbine-powered unit of equivalent weight which would use 17 gallons per hour.

Curtiss-Wright Corporation and Outboard Marine Corporation announced on March 2, 1966, that Outboard Marine had obtained licenses under patents to develop, manufacture and market Wankel engines. In a joint statement, T. Roland Berner, Chairman and President of Curtiss-Wright, and W. C. Scott, President of Outboard Marine, said that their companies had signed a license agreement covering development and manufacture of the new-type powerplant by OMC, involving the payment to Curtiss-Wright of fees and guaranteed royalties in excess of $1,000,000. OMC, the world's largest manufacturer of outboard motors and a leading producer of stern-drive marine engines and fiberglass boats, concluded agreements with NSU Motorenwerke AG, Neckarsulm, Germany, and Wankel G.m.b.H., Lindau, Germany, as well as with Curtiss-Wright. These agreements grant OMC *non-exclusive licenses* to manufacture and sell Wankel engines as powerplants in its marine products throughout the world, and for certain non-marine purposes in North America. OMC also has options for worldwide licenses for a limited number of non-marine applications.

In addition to Evinrude and Johnson outboard motors and stern-drive engines, OMC produces several other products powered by gasoline engines. They include Cushman golf carts and industrial vehicles, Lawn Boy power lawn mowers, Pioneer chain saws, and Evinrude and Johnson snowmobiles. The negotiation of the license agreement between Curtiss-Wright and OMC, Berner and Scott said, marked the culmination of a co-operative development effort carried out over a period of several years. During this time, OMC had extensively tested and eval-

Installation sketch for the RC2-90 Y2 engine.

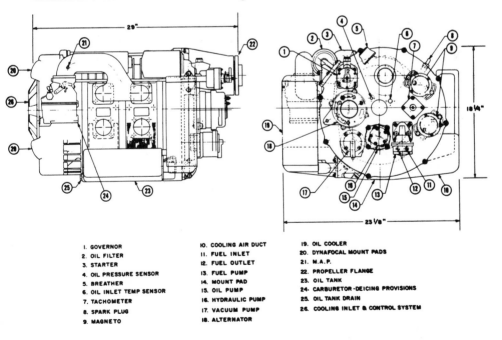

I. GOVERNOR	IO. COOLING AIR DUCT	19. OIL COOLER
2. OIL FILTER	II. FUEL INLET	20. DYNAFOCAL MOUNT PADS
3. STARTER	I2. FUEL OUTLET	21. M.A.P.
4. OIL PRESSURE SENSOR	I3. FUEL PUMP	22. PROPELLER FLANGE
5. BREATHER	I4. MOUNT PAD	23. OIL TANK
6. OIL INLET TEMP SENSOR	I5. OIL PUMP	24. CARBURETOR-DEICING PROVISIONS
7. TACHOMETER	I6. HYDRAULIC PUMP	25. OIL TANK DRAIN
8. SPARK PLUG	I7. VACUUM PUMP	26. COOLING INLET & CONTROL SYSTEM
9. MAGNETO	I8. ALTERNATOR	

uated models of the rotating combustion engine for marine and other uses.

The starting point for the design of Curtiss-Wright's first experimental Wankel engine was the NSU DKM-54. It was redesigned on a vastly enlarged scale, and received the benefit of Dr. Froede's kinematic inversion. It was called the IRC-6 and operated on the KKM (Kreiskolbenmotor) principles. It had a displacement of 60 cubic inches, or almost one liter, and it was eight times bigger in displacement than the NSU KKM-125. The design was not started until an evaluation program of dimensions, R/e ratios, porting configuration, rotor and housing cooling, carburetion and ignition had been concluded.

Bentele had decided to stick with the three-cornered rotor and two-lobe epitrochoidal housing, as developed by NSU. The design pattern of the Curtiss-Wright Wankel engine was closely related to the KKM-125, but it was designed with the same geometrical relationship as the DKM-54. In the IRC-6, rotor radius was 113.4 mm., eccentricity was 16.6 mm., and the R/e ratio was 6.85:1.

The IRC-6 mainshaft ran in two sleeve bearings, one on each side of the eccentric, with two more bearings outside the balance weights. The flywheel was mounted separately and was easily detachable to facilitate experiments to study its effect on engine operation and torsional vibration characteristics. An advanced liquid cooling system was devised for the housing. It worked on a principle they called "multi-pass." The housing structure was made up of double walls interconnected by ribs. The ribs formed passages for the coolant. The outer walls

Air cooling system configuration designed for the twin-rotor RC2-90 Y housing.

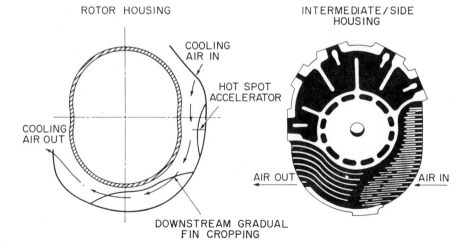

ROTOR HOUSING   INTERMEDIATE/SIDE HOUSING

COOLING AIR IN

HOT SPOT ACCELERATOR

COOLING AIR OUT

AIR OUT   AIR IN

DOWNSTREAM GRADUAL FIN CROPPING

were relatively thick and stayed at low and uniform temperature, so that they helped restrict deformation of the inner walls, which were quite thin.

The coolant passages were designed as a multi-pass forced flow system, flow velocity being dictated by heat transfer requirements. In other words, the coolant was speeded up in the combustion zones to carry away as much heat as possible, and slowed down in other areas with lower heat input. The multi-pass water cooling system was patented by Max Bentele, Charles Jones and Ferdinand P. Sollinger on November 7, 1961. It took full account of the fact that the heat input to the housing is not uniform around its periphery because each of the operational phases always takes place adjacent to the same portion of the housing surface. This was met by designing channels of different size, giving different coolant flow velocity according to requirements. The patent also covered the repeated return of the coolant through different passages between the two end covers before exiting from the housing to the radiator.

With this design, the coolant flow passages, particularly in regions of high heat input, have smooth hydrodynamic contours, that is, they have no abrupt changes in direction or area. This feature serves to prevent the presence of dead spots in the coolant flow passages. Dead spots are areas in the flow passages having little or no flow velocity of the coolant. By avoiding dead spots, any vapor produced in the passages is instantly carried away by the coolant flow, which in turn prevents hot spots resulting from vapor accumulation. The serially con-

Cooling air flow through the RC2-90 Y2 rotor housing.

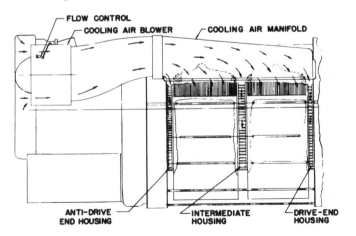

nected groups of passages are circumferentially spaced about the axis of the rotary mechanism and the passage groups are so connected that the passage group to which the liquid coolant is first supplied is located adjacent to one end of a region of relatively high input to the outer body. As the coolant flows through these serially connected groups of passages, it progresses around the axis of the outer body toward the other end of the high heat input region.

The multi-pass, forced-flow cooling system is clog-free, prevents vapor accumulation, and is easily manufactured. The block and end housings are located by dowels and clamped together by 21 bolts, spaced with regard to gas loads and located within the area of the coolant passages and hollow dowels. The next step was a water-cooled housing with a very thin inner wall and non-axially positioned, passage-forming ribs linking the inner and outer walls of the housing. (If the ribs run axially they tend to produce chatter marks on the working surface as each apex

Evolution of RC2-90 performance during 1966, at a constant 5,000 r.p.m. using JP 5 fuel.

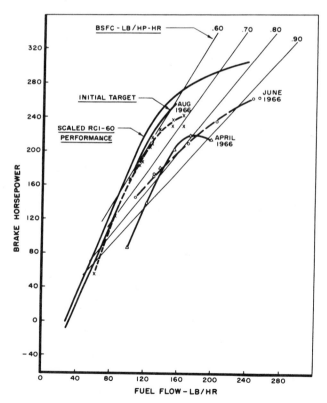

seal sweeps over the trochoidal track and strikes the reinforced portions of the wall in rapid succession.)

Curtiss-Wright followed in NSU's footsteps by deciding on an oil-cooled rotor for the IRC-6. The first type of rotor was an aluminum forging with forced-flow oil cooling. The aluminum alloy provided the important benefits of great weight saving in the principal moving part of the engine and a rotor structure with high thermal conductivity. A high heat dissipation rate was particularly beneficial in preventing the formation of hot spots within the rotor, while lightweight construction greatly reduced the energy losses that resulted from rotor inertia forces.

A rotor constructed of a lightweight metal alloy, however, demands adequate and efficient cooling; such alloys fail from overheating at a considerably lower temperature than materials such as cast-iron or steel. In the IRC-6, oil was fed into the rotor shaft oil intake under pressure. The flow separated at the rotor bearing—most of the oil going inside

High-speed performance of the RC2-90 improved month by month, although target output was never achieved. The curves show output related to fuel flow at a constant 6,000 r.p.m., using JP 5 fuel.

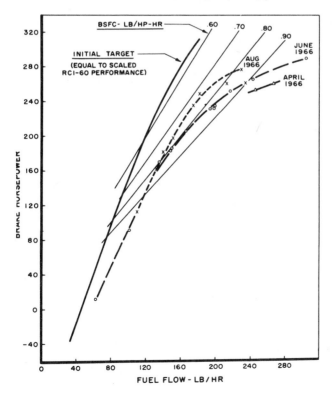

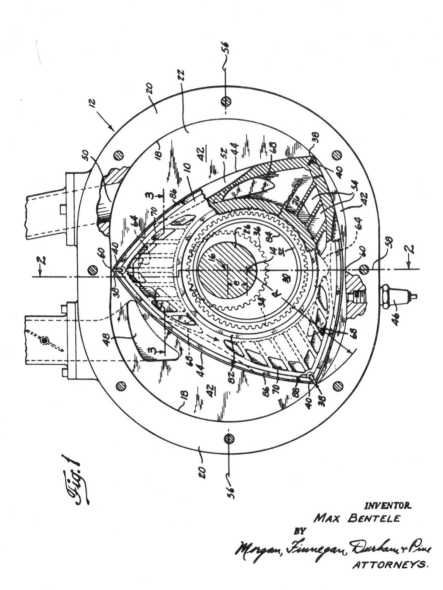

Fig. 1

INVENTOR.
MAX BENTELE
BY
Morgan, Finnegan, Durham & Pine
ATTORNEYS.

Air cooling experiments began with patents covering air cooling of the rotor rather than the housing. This cross-section shows the multiple ribs inside the rotor structure.

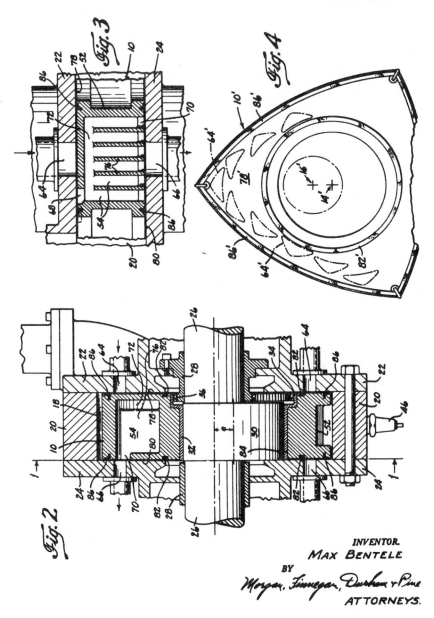

INVENTOR.

MAX BENTELE

BY

Morgan, Finnegan, Durham & Pine

ATTORNEYS.

The inner makeup of the air-cooled rotor, patented by Max Bentele.

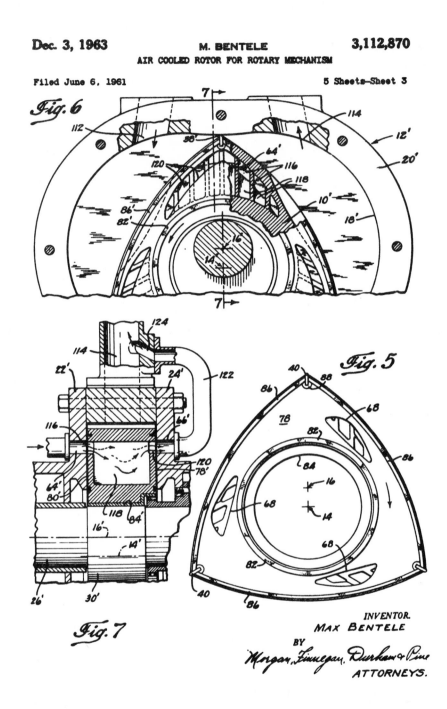

Fig. 6

Fig. 5

Fig. 7

INVENTOR.

MAX BENTELE

BY

Morgan, Finnegan, Durham & Pine

ATTORNEYS.

Air flow through the air-cooled rotor went from end cover to end cover, through ports in both rotor flanks.

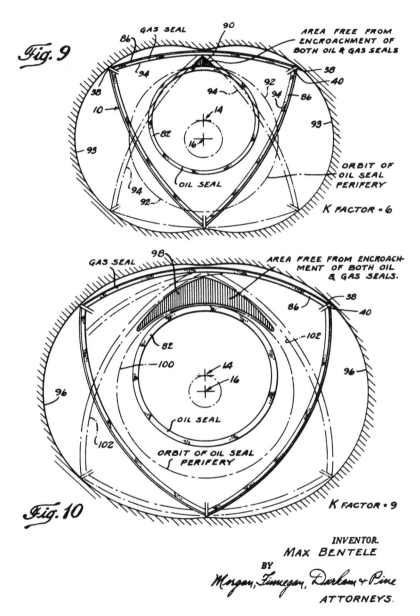

Fig. 9

Fig. 10

INVENTOR.

MAX BENTELE

BY

Morgan, Finnegan, Durham & Pine

ATTORNEYS.

Raising the R/e ratio (K factor) can considerably increase the rotor profile area free from any kind of sealing elements.

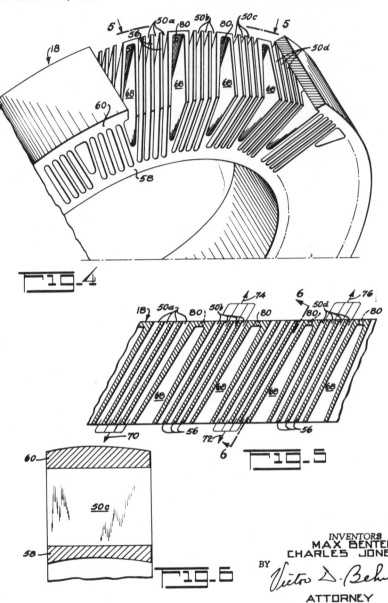

INVENTORS
MAX BENTELE
CHARLES JONES

BY

Victor D. Behm

ATTORNEY

Diagonal fins for cooling the rotor housing were patented by Bentele and Jones in 1964.

to cool the rotor while the rest lubricated the bearing. Instead of a big hollow inside the rotor, there were three separate chambers, each backing up the rotor face cavities which were subject to the highest temperature inputs.

A mechanic adjusts the throttle control of the 4 RC-60 on the test bench.

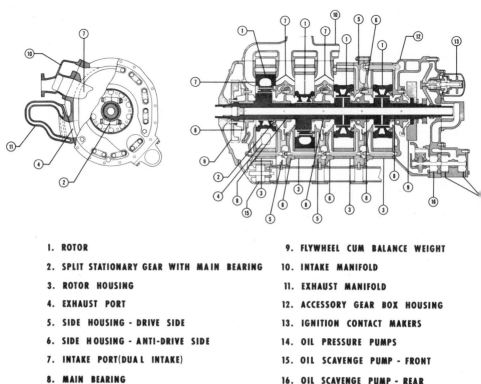

1. ROTOR	9. FLYWHEEL CUM BALANCE WEIGHT
2. SPLIT STATIONARY GEAR WITH MAIN BEARING	10. INTAKE MANIFOLD
3. ROTOR HOUSING	11. EXHAUST MANIFOLD
4. EXHAUST PORT	12. ACCESSORY GEAR BOX HOUSING
5. SIDE HOUSING - DRIVE SIDE	13. IGNITION CONTACT MAKERS
6. SIDE HOUSING - ANTI-DRIVE SIDE	14. OIL PRESSURE PUMPS
7. INTAKE PORT(DUAL INTAKE)	15. OIL SCAVENGE PUMP - FRONT
8. MAIN BEARING	16. OIL SCAVENGE PUMP - REAR

Elevation and end view of the 4 RC-6 engine.

The first IRC-6 was tested in March of 1959. Extensive endurance tests at continuous high power resulted in a gradual pounding of the apex seal slot in the light alloy rotor. Naturally, this bellmouth distortion of the sealing surface caused a power loss. Bentele concluded that if they were to continue development work on light alloy rotors, an insert of higher wear resistance than the basic aluminum was needed. This initiated the development of the cast-iron rotor.

A nodular cast-iron rotor was designed and used experimentally. It performed satisfactorily, under less stringent conditions, with no internal rotor cooling at all. A new cast-iron rotor was designed for splash cooling with scavenge oil. This was a considerably less complex and costly system than that which had been regarded as necessary for the aluminum rotor. Going to a cast-iron rotor solved the apex seal slot wear but introduced other problems. One of them was inconsistent and unpredictable peak power output. The engineers went back to reworking their stress and heat transfer analysis calculations, made new cooling tests and temperature measurements, and came up with the answer that the problem centered on thermal distortions within the elastic range of the rotor. There also were minor contributory problems associated with the machining of cast-iron, after having used aluminum.

The next step was to design a cast-iron rotor with jet impingement cooling. That was calculated to minimize the thermal gradients, and also to give a structure which could better resist the heat distortion threats that remained. Some modifications in the internal ribbing of the rotor were made after tests had shown some differences between the theoretical oil flow, as calculated on a basis of the accelerative forces in the rotor, and the actual oil flow as observed by a high-speed camera. The difference was found to be due to inertia in the oil mass. The end result was the I-beam rotor of 1964.

By that time, the IRC-6 had been thoroughly reworked and carried the designation RC1-60. The internal construction of the rotor was designed to ensure complete circulation of the oil in an entire pass through all cavities prior to ejection into the scupper in the lateral

Test results showing the performance of the 4 RC-60 engine.

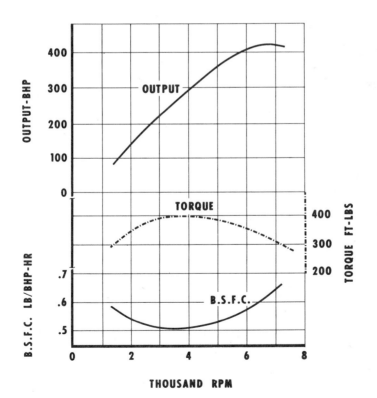

members. Oil motion inside the rotor was ensured by the fact that acceleration in the rotor changed according to rotor position in a cyclical pattern. The I-beam cast-iron rotor was so good that tests were made with higher mean effective pressures and at higher rotational speeds. This brought on a new series of problems, such as molybdenum deterioration and increased rate of seal wear.

The key problem was not gas leakage, but seal tip wear. Naturally, such wear was detrimental to effective sealing; therefore parallel development in both areas was begun. The emphasis, however, was placed on durability rather than efficiency. Apex seal end leakage was most significant at low speeds, they found, because of rotor width variations according to operating temperatures. New self-adjusting apex seals were developed, giving a considerable reduction in minimum starting speed, and also permitting low-speed idling without excessive flywheel inertia.

Bare cast-iron housings soon gave way to cast-iron housings with a soft nitriding on the working surface. Nitriding gave an improvement in seal tip wear, but warped the housing surface and prevented proper sealing under high mean effective pressure conditions. Molybdenum spray on the working surface solved the warping problem by permitting a grinding operation on the surface *after spraying*. This feature, combined with a switch to an aluminum housing, brought development to a new peak in 1960.

Torque is produced far more steadily in a four-rotor engine than in a two-rotor unit. The curve for the single-rotor has a strong negative segment.

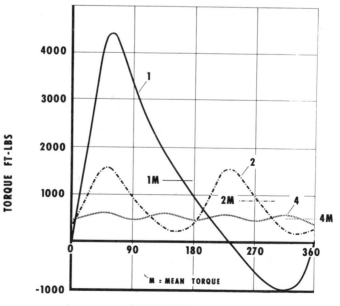

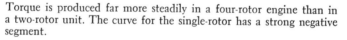

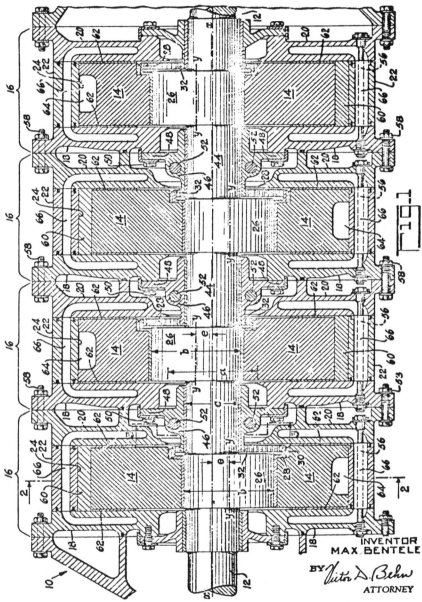

INVENTOR
MAX BENTELE

BY *Victor D. Behn*

ATTORNEY

Patent drawing for the four-rotor Curtiss-Wright Wankel engine, showing the main-shaft arrangement and rotor phasing.

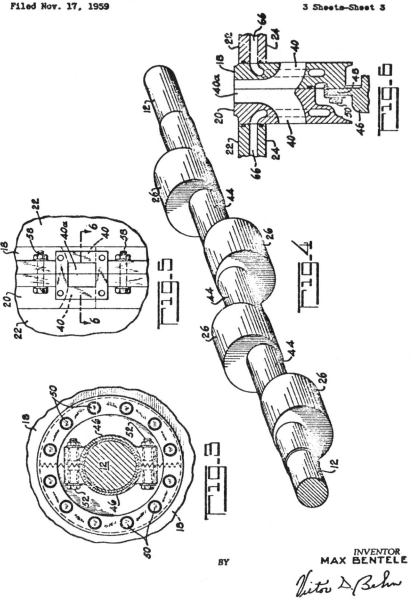

Mainshaft for the four-rotor engine.

INVENTOR
MAX BENTELE

BY

ATTORNEY

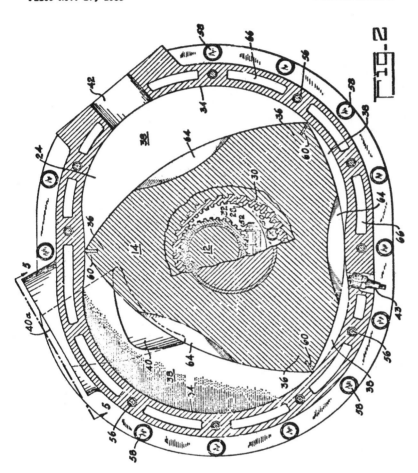

FIG-2

INVENTOR
MAX BENTELE

BY   *Victor D. Behn*

ATTORNEY

Cross-section of the four-rotor engine.

(ITEMS IN DOTTED LINES NOT FURNISHED WITH ENGINE)

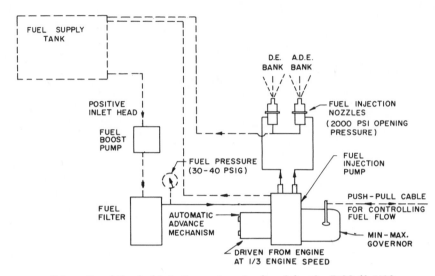

Schematic of the fuel injection system developed for the RC2-60 U10.

Curtiss-Wright then adopted chromium plating of the working surface in the aluminum housing. This chrome surface, with cast-iron
alloy apex seals, permitted long-duration cyclic tests with average wear
rates of 0.0015 inch per 100 hours for the apex seals and less than
0.0005 inch on the working surface. The first type of seal used for oil
control at the eccentric rotor bearing was a cantilevered sheet metal
scraper. This type seal was difficult to make and difficult to maintain.
Worse still, the sealing surface would not remain perfectly level during minute axial displacements of the rotor. As a result, the seal failed
to give consistent oil control. NSU had made better progress and was
using a pivoted Belleville washer type diaphragm seal. Curtiss-Wright
adopted it, but found that, although it was a definite improvement, it
still had some of the same shortcomings as the cantilevered seal. It
also had an additional problem: materials with suitable wear characteristics proved to lack the necessary strength to withstand the high hoop
stresses at the sealing edge. Conversely, materials that could handle the
stress levels at the edge did not have adequate durability. Curtiss-
Wright found an effective and relatively simple solution, evolved
through closer definition of the seal angle at the scraping edge, closer
limits for flatness and surface finish between the seal and the housing,
and more precise definition of the optimum unit loading. It was

patented in 1965 by Max Bentele and William T. Derner and was
based on the fact that just as the orbital movement of the rotor subjects
the oil seal to variations in centrifugal force during the course of every
rotor revolution, the oil seal is also subjected to temperature variations
as the rotor completes its orbit. These factors tend to cause distortions
of the oil seals on the rotor sides. Adding a number of seals of the
same type around the eccentric bearing does not solve the problem
because oil which leaks past one seal will build up a pressure head between the seals—causing the eventual breakthrough of the oil into the
chamber. The new type oil seal was designed to maintain improved
sealing contact with sidewalls and end covers.

The invention consisted of using two seal rings—one inner and one
outer. Both were supported by annular legs, extending inward into the
rotor, axially flexible to ensure sealing effectiveness. The annular legs
of the outer seal were more flexible than the legs of the inner seal;
therefore the outer seal was better able to take up clearance and con-

Estimated fuel consumption of the RC2-60 U10 engine throughout its speed range.

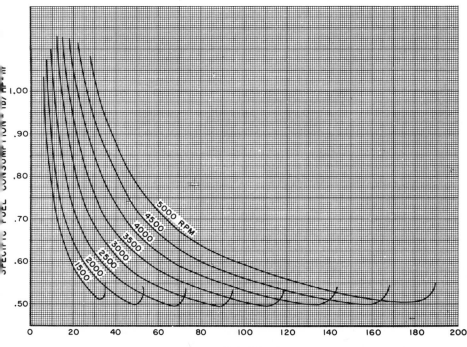

form to the end cover wall. The inner seal had the job of presenting a barrier to the main body of oil inside the rotor, although it was working with oil on both sides. Included was a drain gutter of annular shape between the two seals, into which any oil that leaked past the first seal would flow. This flow was forced by the rotation of the rotor, which acted as a rotary pump. This invention has never come into use, however, and we can assume that the rotor face cavity performs a certain percentage of the same work. The question of compatibility between quench areas and exhaust emissions seems to be the crucial point here.

While NSU remained faithful to the peripheral intake port, Curtiss-Wright designed the IRC-6 with a side intake port. Bentele and Jones were well aware that side ports reduce overlap between exhaust and intake phases, and they chose side ports because they permit earlier intake closing without reduced mixture filling. A power peak increase of 20% was possible by switching from side to peripheral intake ports, Jones admitted in November, 1966. This is analogous to changing valve overlap from essentially zero to over 100 reciprocating-engine degrees.

Peripheral intake ports also increase the power loss due to exhaust back pressure, in similar measure to a high-overlap piston engine. There

Schematic of the oil circulation system of the RC2-60 U10 engine.

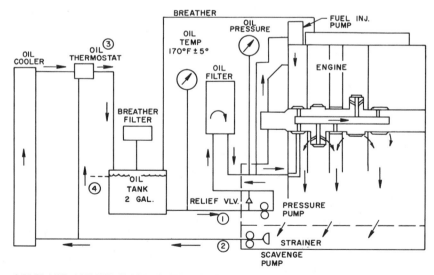

1. INLET LINE CAPACITY IN ACCORDANCE WITH OIL FLOW CURVE.
2. OUTLET LINE MUST HANDLE TOTAL OIL FLOW PLUS ENTRAINED AIR OF 2 CU. FT. / MIN. MAXIMUM.
3. OIL THERMOSTAT MUST CONTROL OIL IN TEMPERATURE TO 170°F ±5° AND LIMIT BACK PRESSURE ON SCAVENGE PUMP TO 30 PSI MAXIMUM.
4. OIL LEVEL IN EXTERNAL SYSTEM MUST BE AT LEAST 4" BELOW SHAFT CENTERLINE UNLESS VALVING IS PROVIDED TO ASSURE AGAINST DRAINBACK THROUGH PUMPS ON SHUTDOWN.

is no middle ground in Wankel engine porting. In the case of side ports, timing is controlled by the rotor contour; with peripheral ports, timing is controlled by the apex seal. It has been suggested that engines combining both types of porting would be a good compromise. Such

Performance of the RC2-60 U10 under wide open throttle (full load).

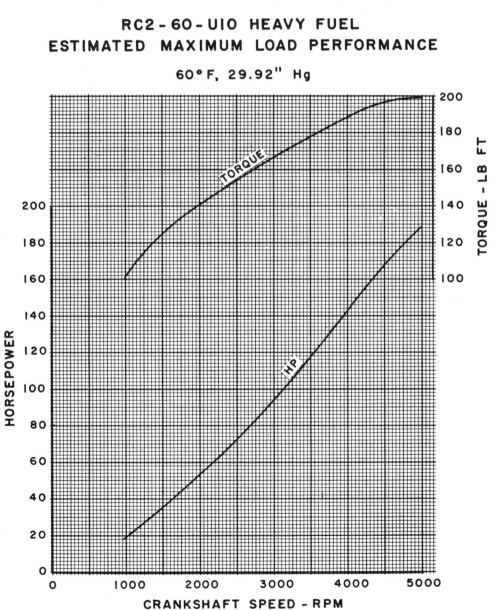

RC2-60-UIO HEAVY FUEL
ESTIMATED MAXIMUM LOAD PERFORMANCE
60°F, 29.92" Hg

The RC2-60 U10 heavy-fuel engine.

experiments have been made in Japan and will be dealt with in a later chapter.

The surface wear on end covers and center separations was a fairly easy problem to solve. They tried molybdenum spray on the surfaces in 1959 and found that it worked—so well that they continued to use it for all subsequent engines. For moderate loads and speeds, Curtiss-Wright said, uncoated iron and aluminum parts could be used. When the IRC-6 engine was first tested in 1959 it developed 100 horsepower at 5,500 r.p.m.

Initial engine tests used standard automotive ignition systems, with spark plugs of normal heat range. With an increase in power output and rotational speeds, the resulting temperature increase made it necessary to use spark plugs having very cold characteristics to ensure plug reliability. Some plug fouling then was encountered. An auxiliary air gap at the spark plug reduced the fouling, but the periods between plug changes due to fouling and gap erosion still were not acceptable. A capacitive discharge system designed by Curtiss-Wright produced a voltage rise rate four times faster than that of the standard automotive system. This system reduced gap erosion and sensitivity to plug fouling, and plug change intervals were increased four to ten times.

Curtiss-Wright's interest in automotive applications grew in 1962. Further research on the Wankel power unit was directed towards low-speed operating characteristics, and programs of part-throttle and varied-throttle operation were instituted. The high-speed potential of the RC1-60 was investigated in 1963. This test engine had large peripheral ports for both intake and exhaust, and characteristically high phase overlap. It developed a peak power of 154 horsepower at 7,000 r.p.m.; torque was above 120 foot pounds from 4,000 to 6,600 r.p.m.

Following the redesign of the RC1-60 to improve low-speed performance, Charles Jones directed the design of a twin-rotor unit incorporating all the improvements. The designing of the RC2-60 U5, the powerplant that was to be used in automotive applications, was started in January of 1963, and it first ran on the test bench in September of that year.

This graph shows nozzle flow rate in relation to pump shaft rotation. The sketch on the right shows how the test was accomplished.

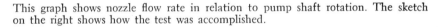

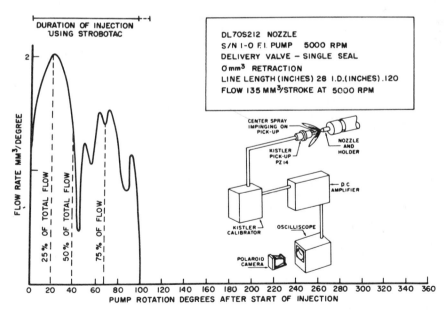

This was the version that was two years later installed in the Ford Mustang that is described in Chapter 9 of the next section.

The durability of the first RC2-60 had reached, in 1962, the point where a 1,500 hour cyclic endurance test was run without adding any oil to the fuel and using regular pump gasoline. Some 40% of this test was run at the maximum rated engine speed of 5,000 r.p.m.; 25% of the time the engine was run at wide open throttle (WOT). The engine recorded a mean effective pressure of 117–130 psi for 600 of the 1,500 hours, and over 100 psi for the remaining 900 hours.

Cranking speed investigations led to an interesting discovery—the fantastic acceleration from first fire to idle speed. The engine went from

This chart shows how fuel flow varied according to engine speed and cam rotation.

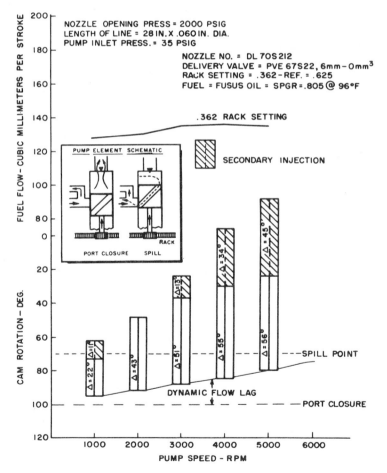

43 r.p.m. to 445 r.p.m. in 0.15 second. Sustained starts were obtained down to 60 r.p.m. shaft speed, and even below. On dynamometer tests, the oil consumption of the RC2-60 U5 was a mere 0.09 pounds per hour, on test cycles simulating high-speed road-load conditions. (This corresponds roughly to 1,000 miles per quart.)

As of June, 1964, the RC2-60 U5 had completed 2,000 hours of bench testing. Performance and reliability were regarded as favorable. No inherent problems came to light, and no new problems were encountered. But many of the old problems persisted: seal tip wear, chatter marks, and cold starting. The overall sealing problem of the Wankel engine, Jones recognized, did not end with the development of effective rotor sealing element configurations and the selection of materials providing a compatible rotor to housing interface. It was equally important, he felt, to achieve adequate rotor and housing cooling by simple engineering means, on the basis that effective cooling was necessary to prevent distortion of the vital parts and their seals. Leakage past the side seals gen-

The RC2-60 U10 would run on a variety of fuels, but best results were obtained with JP 5 jet fuel (after the engine had been modified to run with that fuel). The injector and spark plug configuration is shown in the top sketch.

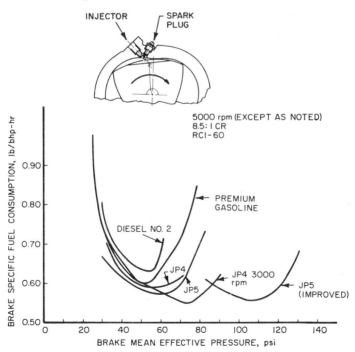

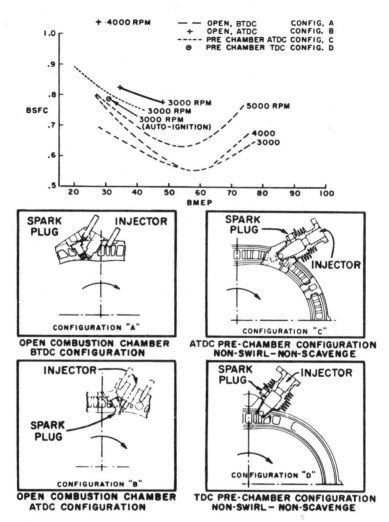

Four different fuel injector and spark plug configurations were tested. The curves in the graph above show that configuration "A," with its open combustion chamber, injector and spark plug slightly ahead of the minor axis, give highest mean effective pressures with lowest specific fuel consumption, throughout the useful speed range.

erally was small, and in the RC2-60 U5, gases that escaped this way were automatically vented to the side intake ports. Leakage past the trailing apex seal was considered similarly unimportant, as unburned gases passed into the chamber behind to be burned on the next cycle. Unburned gases that passed the leading apex seal, however, went directly into the exhaust. Development work therefore was concentrated on improvements in leading seal efficiency under pressure build-up and reversal.

Housing wear was found to occur in a characteristic wave pattern. This wear was severe enough, in early engines, to reduce apex seal life as well as sealing effectiveness. Wear of the working surface was not regarded as a barrier to effective operation. Jones felt that the nature and propagation of such wear must remain a subject for continued study. During 1965, Jones came up with a new type of coating for the working surface. It was a Wolfram-carbide alloy layer which gave an excellent hard-wearing, but very costly, surface. Several thousand hours of testing showed virtually no wear on the surface, even though 30% of the test cycle was run at wide open throttle between 3,600 and 5,000 r.p.m. After 528 hours on a severe-duty cycle the engine showed only 0.0000421 inch of wear on the working surface. This wear rate is just over 2% of the wear recorded in a 1,500 hour test two years earlier. Wear of the working surface had been eliminated.

During 1965, new rotor oil seals also were developed. Tests with the new seals showed leakage rates down to a few hundredths of a pound per hour under road-load operation, and only slightly more at maximum speed. This was followed by a program to determine the minimum rate of controlled leakage necessary to maintain acceptable wear rates on the eccentric seals and rotor side seals.

Rotor spray pattern indicates changes in intensity and direction at various points during the combustion process.

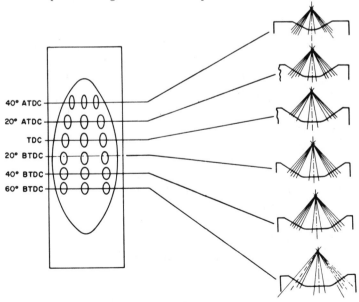

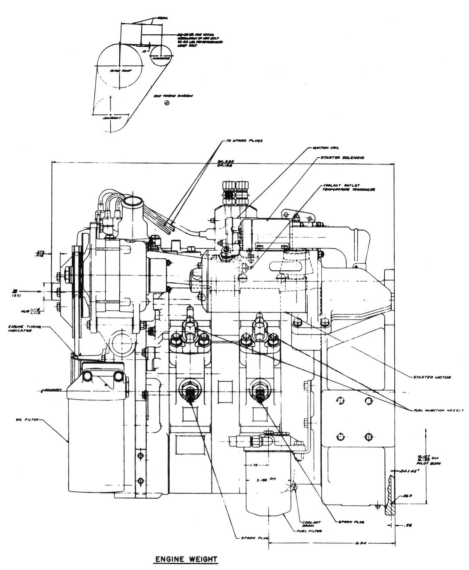

ENGINE WEIGHT

THE CURTISS-WRIGHT MODEL RC2-60-U10 HEAVY FUEL ROTATING COMBUSTION
ENGINE WEIGHT, IN THE CONFIGURATION DEPICTED ABOVE, BUT WITHOUT THE
FLYWHEEL, IS 325 LBS. THE WEIGHT OF THE FLYWHEEL SHOWN IS 50.2 LBS.

Installation diagram for the RC2-60 U10.

The conclusions drawn from their original 1963 experiments with the
single-rotor RC1-60 led Curtiss-Wright to conclude that Wankel engine
design features had to be tailored to the application. They later reversed
this opinion, but before they got to that point they had conducted

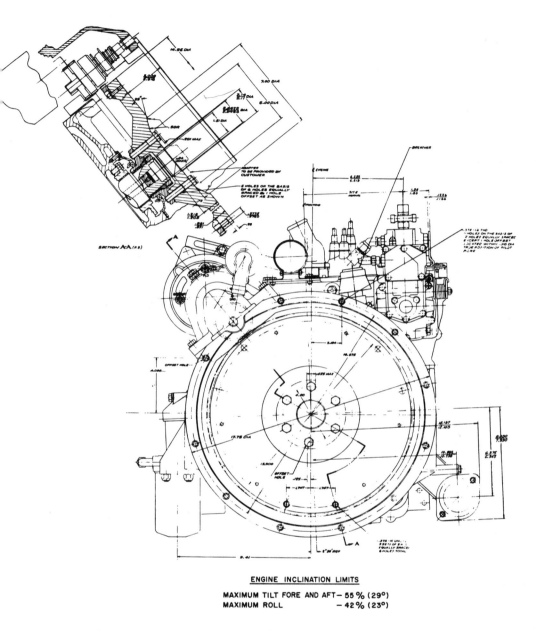

ENGINE INCLINATION LIMITS

MAXIMUM TILT FORE AND AFT— 55% (29°)
MAXIMUM ROLL — 42% (23°)

Installation diagram for the RC2-60 U10.

design, research and experimental work in a number of other directions —such as multi-rotor engines, air-cooling, and multi-fuel operation. After the evaluation of the IRC-6 test results, two new engines had gone on the drawing board. One, a single-rotor unit, was eventually developed

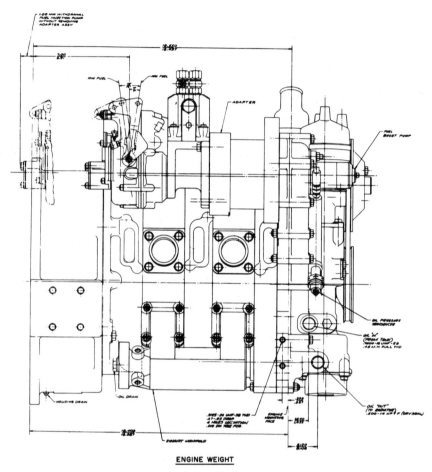

ENGINE WEIGHT

THE CURTISS—WRIGHT MODEL RC2-60-U10 HEAVY FUEL ROTATING COMBUSTION ENGINE WEIGHT, IN THE CONFIGURATION DEPICTED ABOVE, BUT WITHOUT THE FLYWHEEL, IS 325 LBS. THE WEIGHT OF THE FLYWHEEL SHOWN IS 50.2 LBS.

Installation diagram for the RC2-60 U10.

into the RC1-60, while the other, a twin-rotor unit, was used to evaluate the basic twin-rotor configuration. With both engines, chamber dimensions were identical and shaft and bearing arrangements were simplified. The work on these engines encouraged Bentele to investigate the possibilities of a four-rotor engine.

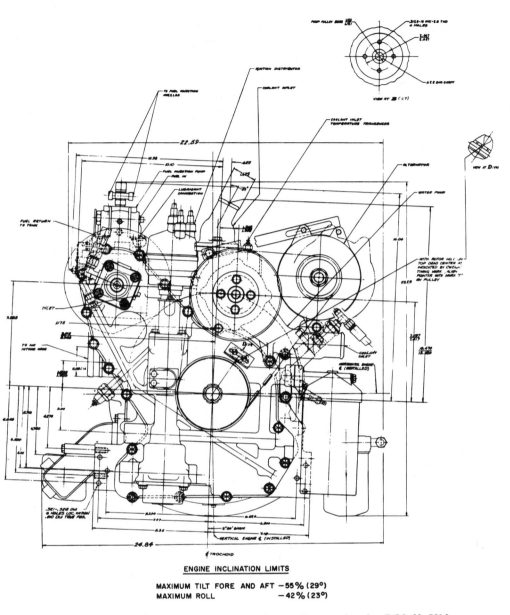

ENGINE INCLINATION LIMITS

MAXIMUM TILT FORE AND AFT —55% (29°)
MAXIMUM ROLL   —42% (23°)

Installation diagram for the RC2-60 U10.

There are only two ways to put more than three rotors on a rotor shaft in a Wankel engine. Either the bearings and reaction gears must be split or the shaft must be a built-up assembly. Bentele decided to use an integral one-piece mainshaft, which meant that the bearings and stationary gears had to be split along the plane of the engine axis. (A

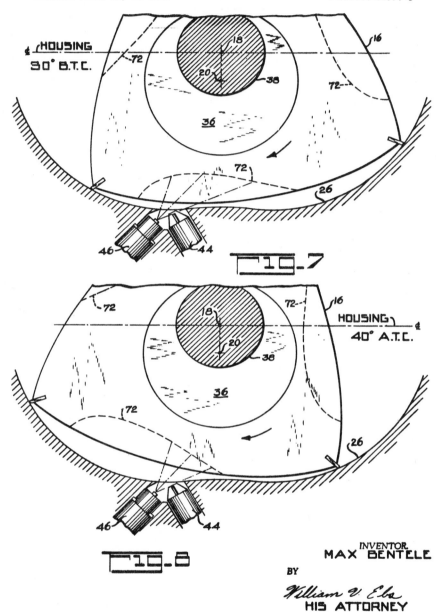

Injection spray pattern proposed for a rotor cavity having maximum depth in the leading portion of the combustion chamber.

INVENTOR.
MAX BENTELE
BY
William V. Els
HIS ATTORNEY

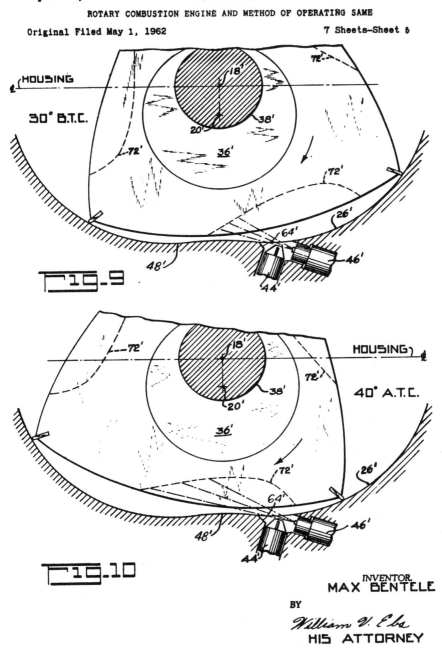

INVENTOR.
MAX BENTELE

BY

William V. Els

HIS ATTORNEY

Injection spray pattern proposed for a rotor cavity with maximum
depth in the trailing portion of the combustion chamber.

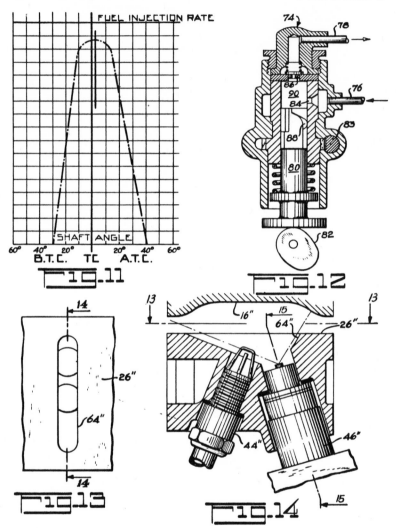

INVENTOR.
MAX BENTELE

BY

William V. Els
HIS ATTORNEY

Injection pump cross section, and fuel injection rate for Bentele's
patented heavy-fuel engine.

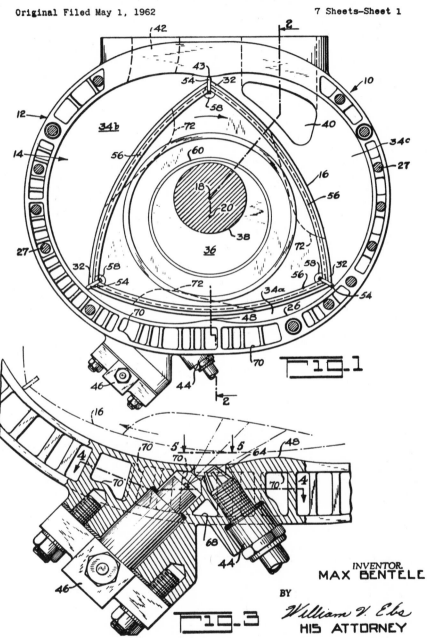

Injector and spark plug configuration for the heavy-fuel Wankel
engine, patented in 1966 by Max Bentele.

INVENTOR.
MAX BENTELE

BY

William V. Ebs
HIS ATTORNEY

split-rotor shaft would introduce structural weakness and make engine assembly difficult.) The plan was to use a one-piece rotor shaft with four eccentrics. Bentele was granted a patent for a four-rotor Wankel engine on November 6, 1962. The rotor shaft was designed like a one-plane crankshaft for a four-cylinder engine, with eccentrics #2 and #4 diametrically opposed to eccentrics #1 and #3. Bentele also devised a novel form of split bearing.

The engine had an integral gear-and-bearing unit, so that each gear half also carried a bearing half. The complete rotor shaft and bearing assembly was held in place within the housing by bolts, and the rotor shaft bearings were split into two semi-circular halves on the line of the lowest bearing loads. The joint faces between the bearing halves had interlifting ribs for accuracy of alignment. Each rotor had its own reaction gear, and the four gears in the four-rotor engine also were split into two semi-circular pieces. The sidewalls were designed with big holes near their centers to allow the engine to be assembled by threading the rotor shaft through a succession of separating walls and working chamber housings.

The four-rotor engine was assembled by mating the mainshaft and the front end cover with its accessory drives. Then, one rotor and rotor hous-

The Wankel-powered Bertram boat at speed.

Installation of the RC2-60 M4 in a Bertram 22 boat.

ing followed, the rotor fitting onto its bearing on the first eccentric. After one separating wall, with its port, the next bearing and gear set followed. They were piloted in the side separating wall, and the split plane was serrated as a further precaution against dislocation. This process was repeated until the final end cover and flywheel assembly could be bolted in place. The eccentric arrangement provided power strokes at 90 degree intervals, with a firing order of 1-4-2-3. There was, however, a small unbalanced couple which made it necessary to add a balance weight at both ends of the shaft.

This four-rotor engine had a maximum output of 425 horsepower at 6,500 r.p.m. and showed fairly flat torque and fuel consumption curves. It was the first multi-rotor Wankel engine designed and built anywhere. But the Japanese were not far behind, and were soon to surpass the achievements of Curtiss-Wright in several areas.

Automobile engineers used to argue among themselves about how many cylinders a 200 horsepower engine should have, due to the fact that breathing and torque characteristics, the combustion process itself, engine balance and transmission requirements change greatly according to the number of cylinders. An engine having too many cylinders tends to be excessively complex and often runs with extremely high friction losses, while too few cylinders leads to enormous bore sizes and inefficient combustion because the flame front fails to reach the far corners in time to produce power. All these arguments were applied to the Wankel engine during its design.

In one experiment, the linear dimensions of the RC-6 engine were literally scaled up about 30 times, to give a displacement of 1,920 cubic inches. Adjustments were made on dimensions and features which affected stress, heat transfer and thermal gradients, but the basic design remained the same. With its dual side ports, the aspiration capacity of the RC-6 proved more than ample. The preliminary tests showed some of the power potential in this engine. However, it suffered from one phenomenon that had not occurred in the smaller RC-6—detonation. This happened under high-load conditions, just as in piston engines.

It seems safe to conclude that the size of the combustion chamber provided opportunities for spontaneous ignition or surface ignition, because flame front velocity failed to increase on the same scale as the physical dimensions of the combustion chamber. Bentele insisted that

Fuel consumption related to engine speed and vessel speed, for the RC2-60 M4 installation in the Bertram boat.

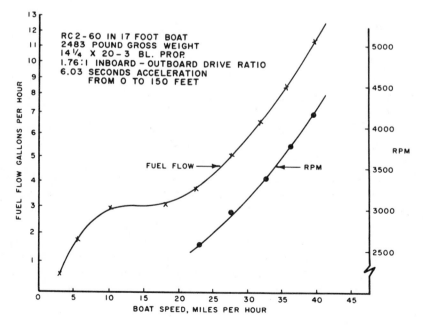

it was too early to say, in January, 1961, that this was in fact the case. The preliminary tests gave readings of 872 horsepower at 1,525 r.p.m., with a brake mean effective pressure as low as 106 psi.

In a continued study of scale effects, Bentele went to the opposite extreme and designed the smallest Wankel engine yet—the RC1-4.3. Design studies indicated that a basic single-unit, die-cast aluminum engine, rated at 10 horsepower for high speed, could be provided in the 25-pound class and within an envelope of .66 cubic feet. This package

Side view of the RC2-60 M4.

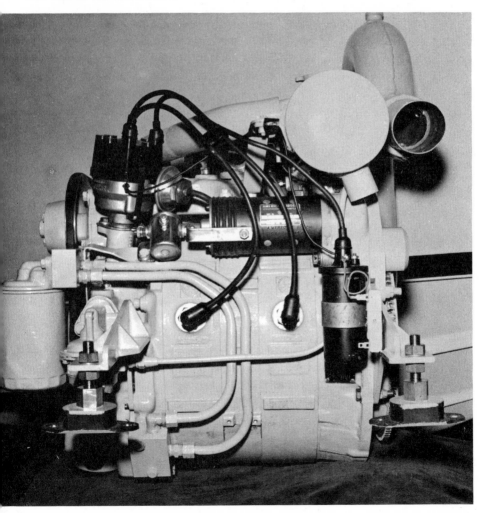

Latest version of the RC2-60 M4. Two of these were installed side by side in an Owens Concorde boat.

included all accessories except the fuel tank. The basic single-unit engine could, of course, be extended by the addition of one, two or three power sections providing 20, 30 and 40 horsepower engines—each rotor addition would add some 10 pounds. The RC1-4.3 was air-cooled and designed for the low speed of 4,000 r.p.m. where it produced the predicted 3.5 horsepower on the test stand.

It was chiefly an interest in air-cooled aircraft engines that led to an analytical feasibility investigation of air-cooling for Wankel engines, small and large, early in 1961. The analysis led to the conclusion that the Wankel engine could be cooled adequately by air within the power output requirements for aircraft applications. However, this conclusion was challenged because some engineers were worried about the localized heat flux peaks of the Wankel engine (much higher than in the four-stroke piston engine because of the frequency of ignition).

A more detailed feasibility study then was started. Ultimate aircraft

engine configurations were thoroughly defined, and test rig capability
worked out for a wider scope of evaluation. Test instrumentation included thermocouple heat flux measurements in both air- and water-
cooled engines. These comparisons showed that rotor housing cooling
was the biggest problem. The first totally air-cooled test engine was designated the RC1-60 J1.

By March of 1963, the RC1-60 J1 had run about 500 test hours. That
month the engine exceeded its target power output of 105 horsepower by
a margin of 10 horsepower, or almost 10%, without overheating. Success
was achieved mainly by the analytical approach prior to testing, although
a considerable number of significant technical refinements had to be redesigned and evaluated in order to match the experimental heat inputs
(which did not conform exactly to those that had been theoretically
predicted). No final conclusions, however, were drawn from this test program.

The air-cooled Wankel engine is attractive for use in light aircraft, fixed
and rotary wing, because of its acceleration, throttle response, and inherent ease of control—all of which are at or above the levels of attain-

The Owens Concorde boat heading for open waters.

ment for a reciprocating engine. In addition, the Wankel offers low inertia for torsional matching, relatively low cost, low fuel consumption, and light weight. Mission studies in this area generally show the engine plus fuel weight to be less than that for a comparable gas turbine, even though the bare Wankel engine weight is higher. This conclusion has held true for all but short duration flight missions.

What actually resulted from the air-cooling experiments was the creation of a definable system for minimization of air pressure drop and quantity flow. This was achieved through close matching of air-side requirements to gas-side inputs, based on a desired metal temperature distribution. Following re-evaluation and re-definition of objectives, Curtiss-Wright went on to design a larger air-cooled aircraft engine, the RC2-75.

This engine was gasoline fueled, with a propeller reduction gear for light aircraft propulsion, and was rated at 275 horsepower. That is equivalent to about one horsepower per cubic inch, giving it a power-to-

Initial test results obtained with the passenger car version of the RC2-60 U5.

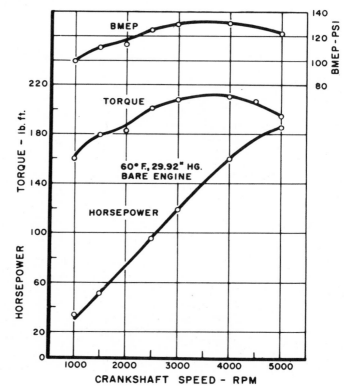

weight ratio comparable to current turboprop engines. The RC2-75 led directly to the design of a somewhat larger air-cooled, twin-rotor Wankel engine—the RC2-90 Y2. It was designed to burn diesel fuel and is described later.

The RC2-75 did not have the disadvantages of the turboprop, such as throttle response lag, high starter motor power requirements and high fuel consumption. As a bonus, it presented no severe need for specialized overhaul and maintenance personnel. The next step was to design a four-rotor, air-cooled aircraft engine. This engine was completed in 1962 and given the designation RC4-60 J2.

The RC4-60 J2 used the same type I-beam rotor that had been developed for the RC1-60. Rotor cooling was by the same inertia-actuated oil circulation method, and sealing elements were of similar design. Air to cool the housing was supplied by a gear-driven, two-stage cast aluminum axial compressor operating at a 1.2:1 pressure ratio and delivering 1.75 cubic feet of air per second. Unfortunately, the military requirements for which this engine was designed did not materialize and it was never developed.

From 1962 onwards, the predominant direction at Curtiss-Wright was the development of a basic unit, adaptable to a variety of applications rather than the exploration of various sizes and specialized versions of the engine for specific applications. Development work centered on the RC2-60 and the YRC-180-2. The RC2-60 is a 185 horsepower, liquid-cooled, twin-rotor engine developed for vehicular, marine, ground support and other military and commercial applications. The other engine, the air-cooled YRC-180-2, being developed under a U.S. Navy contract, weighs 278 pounds and was designed to produce 310 horsepower at 6,000 r.p.m.

Testing with the RC2-60 has been concentrated in the area of demonstrating the feasibility of various configurations. The first tests were made

Comparison between the RC2-60 and a V8 piston engine of comparable performance, showing differences in fuel consumption characteristics.

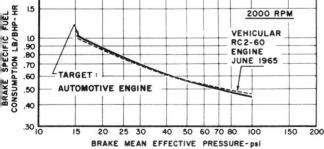

in 1966 with a 17 foot Bertram boat, using an inboard-outboard installation. Next, two RC2-60 engines were installed side by side in an Owens Concorde boat. Results were most encouraging. Jones reported to Curtiss-Wright's management: "The future is promising in the marine powerplant area because of the output smoothness, high specific output and, as compared to two stroke cycle engines, improved fuel consumption on the order of 20% more, freedom from a fuel-oil mix, and a lower noise level."

Working in collaboration with Westinghouse Aerospace Electrical Division, Curtiss-Wright supplied a prototype engine for an RC2-60 generator set of 60 kilowatt capacity. It had a brushless 400 cycle AC generator cantilevered from the engine adapter housing and engine flywheel without additional bearings. Westinghouse released information on how it compared with existing units, using both gasoline and diesel piston engines and gas turbines. On a weight basis, the RC2-60 had a slight advantage over the gas turbine (900 pounds against 1,000), while the gasoline piston engine weighed 1,450 pounds and the diesel carried a terrific weight penalty at 4,880 pounds. The RC2-60 took up less space than the gas turbine (37.6 cubic feet against 45). The gasoline engine used 84 cubic feet, and the bulky diesel 163 cubic feet.

Basic components of the RC2-60 U5 engine.

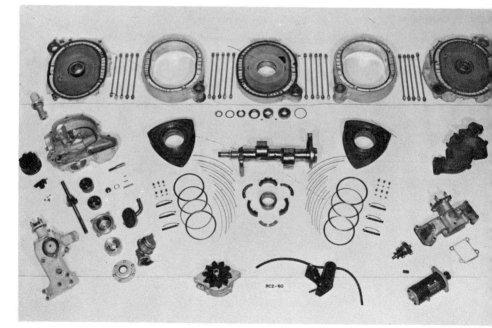

In 1969, Curtiss-Wright reached an agreement with Lockheed aircraft for development of the RC2-60 engine for installation in the Q-star —a civilian version of the U.S. Army's QT-2 reconnaissance plane. The Wankel engine has 85% more power than the original YO-3, as the Army calls an improved version of the Lockheed plane. The YO-3 is basically a Schweizer, two-place, high-performance sailplane with a 100 horsepower piston engine and slow-moving, six-blade propeller. Installation of the RC2-60 could be accomplished at a cost of only 6% in weight.

In order to explore the military market for the Wankel engine, Curtiss-Wright started in 1960 a program to determine its multi-fuel capability. The fuels to be burned covered the range from gasoline to diesel and gas turbine JP type fuels. A comprehensive study conducted by Bentele led in 1961 to the conclusion that the Wankel engine geometry and design parameters favor the low compression spark ignition and not the diesel type high compression approach. High compression would necessitate a radical departure from the proven geometry and construction and would be intrinsically inferior, the penalties being larger and heavier designs for the same power, and severe operating difficulties. Bentele evolved various basic concepts with low compression, spark ignition, direct and pre-chamber fuel injection and was granted patents on them. These schemes were considered feasible and advantageous, and all experience which Curtiss-Wright had accumulated with experimental engines was directly applicable.

The first version of the heavy-fuel Wankel engine was an adaptation of the RC1-60 with water-cooling, spark ignition, and fuel injection directly into the combustion chamber. This meant using diesel-type nozzles and a high-pressure injection pump, consequently the multi-fuel Wankel engine carried the cost penalty of the high-pressure injection equipment. Fuel injection is not the same in the heavy-fuel engine as in a gasoline engine. The gasoline engine has an air intake valve, working as a throttle, to control engine speed. In the diesel engine, crankshaft r.p.m. is controlled by the amount of fuel injected into the compressed intake air.

Fuel injection was dictated by the type of fuel, which has a low octane rating and very low volatility. JP fuel is a hydrocarbon in the kerosene family, having a specific gravity and volatility comparable to paraffin. When burned, it leaves combustion products similar to those of gasoline —carbon dioxide and water vapor, plus the nitrogen from the air; however, it contains no carbon monoxide. With fuel injection, the combustion process is quite different, because the mixing of the particles of fuel and air takes place inside the combustion chamber. The fuel is injected in a fine spray to facilitate this mixing. Because the fuel is

In 1966 Curtiss-Wright installed an RC2-60 U5 engine in this Reo 2½ ton 6 x 6 military truck.

bound to travel in straight lines, as directed by the nozzle, the air must be made to swirl in order to ensure complete vaporization.

In the gasoline engine, combustion takes place so quickly that it almost can be said to be completed at constant volume. That is, the combustion chamber displacement changes relatively little during the combustion process—the burning of the fuel mixture is over in a flash. The result is an extremely high pressure rise rate, followed by an almost equally fast drop in pressure as cylinder displacement is rapidly increased during the power phase. In the diesel engine, there is no mixture in the first place; there is no spark and no flame front. No such combustion phenomena as pre-ignition, knock or rumble exist in a diesel engine. Fuel injection determines the start of combustion. A fine spray of fuel is injected into the compressed air, and atomization occurs simultaneously in several areas of the combustion chamber. However, it takes a while for all the fuel to be injected and atomized, and consequently the combustion process must be stretched out in time. As opposed to the constant volume combustion of gasoline engines, the diesel engine gives what is often called constant pressure combustion. Pressures rise steadily during initial expansion; but displacement is also steadily increased. Instead of a pressure rise, the result is relatively constant pressure with increasing volume. In practice, the pressure rise rate is almost a matter of chance.

With diesel engines, the rate at which fuel is injected into the combustion chamber does not necessarily control the rate of burning, nor can the engine or its injection equipment ensure sufficient time for a degree of mixing that will guarantee clean and efficient combustion. The moment burning starts in some area of the combustion chamber, another area may contain a body of partly vaporized fuel with vapor envelopes approaching the fuel's self-ignition temperature. For this reason, the fuel throughout the combustion chamber may ignite either because it is reached by the progressive spread of the flame from the initial nucleus, or because a body of fuel vapor reaches its self-ignition temperature, or a combination of the two. The diesel engine has three phases of combustion: it starts with a *delay period*, which corresponds to about 15 degrees BTC to about 3 or 5 degrees BTC. In this delay period, ignition is initiated but there is no measurable change in pressure rise rate. Pressure continues to rise under continued upward piston travel just as if no injection had taken place. Next comes the *peak pressure period*, which corresponds to 3 or 5 degrees BTC to approximately 10 degrees ATC. During this period, there is a sharp pressure

The world's first Wankel-powered airplane first flew in 1969. Built by Lockheed, it is powered by Curtiss-Wright's 185 horsepower RC2-60 engine.

rise from about 450 to over 700 psi. This period also represents a phase of rapid ignition and combustion of the whole of the fuel present in the cylinder. The pressure rise rate, usually some 30 psi per degree of crankshaft rotation, is a determining factor in causing diesel knock or rough running. (Diesels are noisier than gasoline engines because the pressure rise during the power stroke induces transient vibrations in the engine structure and causes the outer surface to give off noise. The noise level is determined by the characteristics of the exciting force, by structure response to vibrations and by the ability of the engine surfaces to radiate sound.) The *rising volume period* is the last phase of combustion. It has a duration from about 10 degrees ATC to about 60 degrees BBC. Pressure falls off as volume increases, caused by the piston completing its power stroke. Most of the fuel is mixed with air during this period, injection is completed, and combustion is concluded.

By running with moderate compression ratios and spark ignition, the problems usually associated with heavy fuels and their slow combustion and starting difficulties could be circumvented. Early development centered on finding the best location for the nozzle, the form of

The RC2-60 retains its water-cooled housing and oil-cooled rotor, even when used in aircraft. This is the power unit for the Lockheed Q-Star.

RC-90-Y Aircraft Installation

		YRC-180-2	TURBOSHAFT
MISSION FUEL:		182 LB	348 LB
MISSION WEIGHT ADVANTAGE:		105 LB	—
COST EFFECTIVENESS FACTOR:		0.5	1.0

ENGINE PLUS FUEL LB

3000 / 2000 / 1500 / 1000

CONTEMPORARY TURBOPROPS

RC4-90-Y

0   1   2   3   4

MISSION LOITER TIME—HOURS

A NEW SHAPE IN MILITARY POWER

Curtiss-Wright proposed a four-rotor air-cooled Wankel engine for military planes, and a twin-rotor air-cooled unit for helicopters.

the nozzle spray pattern, the location of the spark plug, ignition and injection timing. As the results of these tests began to roll in, patterns started to emerge.

Four different types of combustion chamber were tested. The first was an "open" combustion chamber with the injector installed very close to the spark plug angle so as to inject fuel against the rotation of the rotor, while the spark plug was angled the other way. The ignition point was about 10 degrees before the minor axis. The second type was similar in the injector/spark plug relationship, but both were moved back to a position about 10 degrees after the minor axis. This engine demonstrated capability of operation on both JP4 and JP5 fuels, as well as on diesel fuel #2 and high-octane gasoline. The third type of combustion chamber was a pre-chamber design with non-swirl and non-scavenge characteristics. Both injector and plug were mounted well after the minor axis. The fourth combustion chamber was also a non-swirl, non-scavenge type, with spark plug and injector advanced to about 6 degrees after the minor axis.

The essential element of all these experiments was coordinated ignition and injection for burning of the injected fuel at a controlled rate to avoid unduly high burning rates with low octane fuels and insensi-

Curtiss-Wright also installed an RC2-60 U5 engine in an F.V. 432 Armored Personnel Carrier in 1966. It had covered almost 10,000 miles by August 1969.

A Ferret Scout Car was equipped with a Curtiss-Wright RC2-60 U5 Wankel engine in 1967. By August 1969 it had covered 16,800 trouble-free miles.

tivity to combustion lag with low cetane fuels. The success of the process was strongly dependent upon controlled wetting and evaporation from the rotor face, combined with the transfer of air past the rotor cavity and minor axis of the trochoid and combustion chamber surface temperatures.

The end result appeared to be a natural outgrowth of this engine's fundamental geometric characteristics. It was designed to sweep all of the charge air, in a reasonably predictable and consistent fashion, past the stationary injector-igniter combination. Superimposed turbulence induced by rotor "squish" action was another design objective. Full advantage was taken of its relatively localized heat input distribution. In effect, "repetitive turbulence" was built into the engine without resorting to shrouded intake ports or swirl pockets. Retaining the low K factor (R/e ratio) of the RC1-60 preserved the advantages of high swept volume per unit of frontal area plus a reasonably flexible combustion chamber shape, as opposed to the situation with high K values required for compression-ignition operation.

Curtiss-Wright's spark-ignition, heavy-fuel injection engine enjoyed the benefits of unthrottled part-load operation at low fuel-air ratios with the attendant savings of: (a) lower pumping losses; (b) reduction of

These were the arguments used by Curtiss-Wright in its negotiations with the U.S. military in order to promote the use of their Wankel engines in the Drone Helicopter.

Cost Comparison

RC2-90-Y2 VS BOEING T-50B10 AND ALLISON 250-C-14 IN THE QH-50D DRONE HELICOPTER

RC2-90-Y2 DESIGN SIMPLICITY RESULTS IN:

- 50-75% LOWER UNIT SELLING PRICE
- 31-64% LOWER INVESTMENT COST
- 26-47% LOWER MISSION FUEL CONSUMPTION
- 16% LOWER (B10) AND 4% HIGHER (C-14) ENGINE PLUS FUEL-WEIGHT
- 83% LOWER TRAINING COST
- 60% FEWER BILL OF MATERIAL LINE ITEMS
- 87-91% FEWER CRITICAL ALLOYING ELEMENTS
- CONVENTIONAL MANUFACTURING TECHNIQUES AND EQUIPMENT
- SIMPLIFIED MAINTENANCE AND HIGH DEGREE OF RELIABILITY
- INHERENT STABILITY
- RAPID RESPONSE AND SIMPLICITY OF CONTROL

variable specific heat effects, dissociation, and heat losses that go along with high temperatures; and (c) lower exhaust temperature and cooling requirements. The anticipated lower level of exhaust carbon monoxide and unburned hydrocarbons at these operating conditions would be another advantage. Partially countering these positive aspects was the fact that controlled burning introduced a slowdown in the combustion process which reduced peak temperatures and pressures for a net reduction in the maximum power output compared to performance with a homogeneous air-fuel charge.

The rotor face cavity influenced, among other things, the rate at which the charge air was brought past the minor axis and the firing location. It was found that an injection duration of some 100 crank angle degrees was required to utilize the air confined by the respective rotor lobe. Because this is equivalent to two-thirds the crank angle for a reciprocating engine, and the RC engine operating speeds were higher, the experimental Wankel's total combustion time was about equal to the present maximum of the highest-speed, two-stroke cycle diesel engines. The limitations to burning time imposed by the mechanics of mixing air with a heavy fuel proved an additional argument for spark ignition. The alternative would have been making geometry changes and adapting compression-ignition, which would have been a deterrent to high-speed operation and exacted a penalty in engine size and weight.

The rotor cavity shape and the injection timing determined the position of the wetting pattern on the rotor surface. This was important because not all of the fuel ignited as it was sprayed. The start of injection usually ranged from 35 degrees to 70 degrees BTC for the 3,000–5,000 r.p.m. band, depending on speed, injection duration, fuel and design configuration. The amount of required advance increased with rising speed and duration. Fuel-air ratio had a small effect; rich mixtures required slightly less advance (ignition timing is generally close to start of fuel injection). All running tests were made with conventional automotive or magneto ignition using automotive type 14 mm. spark plugs, which exhibited a tendency towards auto-ignition at very rich air-fuel ratios (and still do).

After testing and development work on the single-rotor multi-fuel engine had been concluded, Jones and his staff designed a family of twin-rotor and multi-rotor multi-fuel engines, some with liquid-cooling and others with air-cooling. General specifications for this family of engines are given in the following charts.

The ratings and estimated size and weights for the air-cooled engines include the cooling blower, distributor, fuel boost pump, fuel injection pump, starter, oil pumps (dry sump system) and manifolds. Two ratings are given for each engine to allow consideration for application in aircraft, surface vehicle and ground support equipment.

AIR-COOLED FAMILY

Model	RC1-60	RC2-60	RC4-60	RC2-90	RC3-90	RC4-90	RC5-90	RC6-90
HP (aircraft T/O)	103	206	412	310	465	620	775	930
RPM (aircraft T/O)	6,000	6,000	6,000	6,000	6,000	6,000	6,000	6,000
HP (surface equip.)	80	160	320	240	360	480	600	720
RPM (surface equip.)	5,000	5,000	5,000	5,000	5,000	5,000	5,000	5,000
Dry weight—lbs.	192	266	417	317	410	510	605	710
Specific weight—lbs./hp.	1.96	1.29	1.01	1.02	0.882	0.882	0.781	0.764
Length—in.	24.4	29.4	39.4	32.4	38.9	45.4	51.9	58.4
Width—in.	20.7	20.7	27.3	20.7	22.0	23.3	25.2	27.2
Height—in.	18.0	18.0	18.0	18.0	18.5	19.9	21.0	22.0
Volume—cu. ft.	5.26	6.35	8.5	7.0	9.06	12.2	15.9	20.2
Specific volume—cu. ft./hp.	0.052	0.031	0.021	0.023	0.020	0.020	0.020	0.022

The ratings and estimated size and weights for the liquid-cooled engines include the distributor, fuel boost pump, fuel injection pump, starter, oil pumps (dry sump system), coolant pump and manifolds.

LIQUID-COOLED FAMILY

Model	RC2-90	RC3-90	RC4-90	RC5-90	RC6-90
HP (w/o fan)	285	428	570	713	855
HP (w/fan)	276	414	552	690	828
RPM	5,000	5,000	5,000	5,000	5,000
Dry weight—lbs.	345	480	613	740	860
Specific weight—lbs./hp.	1.21	1.12	1.08	1.04	1.01
Length—in.	28	34.5	41	47.5	54.0
Width—in.	23.8	23.8	23.8	23.8	23.8
Height—in.	22.0	22.0	22.0	22.0	22.0
Volume—cu. ft.	8.5	10.5	12.4	14.4	16.4
Specific volume— cu. ft./hp.	0.030	0.025	0.022	0.020	0.019

A separate program on multi-fuel Wankel engines was undertaken on behalf of Rolls-Royce Ltd. under contract with the British Government's *Fighting Vehicle Research and Development Establishment* in England. This joint venture resulted in the creation of the liquid-cooled RC2-60 U10 engine in 1965. It differs from all other Curtiss-Wright Wankel engines in its use of peripheral ports, but has all the typical design features evolved in the multi-fuel capability investigation, such as direct fuel injection and spark ignition. Two different housing designs were prepared, with notable differences in port timing.

INITIAL ROTOR HOUSING DESIGN

Intake port opens	443° ATC	
Intake port closes	840° ATC	Intake port
Exhaust port opens	215° ATC	area—2.00
Exhaust port closes	609° ATC	square inches
Overlap	166°	

ALTERNATE ROTOR HOUSING DESIGN

Intake port opens	441° ATC	
Intake port closes	845° ATC	Intake port
Exhaust port opens	215° ATC	area—2.400
Exhaust port closes	609° ATC	square inches
Overlap	168°	

In tests with this engine, Curtiss-Wright was able to show a power increase at somewhat richer mixtures than originally anticipated, which were assumed to be similar to those of conventional diesel engines. Because the maximum usable fuel flow was limited by the rate at which the injected fuel could find the necessary air, however, cumbustion at high fuel flow was undoubtedly less than complete.

The specific output of engines running on JP5 fuel translated into some 20–25% less than Curtiss-Wright's gasoline RC engines. Higher specific air consumption was, of course, necessary due to the excess-air system. It is interesting that the fuel-injected engine appeared to attain a higher efficiency in the low- and mid-power regions (i.e., low fuel flow), which may have been part of the theoretically predicted gain over the carbureted engine. Based on other "hybrid" engine test results, there were still further improvements possible in specific fuel consumption. The RC2-60 U10 program proved that the high-speed characteristics of the RC engine, together with its inherent compact dimensions and low weight, could be retained even when burning heavy fuels. The resulting attractive power to weight ratio put this engine in a competitive position with other engines burning these fuels, particularly in light of the RC engine's low specific fuel consumption and low noise level.

Development of the Curtiss-Wright Rotating Combustion engine continues. Over 35,000 hours of engine operation have been accumulated to date in test and evaluation programs. The company says it is ready for mass production.

11

Toyo Kogyo

IN DECEMBER, 1959, news of the Wankel rotary combustion engine reached Hiroshima. At that time, the engineering staff at Toyo Kogyo considered it to be nothing more than a pipe dream. Skeptical criticism was voiced from various parts of the world, and the arguments of the engineering community were divided for and against the engine. However, the rotary engine symposium sponsored by the V.D.I. (Verein Deutscher Ingenieure) in January, 1960 made it clear that this engine possessed great future possibilities.

By early 1960, Toyo Kogyo's engineering staff had reached a definite opinion on the rotary engine. They felt that it would be possible to put the Wankel engine into practical automotive use. Toyo Kogyo immediately sounded out NSU on their views concerning a license agreement, but the reply was discouraging. A few months later, on May 21, 1960, the ambassador of the Federal Republic of Germany to Japan, Dr. Wilhelm Haas, paid a visit to Hiroshima. Toyo Kogyo received him, and during the luncheon held after his tour of the plant, Dr. Haas expressed his appreciation for the warm reception and asked if he could do anything to be of assistance to the company. The topic of conversation soon turned to the Wankel engine and Mr. Tsuneji Matsuda, President of Toyo Kogyo, revealed that the company was anxious to sign a license agreement with NSU. Up to that time, NSU had been approached by approximately 100 other companies, including some in Japan, with similar suggestions.

In July of 1960, Toyo Kogyo received unexpected word from NSU, through Ambassador Haas, that NSU was prepared to conclude a license agreement and a delegation was invited to Neckarsulm. On October 3, President Matsuda and a group of five technical men went to visit NSU. They were shown results of bench tests on the KKM-125, KKM-250 and KKM-400 single-rotor engines. Although a slight lack of stability was seen at idling speed, the Japanese party was surprised to see that vibration at high speeds was so small that a coin could be balanced on the engine. At that time, KKM-250 and KKM-400 engines mounted in NSU Prinz cars undergoing field tests had already covered over 25,000 miles. Negotiations for an agreement were based on the policy that such an agreement would not cover engines already completed, but would apply to future joint studies to develop the engine for practical applications at a quicker pace. The agreement was signed on October 12 and officially approved by the Japanese government on July 4, 1961.

In July, 1961, Kohei Matsuda, Executive Vice President of Toyo Kogyo, and a group of six men from the engineering staff went to NSU. The major objective was to find out why the Wankel engine had not yet been put into production. The group left for home after arrange-

Kenichi Yamamoto, Manager of the Rotary Engine Development Division of Toyo Kogyo Company.

Yamamoto attributes the speedy results obtained in his department to the extensive use of computers. Here, test readings are continuously fed in from four dynamometers, placed in a memory bank, and are instantly available for analysis.

The first twin-rotor production engine carried the designation 0813. It was developed for use in the Cosmo car.

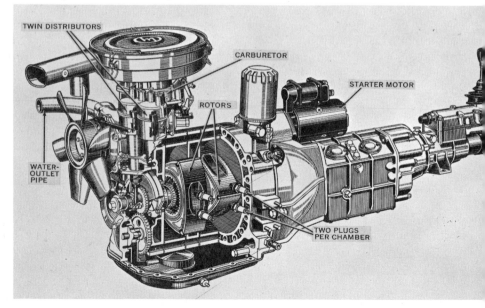

Experimental Mazda four-rotor Wankel engine.

Experimental Mazda three-rotor Wankel engine.

ments had been made for shipment of technical information, drawings, and test engines from NSU. Upon their return, a development committee, composed of members from the company's design, material research, production engineering, manufacturing, and test divisions was organized for the purpose of carrying out full-scale research and development work on the Wankel engine. Toyo Kogyo decided to commence research work on the KKM-400 engine.

The company spent $750,000 to build and equip a special rotary-engine testing center. The test chambers were monitored by closed-circuit television and all test data was immediately reduced to digital form for computer storage and later analysis. A substantial crew, with up to 180 members, was assigned to the project, under chief engineer Yoshio Kono and his rotary engine development chief, Kenichi Yamamoto.

Temperature distribution in the water-cooled rotor housing, under wide open throttle, at 2,000, 3,000, 4,000 and 6,000 r.p.m.

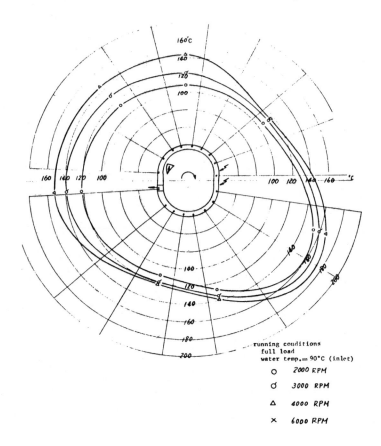

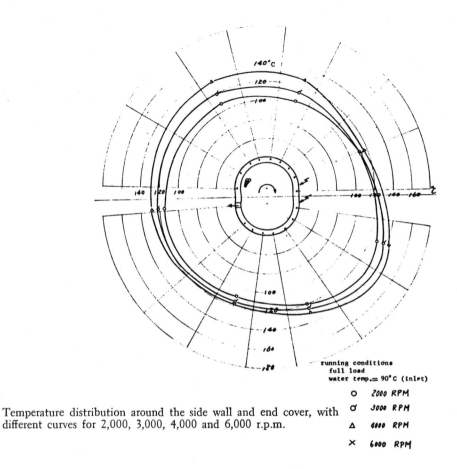

running conditions
full load
water temp. = 90°C (inlet)

○   2000 RPM
ơ   3000 RPM
△   4000 RPM
✕   6000 RPM

Temperature distribution around the side wall and end cover, with different curves for 2,000, 3,000, 4,000 and 6,000 r.p.m.

Temperature distribution in the rotor (with oil cooling), at 5,000 r.p.m. and wide open throttle.

5000 r.p.m. FULL THROTTLE

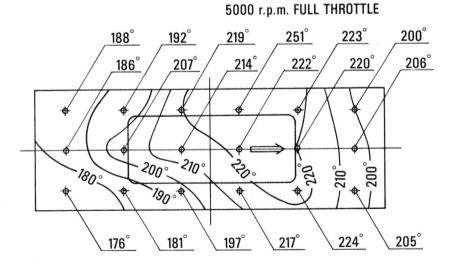

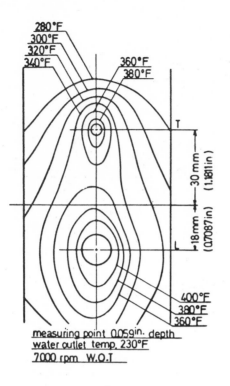

Wall temperatures recorded around the spark plug holes, at 7,000 r.p.m. with wide open throttle.

measuring point 0.059 in. depth
water outlet temp, 230°F
7000 rpm W.O.T

Gas pressure differences between the three working chambers, measured in pounds per square inch (psi).

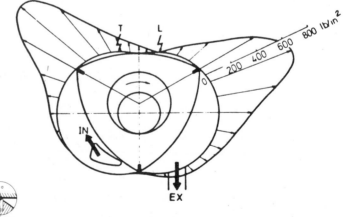

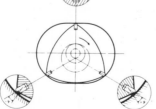

This sketch, and the three insets, illustrate the variations in apex seal leaning angles. When one apex seal has zero leaning angle (top—along minor axis), the other two are at maximum leaning angles, one negative and one positive.

The first NSU KKM-400 engine and its accessories arrived at Toyo Kogyo in November, 1961. The engine was immediately dismantled, examined, reassembled and placed on the test bench. The test showed an output of 43.8 horsepower at 9,000 r.p.m. Toyo Kogyo built their prototype engine No. 1 from KKM-400 drawings supplied by NSU. The test results with prototype engine No. 1 were extremely discouraging. The engine showed excessive vibration at idling speed, emitted large amounts of white smoke, and its oil consumption put it beyond all practical use. When the engine had run for 200 hours power output suddenly dropped. Upon tearing down the engine, it was found that chatter marks on the trochoidal surface had caused the electroplating to fall off.

During the first year, Toyo Kogyo's studies concentrated on the fundamental phenomena of the rotary combustion engine and its problems. In the following two years, emphasis was placed on mapping the basic problems of the Wankel rotary engine. Early in 1962, parallel work was accelerated with the clarification of various problems on the bench. A program was instituted to mount the engine on a test car so that its adaptability as an automotive engine could be tested in its true environment. The prototype engine was installed in a small test car, but new problems were discovered when the car was put into operation. The engine ran very smoothly at high r.p.m., but it became unsteady at slow speeds—when the throttle was closed, strong vibrations followed. Considering the type of passenger cars planned by the company for future production, a decision was made to develop a twin-rotor engine. It must be noted that this was at a time when the single-rotor engine was not yet considered satisfactory.

During the second and third years, all efforts had been directed towards the solution of the basic problems. In the fourth and fifth years, Toyo Kogyo studied the various factors affecting engine performance and concentrated on improving the rotary engine for automotive use.

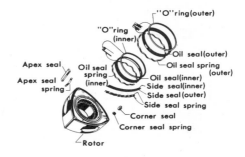

Gas and oil seal configurations for the Cosmo engine.

Test results with the original prototype engine illustrated many un-solved problems. Because of this, the necessity of accumulating new data on the basic operational phenomena of the rotary engine was realized, and a thorough basic research program was started to solve the fundamental problems—Toyo Kogyo wisely saw little or no value in random trial-and-error tests.

Next came the development of special test methods and measuring instruments for this purpose. Full-scale engine tests were devised to verify the results of the basic research. Despite outside criticisms and doubts about the rotary engine, Toyo Kogyo felt that evaluation stand-ards for the rotary engine should be as severe as for the piston engine, which enjoyed the benefits of over 70 years of development. In April, 1963, the rotary engine development division was organized. With Mr. Kenichi Yamamoto as its manager, the division started off with four departments—research, design, testing, and materials research. At first, the division had only five test benches to work on and the build-ing was old and gloomy. The facilities and surroundings could not be called at all suitable for this type of research. Then the company took a decisive step and built a new test laboratory. The construction of the basic test cells in the new laboratory was completed in January, 1964. This was followed by the construction of the endurance test cells, which were put into operation at once.

The new facilities were equipped with the most up-to-date electronic computers and industrial television, which were to play important roles

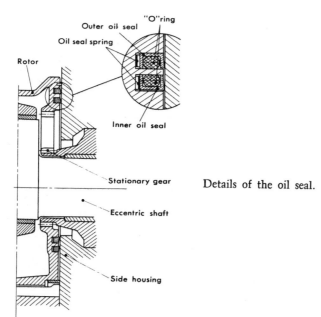

Details of the oil seal.

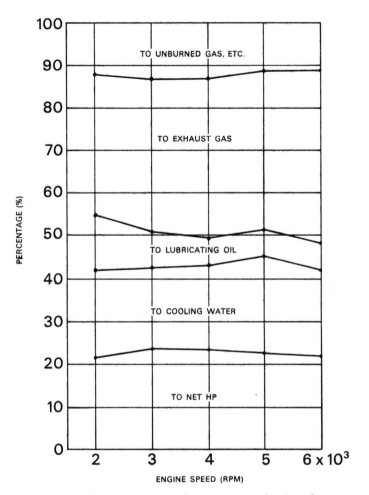

Heat balance chart for the prototype Cosmo engine, showing the percentage of heat given off to water, oil, exhaust gas and unburned gas, throughout its speed range.

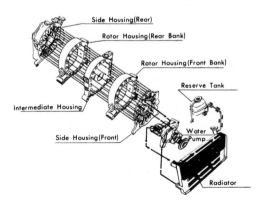

The housing incorporates water cooling by a multi-pass system, designed to provide maximum heat radiation according to the heat load distribution. Water flow is axial so as to give reciprocating balance. The radiator is of the sealed, high-pressure type and is connected to a generous reserve tank.

in expediting research and development. Test benches with centralized controls, capable of running endurance tests 24 hours a day, were installed. The control room was located apart from the test benches, and all the test results were automatically recorded and typed by an electric typewriter. By August, 1964, when construction on the laboratory was completed, there were approximately thirty benches ready for operation.

Yamamoto designed the experimental Toyo Kogyo Wankel engines with light-alloy rotors. The housing and end-covers were made of cast-iron, and the trochoidal surface was given a hard chromium plating. The Wankel engine for the Toyo Kogyo Mazda car contrasts with the NSU units in having apex seals made of carbon rather than cast-iron. Toyo Kogyo now believes in the wear-resistance and self-lubricating

Lubrication and oil circulation system in the 0820 engine. Oil is cooled in a water-oil heat exchanger. Oil and water flow is controlled thermostatically. Oil for lubricating the apex seals and side seals is supplied by a small metering pump, which delivers minute amounts of oil into the working chamber in proportion to engine load. Oil is compressed by the gear-driven oil valves and fed via the thermovalve to the heat exchanger and pressure regulator. Constant-pressure cool oil is passed through a full-flow filter before it enters the main gallery of the housing.

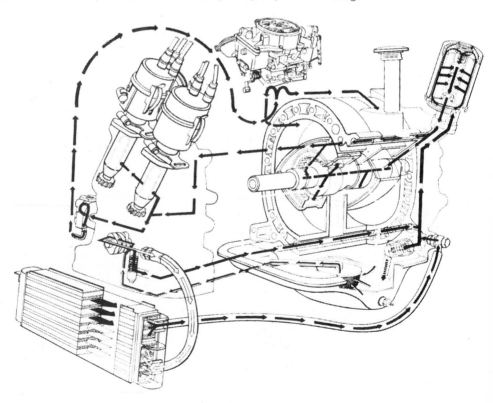

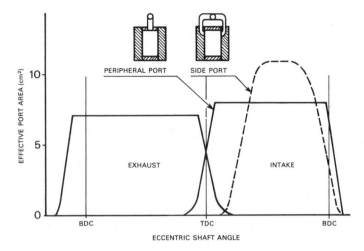

This time-area diagram compares the effective port areas of side ports to peripheral ports, through one full cycle.

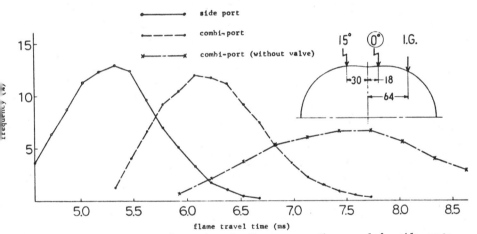

This chart compares the fluctuations in flame propagation speed for side ports (solid line), combi-ports with exhaust heat valve (regular broken line), and combi-ports without exhaust heat valve (irregular broken line). The test engine ran at a constant 1,000 r.p.m., with a mean effective pressure of 186 psi. The sketch shows spark plug location and firing points.

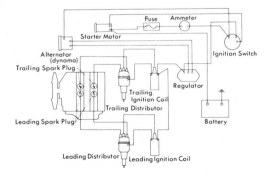

Dual coil four-plug ignition system for the 0813 engine.

properties of carbon, and minimizes the risk of apex seal breakage by detonation through the use of two spark plugs for each rotor, with slightly staggered ignition timing.

As was the case with NSU and Daimler-Benz (about which more will be said in a later chapter), much development work at Toyo Kogyo went into finding compatible materials for apex seals and the trochoidal surface, so as to reduce wear on the seal tips and eliminate chatter marks on the working surface. When compared with a piston ring, which always slides back and forth over the same contact surface, the apex seal always changes its sliding surface; this gives it an added advantage against wear. Moreover, its motion is unidirectional, and the leaning angle of the seal strip changes cyclically along the trochoidal path. The metal apex seals used in the early experimental engines produced chatter marks on the trochoidal surface and wore rapidly.

Retaining iron as the apex seal strip material, the Toyo Kogyo engineers developed the cross-hollow seal. Their proposed method was to drill a hole crosswise near the tip of the apex seal and then drill another hole lengthwise to intersect with the first hole. The results of tests on this new cross-hollow seal proved to be satisfactory. Because the seal

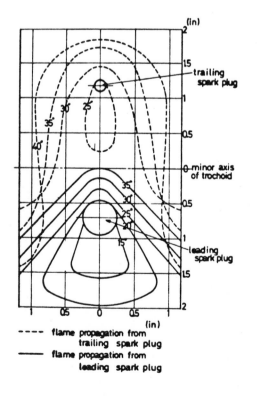

---- flame propagation from
        trailing spark plug
——— flame propagation from
        leading spark plug

Flame propagation along the trochoidal surface (looking directly at the rotor face).

had two longitudinal and several crossing perpendicular holes, the engineers claimed reduced high-frequency vibrations and elimination of the chatter mark phenomenon. Ingenious, and no doubt expensive to produce, the cross-hollow metal seal was soon replaced by a better material—a carbon compound. A special carbon compound impregnated with aluminum was developed for the apex seal material. It had a low coefficient of friction, self-lubricating qualities and high durability.

The strength of this material was much higher than that of ordinary carbon and its friction characteristics did not damage the hard chromium plating. The use of this new seal eliminated the problem of wear and chatter marks on the trochoidal surface. The results of a 60,000-mile car test showed wear on the carbon apex seals of 0.04 inch in

Comparison of two different rotor cavity designs with respect to gas velocity. The evenly shaped cavity produces a later but higher peak velocity before the minor axis, and a very late peak after the minor axis. The cavity with the shallow leading section and deep trailing section showed much more uniform gas velocity, with a peak near the minor axis reaching its maximum before the initial firing, then gradually diminishing.

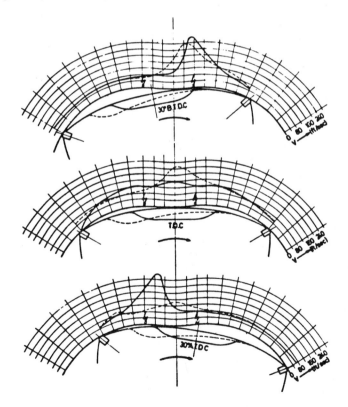

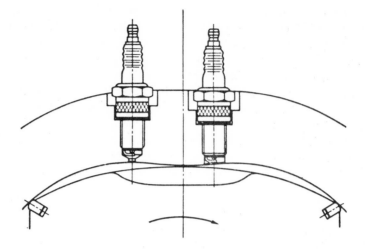

Arrangement of the dual spark plugs.

Performance curves for the 0813 Cosmo engine.

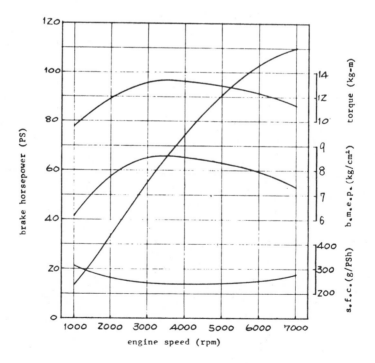

height and of 0.004 inch in width. A power loss of 15% was recorded due to wear. In this case, the wear of the chromium plating of the rotor housing was under 0.0001 mm. and was considered to be negligible.

Lubricating oil was supplied to the working surface on which the apex seals slide by mixing oil into the fuel. This was done by metering the oil with a small, variable-stroke, plunger-type metering pump. The oil was diverted from the main passage into the float chamber of the carburetor in proportion to engine speed and load. Racing engines run continuously at 10,000 r.p.m. Toyo Kogyo technicians say that in these high-speed Wankels, the bearing material is a bigger problem than apex seal material and construction.

Recent reports say that, despite their success with carbon seals and a chromium plated working surface, Toyo Kogyo will soon go to metal seals and a nickel/silicon carbide surface. The reasons are believed to be overwhelmingly economic.

The inner and outer dual side seals, arranged on both flanks of the rotor, were made of the same material as ordinary piston rings. Pin-shaped side seals also were provided at the joint portions of the apex seals and side seals. These seals were pushed against the sliding surface of the housing by spring force and gas pressure behind the seal. The rotor seal construction prevented lubricating oil flowing through the engine from entering the combustion chamber. Near the center of the rotor flanks there were double concentric grooves which housed wave

Performance curves for the 0820 (R-100) engine.

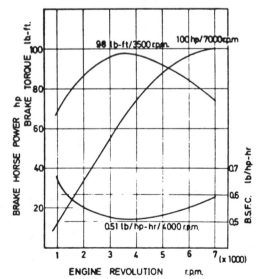

springs and oil seal rings. The tapered lip of the oil seal ring was pushed against the end cover of the side housing and this scraped off excessive oil. Experiments with different materials for the end covers ended with Toyo Kogyo choosing the same material for the newest engines that had been used on the earliest experimental units. The Mazda Cosmo had aluminum end covers with a molybdenum-based coating; on the later R-100, R-130 and RX-2, the end covers were induction-hardened cast-iron.

Although NSU and Daimler-Benz are firm believers in peripheral ports, Toyo Kogyo, along with Curtiss-Wright, chose side ports. The penalty for NSU was irregular running at light loads and low speed (overrun). Side ports gave slower port closing and opening, which meant lack of peak power but better low-speed and light load scavenging. On the earliest test engines, Toyo Kogyo used a peripheral intake port. This induction system was largely responsible for the unsatisfactory performance at low speeds, the shaking in the vehicle on the overrun, and the unsatisfactory idle which was experienced on prototype car No. 1. In the fall of 1963, Toyo Kogyo began to design an engine with a side port induction system. This new system was an induction system with the intake ports located on the side housings, at right angles to the ports of the peripheral system. The reduced overlap lessened the mixing of combusted gas with new mixture, and combustion at low shaft speeds was stabilized. The vehicle with the side port engine showed clear superiority to all earlier versions.

The cycle time of the rotary combustion engine is 1.5 times that of a reciprocating engine (1,080 degrees shaft angle); therefore the Toyo Kogyo engineers felt sure that port-opening time *area* could be ob-

General layout of the 0813 Cosmo engine.

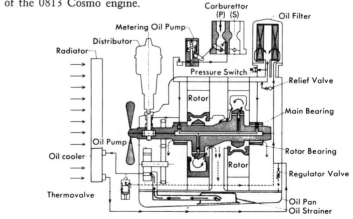

tained regardless of a small overlap. The result was outstanding high-speed performance without any sacrifice of low-speed flexibility. The intake ports were of the double side port type, with mixture intake taking place from both sides to each rotor. The ports on the inter-mediate housing are called the "primary" stage ports and those on the side housing the "secondary" stage ports. The ports were connected to a two-stage, four-barrel carburetor, and each port was fed from a completely independent induction system. The primary stage mani-folds were preheated by exhaust gas in order to speed up engine warm-up, as well as to improve "driveability" and fuel economy.

Cross section of the 0820 engine.

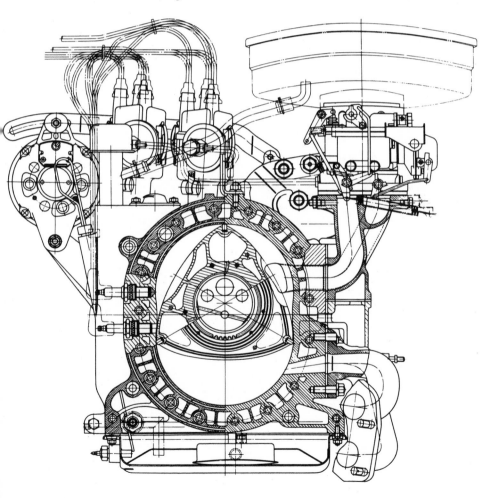

The Cosmo engine had a combination of side and peripheral intake ports. The induction system began with a special triple-throat Stromberg carburetor. The central primary throat, which was in operation at all times, fed a channel in the wall that separated the two rotor chambers. The channel was split in two, giving a side port to each chamber. The other two throats went into action only when the engine load ap-

Elevation of the 0820 engine.

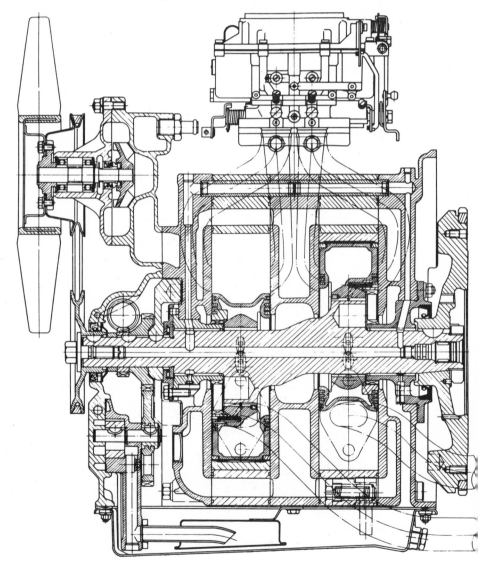

proached wide-open throttle conditions. The two secondary throats delivered fresh mixture to peripheral intake ports. The combi-port was almost equal in volumetric efficiency to the peripheral port, and combustion in the case of the side port was more stable than was the case in the combi-port.

When the 0820 engine for the R-100 was designed, Toyo Kogyo abandoned the compromise combi-port system in favor of dual side ports—one primary and one secondary for each rotor chamber. The carburetor was of the downdraft type with dual throats, one per rotor chamber, and the gas passage from each throat was split in two right below the carburetor mounting flange. One channel led into the dividing wall, the other into the end cover. Each rotor chamber had intake ports on both sides, and the two central ports belonged to separate gas flow systems. It was an important feature of the side ports that their outer edges could not be located very far outboard without causing the apex seal "bolt"—which must have axial freedom for proper sealing—to fall through the aperture, which would be producing, or at best traversing, a wear-inducing depression. This meant that a side inlet port could not open before top dead center, and that a side exhaust port could not remain open after top dead center. Consequently, there could be no overlap between the two side ports. It was also less easy to design side ports with an adequate capacity (which is why Toyo Kogyo reverted to double side intake ports), and the gases they carried had to

Complete R-100 (0820) engine with clutch and gearbox.

Front view of the R-100 engine, complete with accessories.

be turned through at least one right angle. Against these disadvantages must be balanced the improved low-speed torque that followed from the absence of overlap between side ports.

It is interesting to note that peripheral ports suffer from few of the disadvantages of side ports. Although their cross-sectional area may be restricted by the width of the rotor and its housing they have ample capacity because they stay open longer, and the gases which flow through them need not be turned through a right angle. Moreover, a peripheral intake port can open before top dead center and a peripheral exhaust port can close after top dead center, giving better timing than can be obtained from side ports. Overlap is important to ensure complete emptying of the exhaust gases and good filling with fresh mixture. Overlap is theoretically determined by the positions of the rotor apex seals, which open and close the peripheral ports as they sweep over them. Maximum overlap for a rotor of any given size is obtained by moving the two ports as close as possible to the minor axis of the trochoid, so that both simultaneously remain open for a long period as the rotor sweeps through the top dead center position.

0820 rotor with all seals in place.

Exploded view of 0820 rotor and its seals.

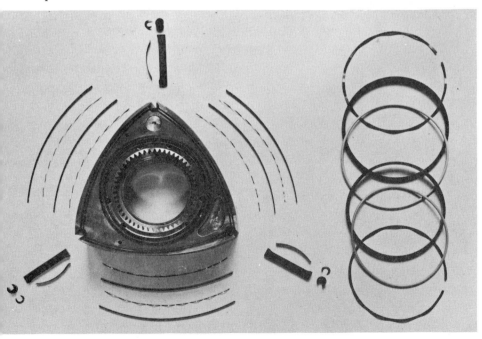

Stable combustion at part throttle is desirable for the Wankel automotive engine, especially at low speeds. The stability of combustion at part throttle is dictated by a large number of interdependent factors, such as induction system design, intake and exhaust port timing, combustion chamber shape, spark plug location, and transfer port configuration. One of Toyo Kogyo's solutions, on the 0813 engine, was the use of two plugs per chamber, 5 degrees apart, one above the other,

The two rotors and their phasing gears, placed in proper relative positions.

Mainshaft of the 0820 engine.

fired by two coils and two distributors. This system was intended to produce more complete combustion regardless of flame front travel patterns. Two spark plugs were installed, one on each side of the minor

Rotors fitted on the mainshaft.

Rotor housings with exhaust ports and spark plugs.

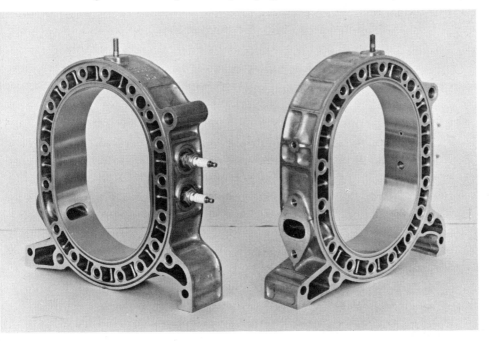

axis. The plug on the leading side in the direction of rotation was located 18 mm. from the axis; the one on the trailing side 30 mm. from the axis. In this arrangement of dual spark plugs, the sizes of the

End covers and center side wall with intake ports.

Two views of the fully assembled 0820 engine.

マツダ10A型エンジン

transfer holes connecting the trochoidal surface with the electrode chambers were not the same on the trailing and leading sides; the one on the trailing side was 4.2 mm., and the one on the leading side 12.4 mm., in diameter. The two sizes were necessary because of the difference in gas pressures between the adjacent chambers separated by the apex seal. It was not possible to make a large hole on the trochoidal surface at the position of the trailing spark plug; therefore the trailing spark plug was mounted in a pocket slightly recessed from the trochoidal surface, with a small transfer hole connecting it with the combustion chamber. At the position of the leading spark plug, the difference in gas pressure was almost zero. The electrode could therefore be brought close to the trochoidal surface, making ignition and resistance to fouling at low speed much more stable as compared to that at the trailing side. The spark plugs were identical for both the leading and trailing sides, and were specially developed for the 0813 engine with due attention to their resistance to fouling at low speed and resistance to preignition at high speed.

Another problem was cooling. The Wankel engine has a "hot lobe," which is never swept by cool unburned gases, as is the combustion chamber of a piston engine, and the temperature rises rapidly as the engine load is increased. In the Mazda engine, the incoming mixture for small throttle openings was preheated so as to warm the cool side of the engine and minimize the temperature differences across it, as well as to improve atomization. The process could be carried a stage further by bleeding exhaust gases through passages in the cool parts of the rotor end housings.

The Mazda engine housing was cooled by water that flowed from the water pump, installed at the upper front end of the engine, through channels in the housing (in the axial direction) to the rear side housing and back into the recirculation cycle. Ribs in the channels were so designed that the water flowed at a rate proportional to the thermal load distribution. After circulating through the engine, the cooling water passed through the bottom bypass-type thermostat to an aluminum radiator. The use of this type thermostat ensured a sufficient amount of coolant circulation inside the engine when the thermostat was closed. This made it possible to hold the maximum wall temperature below 410°F. The cooling fan was driven by a fluid coupling that reduced noise and power loss at high speed. The oil delivered from the oil pump passed through the thermovalve to the oil cooler, which was air-cooled. The oil then went through the oil filter and entered the engine. After lubricating each of the bearings, the oil was ejected into the chambers inside the rotor, where it cooled the rotor. Increased rotor housing temperatures—by up to 70°F. compared to those of the housing formed in the original manner—were a penalty of Toyo Kogyo's basic design. This was due to the relatively low thermal conductivity of the chrome steel layer on the working surface. Tests conducted on the two kinds of rotor housing, however, showed no difference in durability or wear resistance.

Thermal fatigue, as a result of alternate heating and cooling of a given area, was also a problem in early Mazda engines and caused cracking at stress concentration points around the spark plug holes. These cracks, in extreme cases, penetrated to the water jacket. The first approach of the Toyo Kogyo engineers was to improve cooling to cope with the most critical condition—rapid acceleration of a cold engine. In winter the combustion chamber walls could suddenly rise from a low ambient temperature to 450°F. This was the reason for adopting the most modern type of bottom-opening thermostat; one that continuously controlled bypass flow as well as flow through the radiator. A further cause of the cracking trouble was lack of flexibility in the trochoidal wall and conduction of heat from it to the cold end housings through the bolts holding them together. Improved flexibility and reduced heat transfer were simultaneously achieved by separating the side-bolt bosses from the trochoidal wall, and housings made in this way proved completely resistant to thermal fatigue.

As happened with other manufacturers, chatter marks on the trochoid surface of the rotor housing were experienced by Toyo Kogyo during the early stages of development. It was believed that the chatter marks were the result of high-frequency vibrations, and basic research was performed on various subjects such as the vibration characteristics of the apex seal and the rotor housing, friction characteristics as an exciting force, the influence of lubrication, the sliding speed of the apex seal and its relationship with contact pressure and the leaning angle of the apex seal. It was necessary to test and try every possible material available. Great numbers of rotor housings, all with chatter marks, were soon piled high in the laboratory. Then a new method was conceived for measuring vibrations. After six months a proposal was made to alter the seal configuration.

In the course of development, several effective methods of reducing the chatter marks were found, but any methods that would have imposed a limit to the durability of the rotor housing were rejected as undesirable for practical use. At first they used a conventionally cast aluminum alloy rotor housing with hard chromium plating approximately 0.006 inch thick on its trochoidal surface. Their next objectives were to reduce the thickness of this chromium layer (because the rotor spends several expensive hours in the plating bath) and at the same time to make use of the much cheaper die-casting process, despite the consequently poor bond between chromium and aluminum. Their solution was to make a licensing agreement with the Doehler-Jarvis Division of the National Lead Company of America to use a new metal spraying and casting technique known as TCP—Transplant Coating Process.

In this process, a thin layer of steel (which adheres strongly to aluminum) is sprayed on the inner trochoidal core of the die, which is preheated before the aluminum is poured in. The result is a steel-coated, die-cast rotor housing that is machined before being plated with a very thin layer of chrome. After experience with the process it became possible to hold the thickness of the steel layer to 0.015–0.040 inch, after machining, and to reduce the thickness of the chrome layer to only 0.002 inch, thus saving plating time. Because the wear of chrome plating is negligible, there was no problem about the life of the rotor housing. The more important problem concerning the rotor housing was how to obtain an accurate and thin chrome plating on

Cutaway drawing of the 0820 engine and transmission.

the trochoidal surface. There is a fair prospect that chrome-plating with a thickness of several tens of microns can be practically applied, requiring no grinding, in production by using special equipment.

The conventional ordinary carbon seal material could not endure the severe operating conditions required for the apex seal, especially in the areas of strength and wear. For several years, Toyo Kogyo's research group made a study of carbon in close cooperation with a special development team organized by a carbon company. The development of a non-destructive inspection method (NDI) and a carbon which was strong enough to endure the severe operating conditions encountered allowed Toyo Kogyo, by 1967, to run routine tests up to 8,000–8,500 r.p.m. Further improvement in the seal enabled them to raise engine speed further.

The wear of the side seal was negligible, but the wear on the permanent contact section of the oil seal caused an increase in oil consumption and therefore could not be neglected. Toyo Kogyo had adopted the aluminum sidewall for its light weight and high cooling efficiency. However, the sliding surface of the aluminum sidewall, with no surface treatment applied, was subject to serious wear and posed problems in practical use. There were several possible methods of surface treatment. The sliding surface of the sidewall in the production Mazda engine was sprayed with high-carbon steel. However, because all of the seals which slid on the surface of the sidewall were made of metal, some amount of wear naturally occurred on the hardened surface of the housing even with lubrication. The wear in the cylinder bore of the reciprocating piston engine is greater at top dead center. Similarly, the wear on the sidewalls of the Wankel due to *side seal friction* occurred mainly at a position near top dead center. On the other hand, the wear on the sidewalls due to *oil seal friction* occurred within the circular envelope drawn by the outer edge of the oil seal. After considerable mileage, wear on the sidewalls sometimes became so great as to reduce power and increase oil consumption. Because the surface of the side housing was flat and the sprayed-on coating thick enough, the worn out housing could be used again by grinding the surface—the fact that the sidewalls were slightly undersize gave rise to no new problems.

The spark plug in the Wankel engine is exposed to more severe heat conditions than in either the two-stroke or the four-stroke reciprocating piston engine. The problem of high heat input can be solved by using spark plugs with a high heat value. But, taking into consideration cold-starting characteristics and the risk of electrode fouling during continuous running at part throttle, improving only the heat value of the spark plug was not the answer with the Wankel engine. The working conditions of the spark plugs for the Wankel are determined by the relationship of such factors as the construction and heat value of the spark plug itself, the cooling effect of the plug seat, the location of the plug, the volume of the chamber around the electrode and the size of the shooting space between the electrode and the combustion chamber. In conjunction with their studies of engine design, the Toyo Kogyo engineers invented new types of spark plugs in cooperation with the engineers of leading spark plug makers. Spark plugs that lasted between 7,500 and 8,100 miles on the road, and plugs that could endure more than 100 hours of running at full throttle on the test bench, were developed as a result.

After five years' experimental work it was apparent that in order to design an engine that would fully utilize the best of the inherent characteristics of the rotary principle, a twin-rotor side-intake-port engine would be the most desirable for automotive application. In 1964, an engine with 30 cubic inches (491 cc.) chamber displacement was designed. This engine, with an output of 110 horsepower at 7,000 r.p.m., made its first appearance on the market in 1967, mounted in a Mazda 110 S sports car.

12

Daimler-Benz

NSU-BUILT WANKEL TEST ENGINES were running at Daimler-Benz as early as 1959. Wolf-Dieter Bensinger, formerly head of the passenger car engine design office, was in charge of the program. Bensinger had joined Daimler-Benz in 1943, after 12 years at DVL, where he worked mainly on control units for aircraft engines. In 1945, he was assigned to a new passenger car engine program. His work led to the development of the new line of engines with overhead camshafts actuating slightly inclined valves via finger followers. These were the engines that went into production in the 1951 Mercedes-Benz 220 and 300 series.

Bensinger went to Lindau to see Felix Wankel in 1960. They had known each other through DVL contact since 1934, and Bensinger wanted to consult Wankel himself about his ideas and projects. How did Bensinger propose to use the Wankel engine? His first plan was to develop a 1.4 liter Wankel to replace the piston engine in the 220 SE. Felix Wankel was, of course, delighted. Bensinger began an intensive test program, and during its first year of testing, Daimler-Benz spent about $750,000 on Wankel engine experiments. It is significant that Daimler-Benz had not yet acquired a Wankel engine license. There was, however, a tacit understanding between the engineering staffs of the two companies that NSU would be given the benefit of Daimler-Benz test reports and any improvements made, in return for furnishing test engines to Stuttgart free of charge. Bensinger's report to the management of Daimler-Benz, concerning Wankel engine feasibility, was favorable, and the president, Walter Hitzinger, decided to negotiate a contract for Wankel engine rights.

Although Hitzinger was convinced of the engine's good qualities, he did not want to pay too much for the right to make it. The Daimler-Benz attorneys rewrote contract after contract, and the NSU management found Hitzinger to be a real hawk of a negotiator. He also kept finding new reasons for delaying the signing of the contract. This is speculation, but it's possible that Hitzinger stalled because Flick was buying more and more NSU stock, which in turn would have meant that Daimler-Benz, once it owned NSU, could make Wankel engines without paying a license fee or royalties. However, Hitzinger soon realized that ownership of NSU was a long way off, and a contract with NSU was signed in October, 1961. The monetary arrangements involved payment of a flat fee of $750,000 in three annual installments. But, there was also a clause giving a minimum payment to Wankel G.m.b.H. and NSU each year. This was bitterly fought by Hitzinger. The reason this clause was in the NSU contract was to guarantee that license takers would not unduly delay development and production of the Wankel engine for business reasons. In other words, they were forced to push the Wankel engine forward as quickly as possible in their own interest.

Bensinger began design work on a Mercedes-Benz Wankel engine in 1960. His plan was to test and develop both a single-rotor and a twin-rotor version of the same basic engine simultaneously. The single-rotor version, with a chamber displacement of 42.5 cubic inches (700 cc.), was tested first. This engine lost its vibrationless running after 400 to 500 hours under high loads, which corresponded to 25,000 to 30,000 miles on the road. After that period, the seals began to flutter and the casing showed signs of pressure waves and vibrations that robbed the engine of performance. As with the other Wankel developers, chatter marks showed up on the trochoidal surface. Engine life was considered good enough only for minicars and motorcycles, the engines of which usually were overhauled within 25,000–30,000 miles anyway. Following NSU practice, Bensinger then used peripheral intake ports, water-cooled housings, and oil-cooled rotors. Power output from these new engines was remarkably good, but performance fell off at an early stage. Tests with an 85 cubic inch, twin-rotor unit in "as-new" condition showed it to produce 160–170 horsepower at 5,000 r.p.m.

Chatter marks on the working surface was the worst problem during the early experimental period. Dr. Bensinger gave this explanation of the phenomenon: "The apex seals push oil, carbon particles, combustion by-products and even fuel ahead of them. When the resistance of these substances becomes too high, the seals tend to lift off the trochoidal surface. When this happens, the trailing edge of the seal

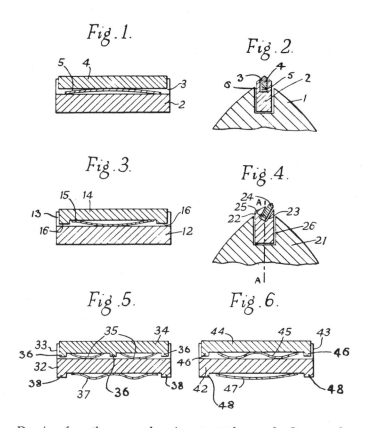

Fig. 1.

Fig. 2.

Fig. 3.

Fig. 4.

Fig. 5.

Fig. 6.

Drawings from the apex seal carrier patent taken out by Goetzewerke. Fig. 1 shows spring loading of the seal strip with a convex spring, without spring-loading of the seal carrier. Fig. 2 is a cross section of the same design. Fig. 3 uses a concave spring to load an inclined seal strip (as shown in Fig. 4). In Fig. 5 and 6 double-concave springs, both one-piece and two-piece, are shown in applications including spring-loading of the seal carrier.

Flame front formations at various points during the combustion phase are shown in the center schematic (looking directly at the rotor face). The sketch at left shows rotor positions before and after the moment of ignition. The diagram at right illustrates gas velocity at the minor axis, during one full rotor revolution.

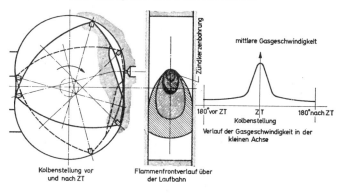

tips tend to break off minute particles from the working surface."
When serious work on the Wankel engine began at Daimler-Benz in
1960, chatter marks could be observed after no more than five hours'
running. Since that time, there has been much improvement. Seal
chatter never may be completely eliminated, said Dr. Bensinger, but the
consequences have, at present, been reduced to levels where they no
longer have any influence on the life of the trochoidal surface. The solution to the problem was Nickasil. This is a layer of nickel, created by
galvanic process, buried in a very fine silicon carbide bearing and used
for the entire working surface. It was originally developed in cooperation
between Daimler-Benz and Mahle Kolbenwerk, although NSU took it
up later, and the process has been made available to Wankel engine
licensees all over the world.

The most serious problem, next to chatter marks, was seal tip wear.
The solution was not to be found by studying materials alone. It was
necessary to analyze the combustion process, pressure levels and direction, and apex seal behavior under the ever-varying conditions along the
trochoidal path. The combustion process in the Wankel engine benefits from one outstanding advantage in terms of sealing, due to the fact
that gas velocity is far higher in the spark plug area at the moment of
ignition than in the other parts of the combustion chamber. Bensinger's tests showed that the gas velocity near the minor axis was
a multiple of the flame front velocity. In his test engine, flame front
travel took place only in the leading direction. The upper flame front
was stationary; ignition occurred only when compressed gas was brought

The loss of some fresh air into the exhaust area, against the direction of rotor rotation, is an advantage in disguise. This air helps
purge the area of exhaust gases that might otherwise have been
recirculated. The air itself does not escape through the exhaust port,
but is swept along into the coming compression phase by the trailing apex. The curve on the right shows the progressivity of variations
in chamber volume.

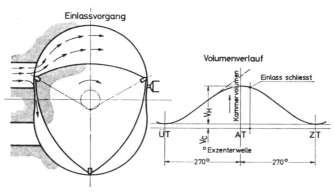

forward to the flame front. By the time the trailing apex seal reached the flame front, temperature and pressure in the combustion chamber were considerably diminished. A leaking apex seal at the trailing end did not show burns—it only resulted in a loss of performance. Ahead of the spark plug, the combustion chamber walls were relatively cold. Because the flame front was chasing the leading apex throughout the combustion process, temperature and pressure both declined from their peak values reached earlier in the process. Since the apex seals operated in a relatively low temperature belt, Bensinger concluded they could be made of materials having low heat resistance, such as carbon or aluminum.

The experimental seals were designed to provide full, or almost full, sealing until 2.5 mm. wear allowed the seals to escape from their grooves. Seals made with carbon materials showed wear characteristics in the range of 0.002 mm. per hour (or 100 kilometers), which would give them a lifespan equal to more than 100,000 kilometers (or about 62,500 miles) on the road. These wear characteristics failed to hold true whenever fuel deposited on the working surface washed off the oil film. Frequent stopping and starting was a severe barrier to long seal life.

The problems were somewhat alleviated by the fact that in case of apex seal failure, such as when they stuck in their grooves or the seal tips were too worn to provide a proper seal, there was no harm in continued operation of the engine. Starting difficulties resulted, but the engine would not stall.

Some engines were tested with a novel type of seal configuration developed and patented by Goetzewerke in Burscheid. Instead of the simple seal strip, a more complex arrangement was invented. The strip was not seated in a groove cut directly into the rotor, but in a channel-section seal carrier that seated in the groove. The strip was spring-loaded radially away from the carrier, and the carrier from the rotor. Two main

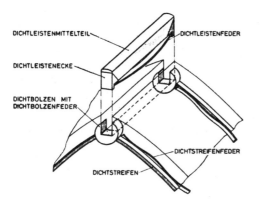

DICHTLEISTENMITTELTEIL

DICHTLEISTENFEDER

DICHTLEISTENECKE

DICHTBOLZEN MIT
DICHTBOLZENFEDER

DICHTSTREIFENFEDER

DICHTSTREIFEN

Final apex seal configuration developed by Daimler-Benz.

advantages were claimed for the new seal configuration. The first stemmed from the fact that the strip could be slimmer and more flexible than was previously possible, with the result that slight distortions in the piston or the track could be more readily accommodated. The second related to the cold clearances between the sidewalls of the working chamber and the ends of the seal strips and seal carriers. This system was not adopted, however, as results failed to warrant the additional cost and complexity.

Nothing has been found that has given superior results to the relatively thick apex seal strip located directly in a groove in the rotor, spring-loaded radially and provided with slots along the bottom leading edge to ensure sealing action by gas pressure, without delay following pressure reversals. For instance, the C-111 apex seal strips are 5 mm. thick and 7.5 mm. high. The trunnions are 10 mm. in diameter and 6 mm. long, while the side seals are 1.5 mm. wide and 2.5 mm. deep. Current engines use cast-iron apex seals, but they are not considered a final solution. Experiments with ceramic edges on the apex seals are being conducted right now. Daimler-Benz continues its search for better apex seal materials, so that with further development they will not be the limiting factor in engine life between overhauls, even under the worst operating conditions.

Apex seal strip, corner seals, and seal spring, as used in the C-111 engines. (*Photo: Ludvigsen*)

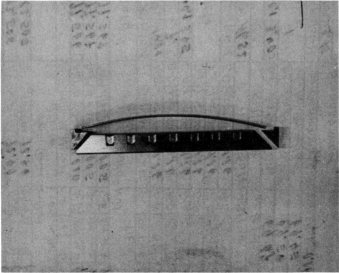

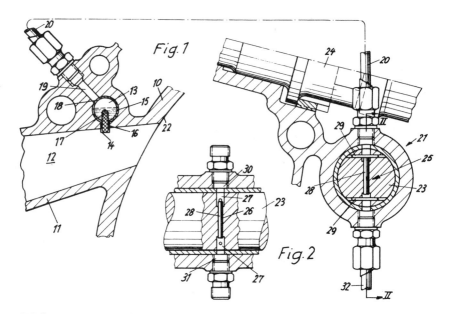

Wick-type metering of lubricating oil into the intake port for apex seal lubrication, was patented by Daimler-Benz in 1965.

The apex seals depend on the presence of an oil film for satisfactory operation. For a long time, NSU simply had mixed lubricating oil with the gasoline, so that the fresh charge would provide the necessary seal lubrication. Although Bensinger adopted this method for the earliest test engines, it was unreliable and suffered from wide variations in oil/gasoline ratios. In addition, it is unsuitable for diesel engines. Daimler-Benz began to study methods to mix accurately metered quantities of oil with the incoming charge. In a patented design, oil was fed from a metering device to a metal-fabric wick projecting into the inlet port. The amount of oil supplied was directly proportional to the speed of the rotor. The metering device was a simple flow control unit housed in a shaft revolving at rotor speed. A short plunger contained in a diametral hole in the shaft had its travel limited by small-diameter pins fitted into drillings that intersected the first bore. The shaft housing had two diametrically opposed ports—one connected to the oil pump and the other to the wick. As the diametral hole in the shaft swept past the port leading from the pump, the plunger was forced away towards the outlet port. After an interval of 180 degrees of rotation the plunger was forced in the other direction, to allow a small quantity of oil through the outlet port. This action was repeated twice per revolution, and the quantity of oil delivered at each stroke was dependent solely on the travel and diameter of the plunger—fluctuations in oil pressure had no effect. The stroke limiting pins also could be made eccentric, so that turning them would alter the stroke of the plunger during running.

The wick device was discarded in favor of a more advanced metering system, using a special drip-feed oil pump that delivered minute quantities of oil to sleeves in the intake ports. The sleeves had a cavity that held an oil film over a large sector. The incoming air carried part of the oil film in with it; not in the middle of the airstream where it would have formed part of the mixture and been wasted, but along the metal surface from where it was deposited on the apex seals. The cavities prevented the oil from getting into the center of the airstream and kept it close to the surface so that the apex seals swept the oil film with them as they passed the intake port. The oil to fuel ratio was estimated to be 1:150, thus the experimental C-111 used no more oil than a piston engine of similar output.

Intake port sleeve with oil film cavity, as used on the C-111 engines. (*Photo: Ludvigsen*)

Pattern of oil flow inside the C-111 rotor.

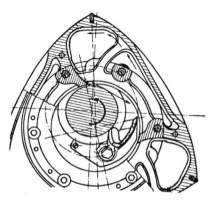

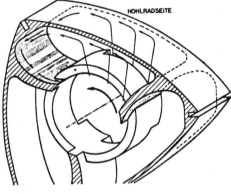

Cutaway sketch showing oil flow towards the rotor face and along the direction of rotation, until arrested at the leading apex and drained back to the eccentric bearing.

Rotor temperatures recorded with and without oil circulation at 5,500 r.p.m. and 135.5 psi mean effective pressure. Black dots = sensor melted. White dots = sensor not yet melted. Kastenkolben ohne oelruecklauf = Without oil circulation. Zellenkolben mit oelfuehrung = With oil circulation. Drehrichtung = Direction of rotation.

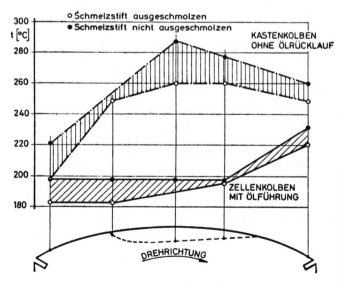

In the course of his development work on the rotor cooling system, Bensinger found that agitation of the oil within the rotor resulted in foaming. Foaming was undesirable because it reduced the oil's cooling effect. It was also found that some of the oil, particularly in the apex cavities, did not circulate and therefore was incapable of providing the intended cooling. One proposed solution was to block off the apex cavities from the oil circulation system and fill them with sodium—a high-conductivity material long used in the exhaust valve stems of Mercedes-Benz piston engines to increase heat dissipation. This has subsequently been found to be unnecessary.

The C-111 engines use a rotor cooling system quite similar to that developed by Curtiss-Wright for the RC2-60. Oil sealing of the rotor bearing was a difficult problem for a long time. A portion of the same oil that cools the rotor had to lubricate the rotor bearing. Because Daimler-Benz chose plain bearings rather than ball or roller bearings for the output shaft, bearing lubrication was highly critical. An ample recirculatory oil supply under pressure would be ideal, but there is always

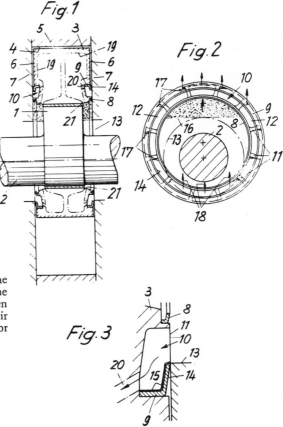

Patent drawings covering the drainage of cooling oil from the rotor, by opening outlets when centrifugal loads force oil in their direction. A scoop is used for trapping excess oil.

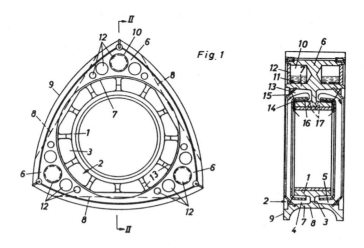

Drawings from Daimler-Benz's patent for sodium-filled holes in the rotor casting, behind each apex.

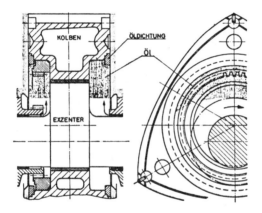

Oil sealing method, as applied to the C-111 engines.

Detail of the oil seal configuration used on the C-111 engines.

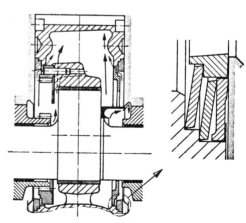

the risk of excess oil leaking through to the combustion chambers. Oil would accumulate behind the seals and remain pressed against the seals by centrifugal forces of increasing magnitude at increased r.p.m., until the oil would force its way through to the sides of the working chambers. Daimler-Benz found a very elegant solution by forming a chamber inside the rotor. The rotor turns more slowly than the eccentric shaft, centrifugal forces act in the direction of the eccentric, and the oil contained in the rotor is kept in constant circulation. Oil inflow and outflow are controlled by centrifugal force action combined with a new dual rotor bearing seal. As the rotor revolves, oil can flow into the rotor cavity towards one apex from both sides of the eccentric. It can only flow out on the opposite side (i.e., the rotor face opposite the apex that is receiving the oil). If too large a proportion of the hot oil from the rotor cavity fails to return to the mainshaft bearing, and instead recirculates back into the rotor, a shield on the eccentric catches it and throws it into the outlets. Because the shield revolves faster than the rotor, it easily scoops up excess oil. This is backed up by a dual rotor bearing seal, which also has to function as a gas seal to keep combustion fumes from contaminating the lubricating and cooling oil. These seals are fitted on both sides of the eccentrics that act as rotor bearings. Each has an inner ring and an outer ring. The oil that passes the inner one is then contained in a chamber formed by radial ribs between the two seals. The ribs have a slight clearance from the resilient lip of the outer ring. This lip extends radially inward and effectively prevents the oil from escaping into the combustion space, while staying at surface level and providing lubrication where needed.

By the use of peripheral intake ports, Daimler-Benz is able to obtain volumetric efficiency between 110 and 115% at 2,000 r.p.m. This is possible because gas flow continues even after the onset of compression, and gas velocity may even increase as the port area is restricted by the passing of the rotor's trailing apex. Gas flow is split by this apex as it travels across the port area, and fresh gas begins to fill the next working chamber, which is not yet halfway through its exhaust phase. This reversed flow helps exhaust gas evacuation and does not cause any notable loss of fresh mixture, simply because the chamber advances so quickly into the intake and compression areas. Tests with side intake ports showed that they could not give equal volumetric efficiency for two reasons. First, the overlap between intake and exhaust was missing and second, the absolute time available for intake was smaller.

The first Wankel engines designed and built at Daimler-Benz used carburetors. There were multiple reasons why fuel injection was eventually adopted, and, again, the change was only made after exhaustive examination of the full combustion process. Combustion is, of course, slower in a Wankel engine than in a piston engine. This is unfavorable in terms of thermal efficiency but does result in favorable part-load operational characteristics. The use of twin spark plugs to shorten flame front travel distance was rejected by Daimler-Benz, first of all because it didn't help much and second because of the faster pressure rise and related problems. The engineers chose to use one surface-gap spark plug per chamber. The Wankel engine does not develop the same compression pressures as does a reciprocating piston engine because more heat is given off to the surfaces of the combustion chamber. This disadvantage does, however, diminish as rotational speed rises. Fuel consumption of the Wankel engine tends to be higher than that of piston engines of similar output, particularly at part load and low r.p.m. Daimler-Benz ultimately chose fuel injection for the C-111 for these reasons, in addition to the following particular disadvantage of carburetors as used on Wankel engines. Condensed fuel droplets in the intake manifold of a piston engine usually are vaporized and mixed with the air due to the warming effect of the hot valve head and the fluctuating gas velocity in the induction system. In the Wankel engine, however, everything is cold and the areas are so large that gas velocity tends to be low. With side intake ports, some turbulence can be created, but with disappointing results. Fuel injection overcomes all these objections. On a closed throttle, the carburetor-equipped Wankel engine continues to aspirate fresh mixture and, because the ports are always open, the engine runs on exhaust gas, or unburned gas mixed with fresh air. Depending on r.p.m., there will be a time lag between the closing of the throttle and the first ignition of these gases, which have had time to mix with fresh

The use of direct fuel injection modifies the operating cycle of the Wankel engine to some extent. Injection begins well before the intake port is closed off and the beginning of compression.

How the Rotary Piston Engine Works

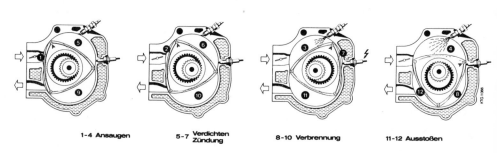

1-4 Ansaugen   5-7 Verdichten   8-10 Verbrennung   11-12 Ausstoßen
                        Zündung

oxygen—it is this ignition lag that produces a bucking sensation, commonly called the "snatch" phenomenon. Fuel injection solves this problem, because fuel delivery is shut off whenever the throttle is closed. Only when the throttle is re-opened suddenly will abnormal combustion take place. Similar problems occur during idling, although they are easily overcome because of the narrow r.p.m. band.

June 14, 1966   W. SPRINGER ET AL   3,255,738

ROTARY-PISTON INTERNAL COMBUSTION ENGINE

Filed March 19, 1963

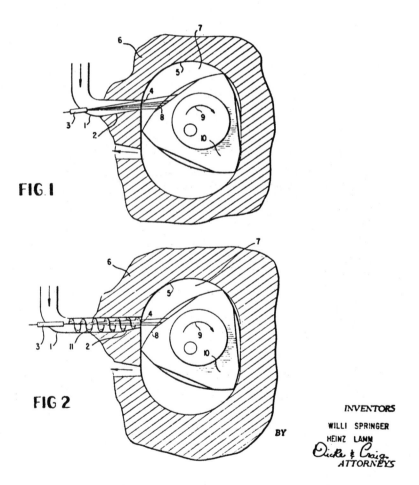

FIG. 1

FIG 2

INVENTORS

WILLI SPRINGER
HEINZ LAMM

BY

Dicke & Craig
ATTORNEYS

Swirl injection in the intake port, as patented by Springer and Lamm.

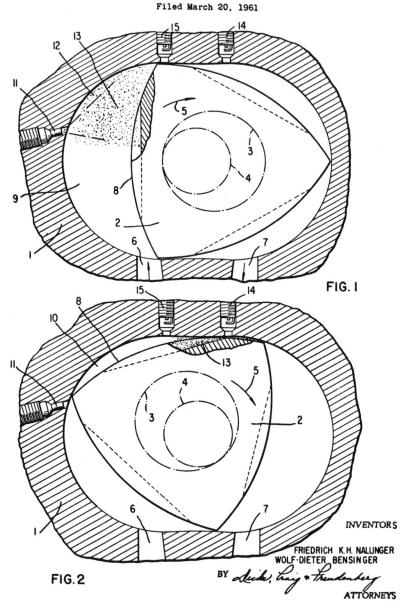

FIG. I

FIG. 2

INVENTORS

FRIEDRICH K.H. NALLINGER
WOLF-DIETER BENSINGER

BY *Luck, Leig & Freulenberg*

ATTORNEYS

Stratified charge fuel injection and dual ignition, according to a patent by Nallinger and Bensinger in 1964.

Several important patents were taken out by Daimler-Benz engineers in the course of developing an advanced gasoline injection system for the Wankel engine. Heinz Lamm and Willi Springer of Daimler-Benz AG received German patent D 38,450 on March 23, 1962, covering port injection. The method of injection was to direct the jet against the cavity in the rotor face. They claimed this would give the effect of direct injection without any loss of compression when the radial apex seals passed the nozzle bores in the working surface. Fuel was injected into the area of the highest air velocity, which promoted better mixing through turbulence. The intake port diameter was narrower at the working surface than back at the elbow where the injection nozzle was positioned. The spray was so directed that the fuel did not hit the port wall, but progressed into the working chamber without changing course. Lamm and Springer also patented a helically wound sheet metal guide to fit inside the intake port. This would have the effect of starting turbulence well back in the port area while still confining the spread of the injection spray to a small area on the rotor face. This invention was intended to improve the mixture preparation, to provide additional cooling of the rotor faces, and allow stratification of the charge within the combustion space. However, port-type injection was soon rejected in favor of direct injection, which promised more controllable stratification and combustion characteristics.

Nallinger and Bensinger took out U.S. patent number 3,136,302 on June 9, 1964, concerning fuel injection and the stratified charge. The injection nozzle was described as injecting in a circumferential direction of the rotor directly into the working chamber. The patent covered methods to ensure advantageous stratification of both rich and lean mixtures within the combustion space during part-load operation of the engine. Stratification was achieved by the nozzle arrangement and positioning. The basic idea was to use two spark plugs per chamber. The rich mixture was to be ignited by the plug—the lean portion of the mixture was to be ignited by the heat in the previously ignited mixture. The first spark plug then could be placed only with regard to initial ignition to provide proper and safe ignition of rich mixtures such as would exist during full-throttle operation. The second plug would be positioned so as to take care of end gas—it would be fired fairly late in the combustion phase. The injection nozzle was to be so inclined that its spray would be aimed at the plugs in the combustion space, where compression and gas transfer take place.

Nallinger and Bensinger reasoned that the rich fuel mixture would fill the head of the combustion space, right up against the plugs, at the moment of firing. The remainder of the chamber then would contain

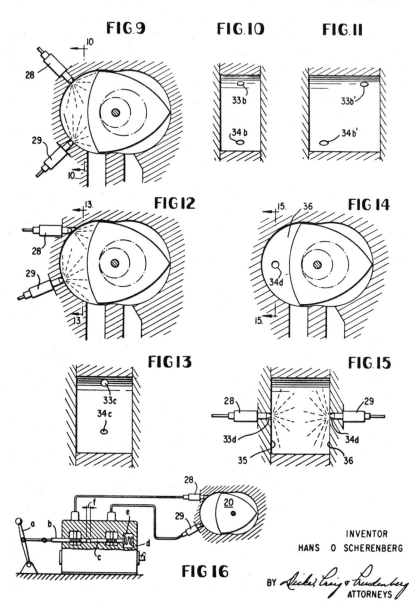

Various suggestions by Scherenberg for use of dual injection nozzles in each chamber.

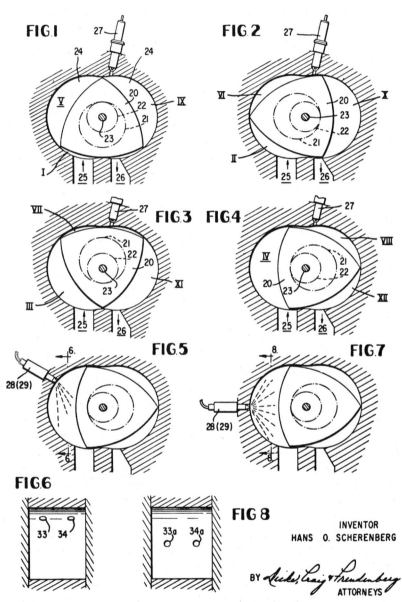

March 23, 1965   H. O. SCHERENBERG   3,174,466

INTERNAL COMBUSTION ENGINE

Filed Feb. 2, 1961   2 Sheets—Sheet 1

FIG.I

FIG.2

FIG.3   FIG.4

FIG.5

FIG.7

FIG.6

FIG.8

INVENTOR
HANS O. SCHERENBERG

BY *Hicks, Craig & Freudenberg*
ATTORNEYS

Scherenberg also proposed stratified-charge injection, with single injectors and a spark plug positioned just after the minor axis.

a relatively lean mixture, to be ignited by the burning of the rich mixture ahead of it. Rotor rotation would, of course, bring the lean mixture to the area where combustion had been started. They worked on the assumption that the engine would operate without turbulence in the combustion space. They also envisaged the flame front travelling against rotor rotation, a concept which later tests proved to be wrong.

Progress came fast when Hans Scherenberg began to take an active interest in the development of the Wankel engine and applied his inventive talent to the fuel injection system. His work was directed towards high-speed operation with minimized fuel losses. Another main goal was better part-load operation. The result was a multi-nozzle fuel injection system, which would render stratification possible and improve operating characteristics. Scherenberg received U.S. patent number 3,174,466 on March 23, 1965. His attention had focused on the use of two nozzles per chamber, arranged in various configurations to produce best results under a variety of conditions. Scherenberg was first attracted to fuel injection because it would not involve the flow losses of fresh fuel that he regarded as inevitable with carburetors. The twin-injector idea came to him when he looked at the elongated shape of the combustion chamber. He pointed out that twin nozzles could be arranged in such a fashion that satisfactory mixtures could be obtained regardless of the unfavorable shape of the combustion space. Depending on spatial requirements, the nozzles could be positioned either one behind the other, as viewed in the circumferential direction of rotor rotation, or axially opposite each other in the sidewalls.

A staggered arrangement was recommended to achieve the best distribution of fuel throughout the entire working air. It was felt that if one or both of the nozzles were arranged in the end walls, mixing would be good because the jets from the individual nozzles would cross each other or be directed against each other so as to impinge against each other. The use of two nozzles could also improve the operating characteristics during partial loads. In order to ensure proper atomization at the nozzles under part-throttle conditions, Scherenberg proposed to shut off one nozzle. Then the relatively small fuel quantity to be injected would have to pass through a single nozzle and thereby maintain pressure and stratification characteristics. Stratification of the charge during partial load operation could be achieved using two injectors by so arranging and constructing the nozzles that the one which is turned off during part-throttle running injects the fuel charge into that region of working air which moves closest to the spark plug(s). This would make it possible to operate the engine on very lean air/fuel ratios.

The geometry of the C-111 engine evolved from a twin-rotor unit with

27 cubic inches (450 cc.) chamber displacement designed in 1965 and 1966. This was the first Mercedes-Benz Wankel engine designed specifically with passenger car installation in mind. A three-rotor version followed almost immediately. Bensinger wanted to explore the multirotor concept because more rotors mean more firings per output shaft revolution, which results in reduced torque fluctuations. That, in turn, cuts noise and vibration. Beyond that, it was soon found in testing that the hexane content of the exhaust gas was greater when the displacement of each combustion chamber was greater. With more rotors, each chamber is smaller, and exhaust emissions are reduced.

Elevation of the three-rotor C-111 engine.

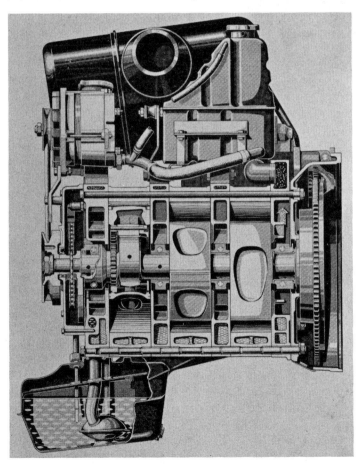

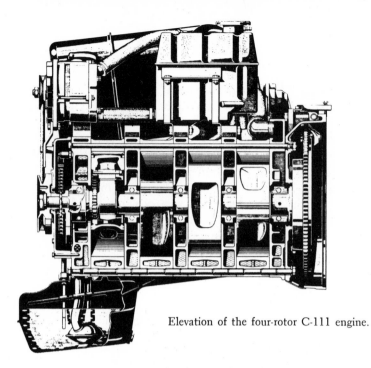

Elevation of the four-rotor C-111 engine.

Horsepower, torque, and mean effective pressure curves for the three-rotor C-111 engine.

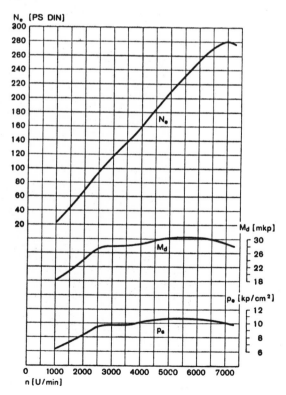

In 1967, a new three-rotor engine was designed to meet U.S. exhaust emission control standards. It had 560 cc. displacement per chamber, and its chief geometrical features were:

Radius   103 mm.
Eccentricity   15 mm.
Width   70 mm.
R/e ratio   6.87:1
R + e   118 mm.

This experimental unit was the direct basis for the C-111 design, which has wider rotors but retains the same radius and eccentricity. Rotor width went from 70 mm. to 75 mm., and displacement per chamber rose to 600 cc. The transistorized ignition system uses one surface gap spark plug per chamber, which gives a strong spark at extremely high r.p.m. and at very high operating temperatures, even with deposits on the electrodes. The concentric electrode plugs were developed by Beru for Daimler-Benz. They have a heat value of 320 and are especially resistant to high temperature and are able to fire despite fouling or lead deposits.

The front cover on the C-111 engine hides the rotor shaft sprocket and chain drive to the accessories. On the other side, three sprockets are stacked, the chain threading its way on a slalom course through from top to bottom. The top sprocket is carried on a shaft that drives both the fuel injection pump and the lubrication oil metering pump. The

Comparison of three and four-rotor C-111 engines.

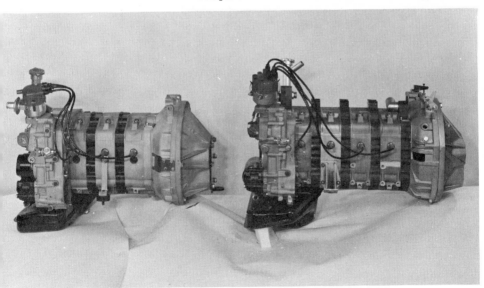

Ignition side of the four-rotor C-111 engine.

Injection side of the four-rotor C-111 engine.

same shaft also carries a gear which has a worm gear attachment to the ignition distributor shaft, which is mounted vertically. The center sprocket drives the shaft for the cooling oil pump, and the bottom sprocket drives the water pump shaft. The main oil pump supplies oil for rotor cooling, shaft bearings and bearing seals.

The Bosch mechanical injection pump is driven by a single chain and runs at one-half mainshaft speed. The injection pump is fed by two electric fuel pumps mounted outside the tank. Exhaust noise and temperature created a muffler design problem. The exhaust gas flow is noisy simply because the exhaust port opens suddenly. The C-111 engine normally is run on 95–100 octane European premium gasoline. Tests with regular gasoline have shown that the engine is remarkably uncritical of fuel grade. Drop in performance is insignificant, and no cases of abnormal combustion have been observed.

The big question today is: will Daimler-Benz produce the C-111 engines or derivatives of them in quantity? Uhlenhaut said in September, 1969, that he did not want to go on record as saying that all problems of the Wankel engine had been solved. Development work at Daimler-

End cover for the C-111 engine. (*Photo: Ludvigsen*)

Accessory drive at the front of the C-111 engine. (*Photo: Ludvigsen*)

Author's face registers amazement when confronted with the torn-down three-rotor C-111. Housings, side walls, and end covers are placed in their proper relative positions, but separated by wooden blocks. (*Photo: Ludvigsen*)

Mainshafts for the three-rotor C-111 engine. The eccentrics are not equidistant, while on the four-rotor unit they are. (*Photo: Ludvigsen*)

Rotor installed in its trochoidal chamber on the C-111 engine. (*Photo: Ludvigsen*)

Benz had been successful in that the engine was far, far better than it was only a few years before. But only when he (Uhlenhaut) had a fleet of a few thousand cars in service with Wankel engines, running satisfactorily, would he feel that all problems were, in effect, solved. One is never sure until production is a fact. The worst problem today is exhaust emissions—there is no immediate and satisfactory solution to that. Uhlenhaut thinks the Wankel engine can be produced more cheaply than conventional piston engines, on a dollar per horsepower basis, when it is "fully developed." The machining of the epitrochoidal working surface was a big headache, but the time involved has now been reduced to an economically acceptable level. Daimler-Benz C-111 test engines run over 60,000 miles or the equivalent between overhauls, which means that durability is fully comparable with piston engines. The C-111 engines now are running in a variety of Mercedes-Benz passenger cars and trucks, around the clock, to amass the data needed for Dr. Zahn and his colleagues to justify quantity production of Wankel-powered cars. How soon? Factory spokesmen decline to mention any number of years or months, but my impression from numerous interviews is that a Wankel-powered Mercedes-Benz production car can be no more than two years away.

13

Citroën

IT MAY HAVE STRUCK YOU in reviewing the license agreements made by NSU for the Wankel engine rights that there was no mention of Citroën, and yet the company is building the Wankel-powered M-35 (covered in a later chapter). Citroën found a different way to gain experience with the Wankel engine. In exploratory talks with NSU in 1962 and 1963, the German company proved willing to consider the formation of joint subsidiaries in partnership with Citroën to be active in areas that promised to be mutually beneficial. In June, 1964, representatives of Citroën and NSU met in Geneva to form a local company—Societe d'Etude Comobil. Comobil was capitalized at $300,000, half of which was held by NSU and half by Citroën. The purpose of the company was to prepare, and evaluate the marketing of, an automobile equipped with an NSU Wankel rotating combustion engine. The technical and commercial staffs of the two companies went to work to complete the assignment.

It was planned that this work would take a period of years. Comobil was not only to make proposals for future Wankel-powered cars, but also was to produce designs for one such vehicle. On March 30, 1965, Comobil began a program for the preparation of manufacturing and sales for a new car equipped with an NSU Wankel engine. According to the understanding between Citroën and NSU, the car was to be made by Citroën in France, except for the power unit which would be manufactured by NSU in Germany. Assembly plants were to operate in both France and Germany.

The next phase of collaboration between Citroën and NSU was the creation of another joint subsidiary, Comotor SA, with headquarters in

Luxembourg. Its purpose was much more ambitious than that of Comobil. Comotor actually was to manufacture and distribute Wankel engines for all applications and also make and sell all the accessories required for these purposes. The Compagnie Europeenne de Construction de Moteurs Automobiles (Comotor S.A.) was organized on May 9, 1967 and was capitalized at $1,000,000. Comotor was authorized by its founders to participate in other corporations to conclude any commercial, industrial, technical or financial transactions, including real and personal property operations, directly or indirectly related to the purpose of its charter. The six members of Comotor's board of directors are Pierre Bercot of Citroën, Gerd Stieler von Heydekampf of NSU, M. A. Bunford, A. Noel, M. Geoffroy, and J. J. Baumann. During its first meeting, the board elected Pierre Bercot as its chairman and president of Comotor S.A. On October 22, 1968, Comotor's board of directors de-

Flywheel end of the M-35 engine, showing the two spark plugs.

cided to increase its capital to $1,620,000 and in February, 1969, a further increase was made, raising capitalization to $2,020,000. In July, 1969, the board agreed to double its capitalization—to $4,040,000. Comotor was at that time buying a future engine plant site in the Saar, which was scheduled for completion by the end of 1970. At the end of 1969, the Citroën M-35 was announced. The vehicle is built by Citroën, and the engine is made by NSU in Neckarsulm.

This NSU engine used the latest technology developed during the first two years of Ro-80 production. For instance, the working surface in the old single rotor KKM-502 engine was chrome plated, but the working surface for the M-35 engine has a special nickel-silicon alloy coating known under the trade name Einisil. The NSU Spider engine often developed cracks around the spark plug holes, but on the M-35 engine the housing has a copper sleeve into which the spark plugs are screwed. The KKM-502 carbon seals had a short life, which led to high fuel and oil consumption. The seals in the M-35 engine are made from cast-iron and have proved to have good durability. The corner segments of the apex seals have a new configuration. The curved side seals, pressed against the sidewalls by springs, are no longer a problem. The oil tem-

Front view of the M-35 engine, complete with accessories.

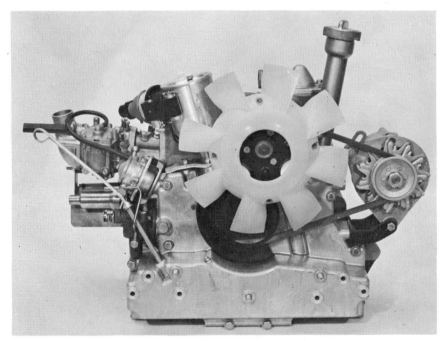

perature in the NSU Spider engine normally hovered around the 360°F. level, which put great stress on the bearings. Through the use of a heat exchanger, the normal temperature has been lowered to around 250°F. It was not until June, 1969, that NSU installed the M-35 type heat exchangers on the Ro-80. In this system, the oil runs in two circuits. The first feeds the rotor bearings and cools the rotor internally; the oil filter and the heat exchanger are incorporated in this circuit. The second circuit takes oil from the sump to lubricate the seal tips and mixes oil with the fuel through a special metering device built into the fuel pump. Lubricating the seals by this method means that a certain amount of oil consumption is inevitable, but it is generally no higher than one quart per 600–650 miles.

NSU found out that both single- and twin-rotor Wankel engines had a problem with bucking or "snatching" at low speeds and under light

Exploded view of the M-35 engine.

loads. This condition comes about because of exhaust gas backpressure when the throttle is closed. Under such conditions there is often a partial vacuum in one or two of the working chambers, and only small quantities of exhaust gas are let in. As a result, large quantities of exhaust gas are mixed with small quantities of fresh mixture—the resulting mixture will not burn. As the rotor follows its course, this mixture flows out of the exhaust port, at which point it finds little or no backpressure. With more fresh mixture added as the chamber passes the intake port, the mixture finally becomes combustible and is ignited after compression. The combustion exerts a force on the rotor that is detected as "rocking." Because there are three ignitions per rotor revolution, the engine loses three power phases every second revolution. Side ports would eliminate the problem, but NSU relies on peripheral ports because of the higher volumetric efficiency. On the Ro-80, the bucking is cushioned by the hydraulic torque converter. With the M-35's conventional clutch and transmission, the technically best solution was to go to fuel-injection.

View of the rotor motion in the M-35 engine.

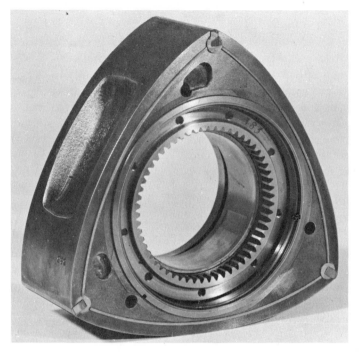

M-35 rotor, complete with seals.

M-35 rotor, with eccentric and mainshaft in place.

One engine with peripheral ports that does not have this problem is the C-111. As has been mentioned, Mercedes-Benz completely shuts off the fuel supply when the throttle is closed. But injection is expensive, and Citroën wanted something simpler. Citroën has tried to overcome the problem with a revised ignition system. However, this does not entirely eliminate bucking, although it is not noticeable inside the car. The ignition system includes two coils, two breakers, and two spark plugs to ensure continued operation in case of failure in any one unit. A special vacuum advance mechanism also is incorporated, which severely retards the spark on the overrun. The resulting late firing effectively limits the effects of the conditions that were such a problem with the original NSU Spider engine.

14

Other Wankel
License Holders

OF THE OTHER Wankel license holders, among which are M.A.N., Krupp, Hanomag-Henschel, Deutz, Porsche, Alfa Romeo, Rolls-Royce, Perkins, and Fichtel & Sachs, only one has done any really serious work in developing the powerplant for commercial application—Fichtel & Sachs.

Hanomag had, in 1961, a Wankel test program, but the entire project has not amounted to very much. At any rate, Hanomag is now closely allied with Daimler-Benz, and if Daimler-Benz ever decides to go into production of a Wankel-powered truck they may well utilize Hanomag's license to do so.

Krupp, the organization that controls, among other things, Europe's largest iron and steel works, was at one time interested in developing a Wankel engine in diesel form for its line of trucks. Krupp has since sold its truck interests to Daimler-Benz, and although it still retains its Wankel license no research and development work is now taking place. M.A.N., another German truck manufacturer, was also interested in the diesel Wankel. Early tests were, however, disappointing and research, although still in progress, is not intensive.

Both Porsche and Deutz have not done too much work on the Wankel. Porsche has held a license since March 2, 1965, but so far has done little but evaluate existing engines to determine their suitability for various applications. It is interesting to note that neither Porsche nor any of its engineering staff has taken out any patents in connection with the Wankel. No data on Deutz test engines has been made available up to this point, but it is worth noting that the company has the industrial strength necessary to become a major factor in Wankel production should a firm decision be made to produce the engine.

Alfa Romeo, the famous Italian builder of high-performance cars, was among the first to recognize the potential of the Wankel engine. As early as 1959, it was rumored that Alfa was negotiating a take-over of NSU. Although this never transpired (Volkswagen eventually bought NSU), Alfa was among the first to conclude a license agreement. Several prototype engines were built, having 500 cc. chamber displacement. Both single- and twin-rotor engines were investigated, and studies were conducted on porting, basic geometry, and combustion, as well as on sealing and lubrication. No test reports have been made available, however, so no real idea of their progress in the field can be determined. If Alfa Romeo were to make a wholesale conversion to Wankel engines in its cars at some future date, they could revolutionize the Italian automobile industry, now dominated by Fiat. Fiat's attitude towards the Wankel engine was skeptical until reports from Citroën were studied in Torino, but they now seem prepared to admit the engine's potential. Alfa Romeo is located in the rival city of Milano. But differences do not end there. Fiat is strictly a private enterprise (publicly owned) while Alfa Romeo is part of a government-owned subsidiary.

Of the two British companies that have conducted Wankel engine research, only Rolls-Royce has conducted an intensive investigation. Perkins, the world's largest builder of diesel engines, has been more concerned with the development of their differential diesel engine, while for years Rolls-Royce has been engaged in research and development of diesel and multi-fuel versions of the Wankel. The reason Rolls-Royce decided to obtain a Wankel license is that the company had the promise of a research contract for a diesel-fuel rotating combustion engine from the United Kingdom Ministry of Defence. Britain's military was looking for a new powerplant for tanks and armored vehicles that would use less flammable fuel than high-octane gasoline. The contract was duly signed, and the Rolls-Royce technicians went to work. The technical details of their experimental engines have been classified information under the terms of British government contracts. The main design features of the prototype engine have leaked out, however.

The Rolls-Royce unit differs from all other Wankel engines in that it works on the compression-ignition principle. Curtiss-Wright's heavy-fuel engines still rely on spark ignition, but Rolls-Royce decided to raise compression to the point where an electrically triggered spark no longer would be required. This, of course, became something of a geometrical problem in that the maximum compression ratio is dictated by the size of the rotor and its eccentricity (R/e ratio and e). A rotor and combustion chamber configuration that permits a high enough compression ratio to ignite the air-fuel mixture tends to have very poor burning characteristics. A study of all possibilities led Rolls-Royce to decide on a supercharged Wankel engine, using a Wankel-type compressor.

Actually, the compressor is bigger than the engine. It has a larger rotor and larger chamber volume. The compressor has an outlet port near its minor axis that feeds into the engine's intake port, and the air is further compressed. Fuel is injected from a nozzle placed approximately where

Operational cycle of the Rolls-Royce engine. The rotors are geared together in a 1:1 ratio and revolve in the same direction. 1 = Intake port. 2/3/4 = Compression chamber. 5/6 = Secondary compression chamber. 7/8/9/10 = Combustion chamber. 11/12 = Exhaust from engine to compressor. 13/14 = Exhaust. 15/16 = Exhaust ends.

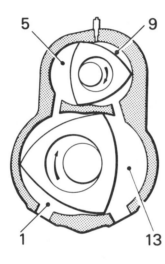

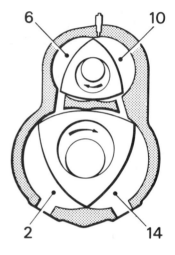

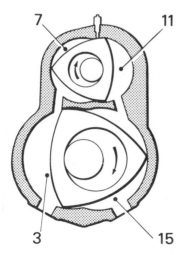

the spark plug would be normally—a few degrees before the minor axis. The combustion process is similar to that obtained in heavy-fuel Curtiss-Wright Wankel engines, but something new is added. When the burned gas leaves the engine through the exhaust port, it is routed back into the compressor, entering through a separate port, to complete its expansion and combustion against the compressor rotor. The compressor is thus both gear-driven and gas-driven.

The Rolls-Royce Wankel engine is said to have a peak operating speed of 4,400 r.p.m., which is considerably below that of spark-ignition Wankel engines, but about twice as fast as reciprocating piston-type diesel engines. The Wankel is also reported to be one-quarter the size of a piston-type diesel engine of comparable power—and far lighter. It is also half the size of a gas turbine of the same power, says one report.

Fichtel & Sachs A.G. is Germany's largest producer of small two-stroke engines for outdoor power equipment, clutches and automatic transmissions, and ball and roller bearings. It was also one of the first German industries to buy a Wankel engine license. This agreement covered a range of power units far smaller in size than any considered by the other license takers. The Fichtel & Sachs Wankel engines are too small to be considered for passenger car applications, but they are used in some snowmobiles and all terrain vehicles (ATV's).

Production of the Fichtel & Sachs single-rotor KM37 engine began in 1965. The design contrasted with NSU practice in several areas, notably in cooling of the housing and rotor. The housing was air-cooled and the rotor had no provision for oil cooling. Fichtel & Sachs relied on mixing

Dimensional sketches of the Rolls-Royce 2-R6 engine.

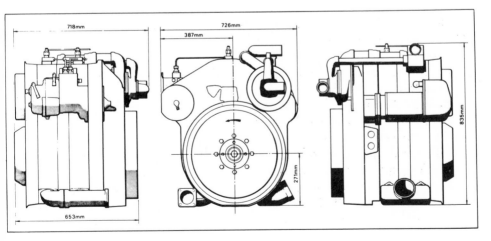

oil with the gasoline in normal two-stroke fashion for apex seal lubrication—the specified oil quantity was 2% by volume. The KM37 compared most favorably with conventional single-cylinder, two- and four-stroke piston engines within the same performance range. The KM37 specifications were:

Displacement	108 cc.
Compression ratio	8.5:1
Power output	6.5 horsepower
	at 5,500 r.p.m.
Weight	30.5 lbs.

The installation dimensions were 16.5 inches long by 8.66 inches wide by 11.22 inches high. This small size was directly attributable to the air-cooling and the absence of an oil cooler. The KM37 unit was extremely smooth, and the vibrations were estimated to be of the same frequency range as those of a 200 cc. single-cylinder two-stroke unit. The single-rotor Wankel engine is, of course, the equivalent of a twin-cylinder four-stroke cycle engine, and the resulting torque fluctuations are considerably lower than those of single-cylinder piston engines tuned to deliver comparable specific power. The vibration amplitudes and the resulting mass accelerations and inertia forces in the KM37 were only 15% of those for a 200 cc. two-stroke engine.

Engines with slightly greater displacement soon followed. They were the 9.8 cubic inch (160 cc.) KM48, and the 18.5 cubic inch (300 cc.) KM914. But before discussing the existing engines in detail, we should look at the overall approach to the Wankel engine taken by Fichtel & Sachs and follow the development work performed by the Schweinfurt engineers.

Fichtel & Sachs ran into the same difficulties that plagued NSU—chatter marks on the trochoidal track and high seal tip wear. At first the surface was electroplated with hard chrome. After a longer running period, chatter marks on the chromium layer developed to the point where the lifespan of the engine was impaired. Chatter phenomena were reduced considerably by increasing the thickness of the apex seals from 1.5 to 3.0 mm. and tests with carbon-type sealing elements completely eliminated them, even on the chromium layer. Unfortunately, the wear of the carbon-type seals proved to be too high, especially at high speeds and when the engine ran at high load. In addition to this problem, the sealing elements tended to break, and erosion occurred at the edges. When the output shaft speed was held down to 3,000 r.p.m., the unit ran for 1,500 hours before the appearance of chatter marks on the bore. At a steady speed of 4,500 r.p.m., however, they appeared after

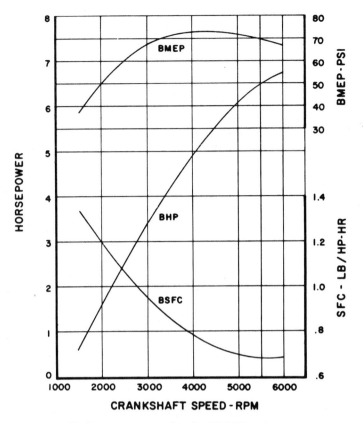

Performance curves for the KM-37 engine.

Specific fuel consumption of the KM-37 engine is plotted against mean effective pressure (psi).

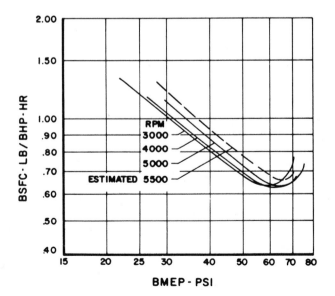

1,000 hours. Fichtel & Sachs found that the matching of high quality gray cast-iron sealing elements with a chromium-plated working surface extended the service life an additional 1,000 to 1,200 hours between overhauls. The chromium layer now has been replaced by a mixture of steel and bronze. With this surface, which has a very long lifespan, no chatter marks are produced. The surface wear amounts to slightly more than 0.001 mm. per hour; the radial length of the apex seal is only worn off about 0.008 mm. per hour.

Fichtel & Sachs engineers established economically acceptable tolerances and clearances for the individual elements of the sealing grid. The grid of the KM37 differed in detail rather than in concept from the original Wankel arrangement. Light spring-loading behind every seal ensured adequate functioning under normal operating conditions, with gas pressure utilized to effect sealing. Experiments with other types of seals did not produce any improvement. One Fichtel & Sachs engineer, Ernst Ansorg, took out a patent for a V-shaped and T-shaped apex seal. He claimed that his seal system would ensure a permanent joint between the rotor and the housing, regardless of their relative movements, and would not allow any undesirable gas flow in the working chamber. The V-shaped Ansorg seal was lodged in a dual slot and provided with spring-loading against the working surface in the radial direction by a helical spring positioned on the apex radius. The seal could not be installed if it were of one-piece construction, so Ansorg devised a built-up seal consisting of two narrow elongated plates integrally connected at right angles to form a sealing edge. The seal plates were adapted to engage certain portions of the seal groove walls and thereby direct gas pressure behind the seal plates when desirable. Because of the rounded shape of the inside edges of the seal plates and the flat shape of the engaging portions of the groove wall, there was to be line contact between them when in sealing engagement. A variant of this system has individual seal strips, each with its own sealing edge, both held in contact with the working surface by pressure from a spring-loaded ball located in the rotor. Ansorg claimed that his seals offered larger available surfaces to receive gas pressure and that this would give greater sealing effectiveness.

Another engineer, Franz Rottmann, proposed a different approach to eliminating the chatter marks on the working surface. He took out a patent for more secure mounting of the rotor on its eccentric bearing. Rottmann claimed that an engine built to his specifications would operate with reduced vibration at lower noise levels. In addition, there would be no rotor tilt that could cause wear marks on the housing. The heart of his device was a huge roller bearing imbedded in the cylindrical face of the eccentric bearing, combined with phasing gears and a sleeve arrangement to ensure rotor alignment. Sleeves on the mainshaft extended into the rotor from both sides and lined up with internal abutments on the rotor. The rotor was guided on its course by four contact surfaces, plus the phasing gears to some extent. The roller bearing on the

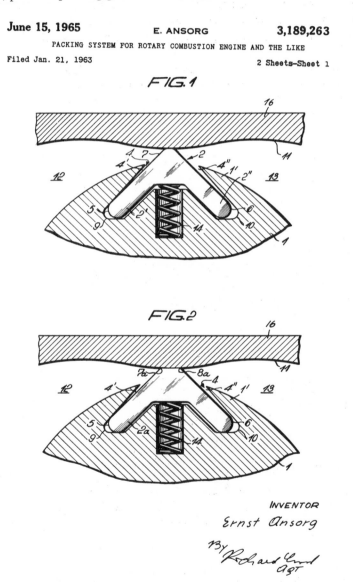

June 15, 1965   E. ANSORG   3,189,263

PACKING SYSTEM FOR ROTARY COMBUSTION ENGINE AND THE LIKE

Filed Jan. 21, 1963   2 Sheets—Sheet 1

FIG. 1

FIG. 2

INVENTOR

Ernst Ansorg

By Richard Ernd
Agt

V-shaped apex seal, patented by Ernst Ansorg.

eccentric was concerned only with transmitting the rotary motion to the shaft.

But, the true solution to the chatter mark problem did not lie simply in the choice of compatible materials, apex seal configuration, or rotor guidance. The key was temperature control, to reduce heat distortions of both rotor and housing. There is an unavoidable temperature variation between the various areas around the rotor housing. The housing is subject to the least heat input in the area surrounding the intake port,

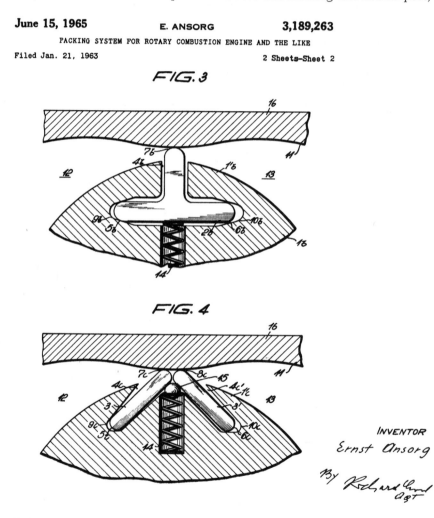

June 15, 1965   E. ANSORG   3,189,263

PACKING SYSTEM FOR ROTARY COMBUSTION ENGINE AND THE LIKE

Filed Jan. 21, 1963   2 Sheets-Sheet 2

FIG. 3

FIG. 4

INVENTOR

Ernst Ansorg

By Richard Ernst
Agt

T-shaped apex seal, as patented by Ansorg (above). Below, a variation on the V-shaped seal, with a spring-loaded ball at the intersection of the two blades.

and to the highest heat input in the area surrounding the spark plug. Fichtel & Sachs managed to keep temperature differences within reasonable bounds by positioning a primary intake port within the hot region. The decision to adopt this arrangement was made because it was felt that for industrial engines the advantage of simplicity was more important than the higher performance potential of an engine with an oil-cooled rotor, especially when the cost and complication of an oil cooler and oil circulation pump were considered.

Feb. 20, 1968   F. ROTTMANN   3,369,738

ROTOR GUIDE ARRANGEMENT FOR A ROTARY INTERNAL COMBUSTION ENGINE

Filed March 16, 1966

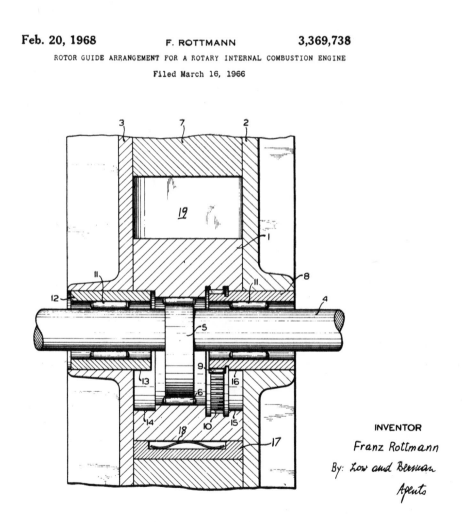

INVENTOR

Franz Rottmann

By: Low and Berman

Agents

Rottmann's patent for rotor guidance.

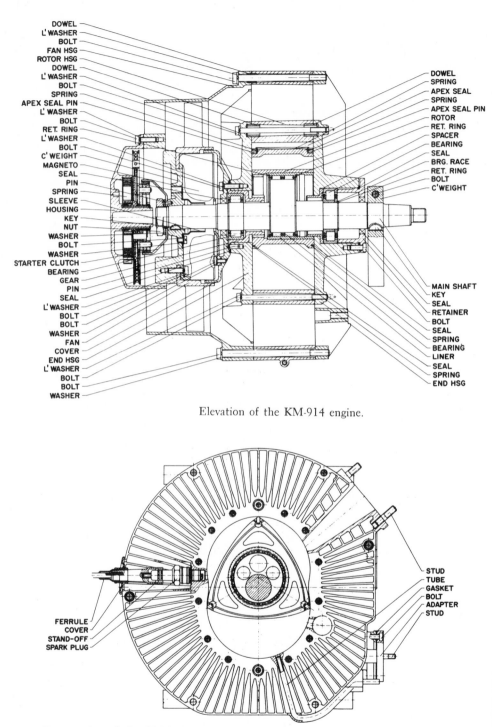

DOWEL
L' WASHER
BOLT
FAN HSG
ROTOR HSG
DOWEL
L' WASHER
BOLT
SPRING
APEX SEAL PIN
L' WASHER
BOLT
RET. RING
L' WASHER
BOLT
C' WEIGHT
MAGNETO
SEAL
PIN
SPRING
SLEEVE
HOUSING
KEY
NUT
WASHER
BOLT
WASHER
STARTER CLUTCH
BEARING
GEAR
PIN
SEAL
L' WASHER
BOLT
BOLT
WASHER
FAN
COVER
END HSG
L' WASHER
BOLT
BOLT
WASHER

DOWEL
SPRING
APEX SEAL
SPRING
APEX SEAL PIN
ROTOR
RET. RING
SPACER
BEARING
SEAL
BRG. RACE
RET. RING
BOLT
C'WEIGHT

MAIN SHAFT
KEY
SEAL
RETAINER
BOLT
SEAL
SPRING
BEARING
LINER
SEAL
SPRING
END HSG

Elevation of the KM-914 engine.

STUD
TUBE
GASKET
BOLT
ADAPTER
STUD

FERRULE
COVER
STAND-OFF
SPARK PLUG

Cross-section of the KM-914 engine.

Surface temperature in the trochoidal chamber generally was kept below 400°F. The working surface was chrome-plated and ground to a fine finish and the end covers were sprayed with a special bronze coating. Earlier test units used a molybdenum coating, which was very expensive. This led to the development of the bronze coating, which offered similar wear characteristics at a much reduced cost. The housing and end covers were light alloy castings with a number of cooling ribs on their external surfaces. Air-cooling was provided by an axial-flow fan working through full ducting. The light alloy axial-flow fan, with its airfoil-section blades, absorbed very little horsepower.

Performance curves for the KM-914 engine.

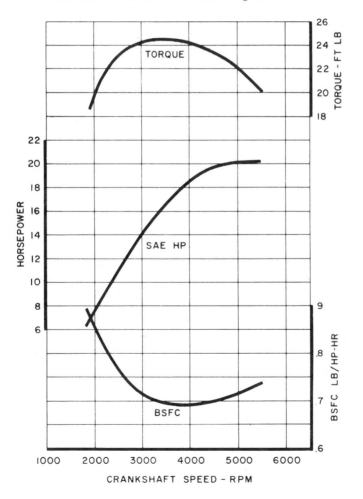

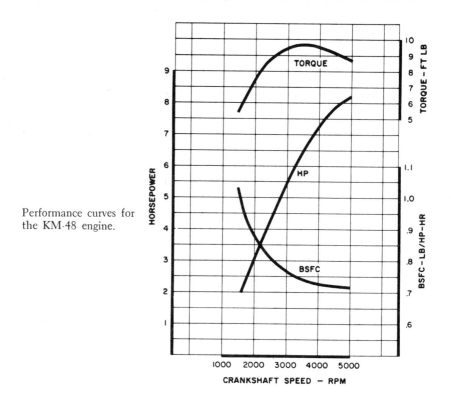

Performance curves for the KM-48 engine.

Specific fuel consumption of the KM-914 engine, plotted against r.p.m. (mainshaft speed) and mean effective pressure (psi).

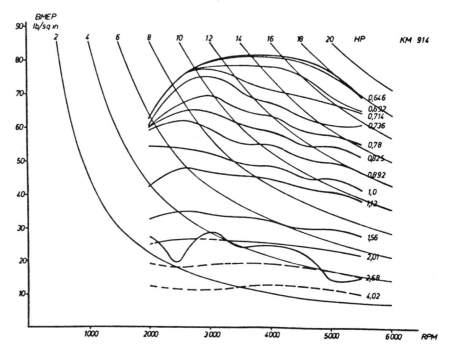

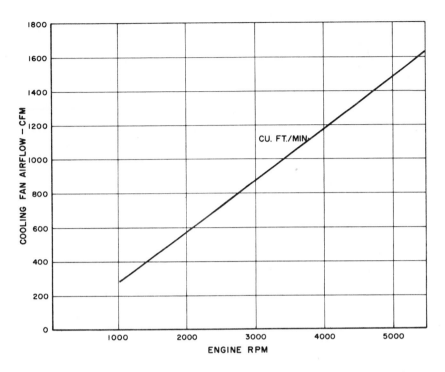

Cooling fan air flow (cubic feet per minute) plotted against mainshaft r.p.m. for the KM-914 engine.

Carburetor air flow in the KM-914 engine.

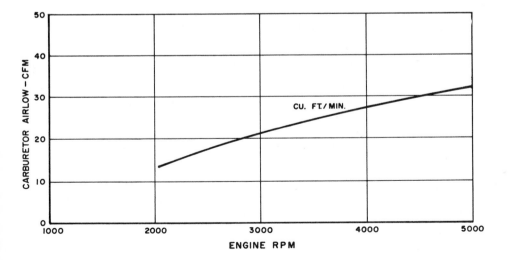

In May, 1968, Helmut Keller delivered a paper to the Society of Automotive Engineers on his cooling system development work at Fichtel &
Sachs and other features of the Wankel engines from Schweinfurt. The
cooling ribs on the latest engines were longer on particularly hot areas
of the housing than they were on the cooler parts. The power expendi-
ture for the blower remained within the normal limits for air-cooled,
reciprocating piston engines, using axial fans made of light alloy. The
cooling system was based on the use of pressure fans, by which cold air
was taken in and blown over the cooling ribs of the side housings and
trochoid housing. The maximum speed of the engine was automatically
limited to 6,000 r.p.m. by using a simple throttle control device sensitive
to gas velocity.

Cutaway model of the KM-914 engine.

Closeup of the combustion chamber in the KM-914 engine.

Exhaust gas temperatures are somewhat higher in the Fichtel & Sachs rotating combustion engines than in conventional engines. The induction system on these engines is complicated and no doubt derived from two-stroke piston engine experience; for, although the KM37 operated on the four-stroke cycle, the induction system relied on two-stroke principles. Fresh mixture from the carburetor was led to a passage in the eccentric. At certain rotor positions, ports in the eccentric passage lined up with ports inside the rotor, while ports in the rotor sides, one inboard of each apex, lined up with transfer ports that channeled the incoming charge to the normal intake port position. This was done to obtain pre-heating of the fresh charge and to improve temperature uniformity in the housing. Gas temperature increased by about 120°F. on the way through the rotor. This meant a loss in volumetric efficiency, but did provide some cooling effect for the rotor.

Rotor bearing lubrication never was a problem, because a gas mixture containing oil flowed though the eccentric bearing always in the same direction, and it was the experience of Fichtel & Sachs that this always

Dual Fichtel & Sachs Wankel engines installed in an experimental Delta-Hover air cushion vehicle. The upper engine provides propulsion by driving an airscrew, while the lower provides a ground-effect lift force by pumping air under the vehicle.

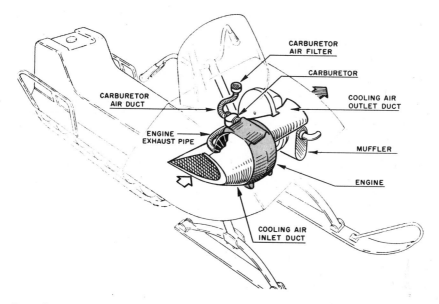

One of America's largest manufacturers of snowmobiles, Polaris, added a Wankel engine option in 1968. The unit is Fichtel & Sachs' RC1-18.5.

Here is Curtiss-Wright's suggested installation of the 18.5 cubic inch Fichtel & Sachs single-rotor, air-cooled Wankel engine in an all-terrain vehicle.

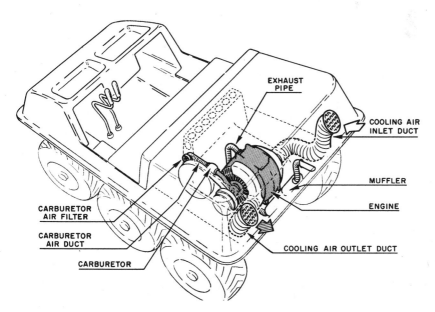

supplied a sufficient lubricating film. This applied also to both main bearings, which were joined with the admission chamber by means of a channel and thereby received sufficient oil. Nevertheless, it was eventually decided to introduce a small oil pump to feed the requisite minute quantities of lubricant (varying with engine speed) into the induction port. This required separate tanks for oil and gasoline. Metering the lubricant relative to shaft speed proved entirely satisfactory.

The first test engines used special spark plugs, the electrodes of which were connected to the combustion chamber by a channel. These spark plugs were very expensive and not readily available; therefore a way was sought to utilize standard spark plugs without a connecting channel. The solution made use of the fact that the spark plug was located exactly at that point of the trochoid at which equal pressures existed on both sides of the apex seal. One chamber has compression pressure, the other has a similar pressure from the burning gases. By moving the spark plug, somewhat better consumption values could have been obtained, but the advantage of being able to use an inexpensive standard spark plug, available everywhere, was regarded as being more important. The Fichtel & Sachs Wankel engine now uses standard Bosch W 150 M11S spark plugs. They are quite satisfactory for 150 to 200 hours service, which is comparable to the life of spark plugs used on the smaller F & S two-stroke engines.

KM37 SPECIFICATIONS

Direction of rotation	Counterclockwise (when looking at drive end)
Displacement	6.6 cu. in. (108 cc.)
Cooling	Air-cooled (blower)
Compression ratio	8.5:1
Performance	6.55 horsepower @ 5,500 r.p.m. (tolerance range +5%)
Ignition	Bosch flywheel magneto MZ/URB 1/116 R 2 or Bosch 15 W
Spark plug	Champion L-85 or Bosch W 190 M11S
Ignition timing	10 degrees before TDC
Breaker gap	0.014–0.018"
Carburetor	Bing 8/14/1
Air filter	Wet type
Starting method	Hand start
Control	Top speed limit (governor)
Weight	34 lbs., engine including starter, governor, carburetor and muffler
Lubrication oil	Shell Rotella SAE 30 or 40
Pre mix ratio	50:1 fuel to oil
Fuel	Gasoline, regular grade

KM48 SPECIFICATIONS

Direction of rotation	Counterclockwise (when looking at drive end)
Displacement	9.8 cu. in. (160 cc.)
Cooling	Air-cooled (blower)
Compression ratio	8:1
Performance	5 horsepower @ 3,000 r.p.m.
	8 horsepower @ 4,700 r.p.m.
	(tolerance range +5%)
Ignition	Bosch flywheel magneto—lighting coil on request
Spark plug	Champion L-85 or Bosch W 190 M11S
Ignition timing	10 degrees before TDC
Breaker gap	0.014–0.018″
Carburetor	Bing 8/14/1
Air filter	Wet type
Starting method	Recoil—hand start
Control	Top speed limit (governor)
Weight	37 lbs., engine including starter, governor, carburetor and muffler
Lubrication oil	Shell Rotella SAE 30
	Essolub HD30
	BP Outboard-Motor Oil
	Mobiloil Outboard
	Mobiloil TT
	Mobilmix TT
Pre mix ratio	40:1 fuel to oil
Fuel	Gasoline, regular grade

KM914 SPECIFICATIONS

Direction of rotation	Counterclockwise (when looking at drive end)
Displacement	18.5 cu. in. (303 cc.)
Cooling	Air-cooled (integral blower)
Compression ratio	8:1
Performance	20.0 horsepower @ 5,000 r.p.m. (tolerance range +5%)
Ignition	Bosch flywheel magneto (with lighting coil) 40-watt, 12-volt
Spark plug	Champion L-86 or L-90 or Bosch W 150 M11S
Ignition timing	10 degrees before TDC
Breaker gap	0.014–0.018″
Carburetor	Tillotson HL242A with fixed main jet or HL252A adj. main jet
Air filter	Inlet mesh filter
Starting method	Hand recoil starter
Weight	56 lbs., engine including starter, carburetor and muffler

Muffler	Flat
Lubrication oil	Shell Rotella SAE 30
	Essolub HD30
	BP Outboard-Motor Oil
	Mobiloil Outboard
	Mobiloil TT
	Mobilmix TT
Pre mix ratio	40:1 fuel to oil
Fuel	Gasoline, regular grade
Fuel pump	Diaphragm-type pump fitted to carburetor, operated by impulse connection with engine

15

Exhaust Emissions

IN THE PREVIOUS CHAPTERS, I have described in considerable detail how the various companies engaged in Wankel engine research and development have attacked and overcome all problems affecting the successful and efficient operation of the rotary combustion engine. But the achievement of high thermal and mechanical efficiency is not enough to guarantee the mass production of any power unit in today's environment-conscious world. The engine, no matter how advanced technologically, how inexpensive to produce, how attractive for a variety of installations, still must fill one all-important requirement—it must not pollute the atmosphere.

The matter of the Wankel engine's emissions has not been fully mapped as yet. It is an area that was ignored for the first six or seven years of development work, but much has been done recently to identify the problems and to find solutions. Solutions had to be sought, not in absolute terms, but within the framework of the laws enacted by the government of the state of California and the U.S. federal government. The automobile industry is confident that the reciprocating piston engine can be "cleaned up" well enough to meet the 1975 standards, and the leading Wankel engine manufacturers will not concede that the rotating combustion engine has worse problems. Beyond that, unconventional powerplants may become necessary. For the time being, two Wankel-powered cars have met the current 1970 standards.

EXHAUST EMISSION STANDARDS

Year	Exhaust HC	Exhaust CO	Exhaust NO$_x$	Partic-ulates	Crank-case HC	Evapora-tion HC
1963 (uncontrolled vehicles)	5.7	87.2	5.8	0.3	3.2	2.8
1966 Fed.	N.S.	N.S.	N.S.	N.S.	N.S.	N.S.
1966 Cal.	3.4	34.0	N.S.	N.S.	0.0	N.S.
1968 Fed.	3.3	34.0	N.S.	N.S.	0.0	N.S.
1968 Cal.	3.4	34.0	N.S.	N.S.	0.0	N.S.
1969 Fed.	3.3	34.0	N.S.	N.S.	0.0	N.S.
1969 Cal.	2.2	23.0	N.S.	N.S.	0.0	N.S.
1970 Fed.	2.2	23.0	N.S.	N.S.	0.0	N.S.
1970 Cal.	2.2	23.0	N.S.	N.S.	0.0	6.0 G
1971 Fed.	2.2	23.0	N.S.	N.S.	0.0	0.5
1971 Cal.	2.2	23.0	4.0	N.S.	0.0	6.0 G
1972 Fed.	2.2	23.0	N.S.	N.S.	0.0	0.5
1972 Cal.	1.5	23.0	3.0	N.S.	0.0	6.0 G
1973 Fed.	2.2	23.0	3.0*	N.S.	0.0	0.5
1973 Cal.	1.5	23.0	3.0	N.S.	0.0	6.0 G
1974 Fed.	2.2	23.0	3.0*	N.S.	0.0	0.5
1974 Cal.	1.5	23.0	1.3	N.S.	0.0	6.0 G
1975 Fed.	0.5*	11.0*	0.9*	0.1*	0.0	0.5
1975 Cal.	0.5*	12.0*	1.0*	N.S.	0.0	6.0 G
1980 Fed.	0.25*	4.7*	0.4*	0.03*	0.0	

HC = Hydrocarbons (grams per mile)
CO = Carbon monoxide (grams per mile)
NO$_x$ = Oxides of nitrogen (grams per mile)
N.S. = No standard
* = Proposed
G = Grams per test (California specifications).

In 1969, Toyo Kogyo successfully subjected the R-100 production car to the American test conditions. This removed the obstacle to importing cars fitted with Wankel engines into the U.S.A. under 1970 rules. The federal government's test consists of a primary 4,000 mile test and a secondary 50,000 mile test. The primary test was conducted during the period mid-June to mid-July, 1969 and the secondary test was completed on October 13, 1969. The prescribed limit of exhaust gas depends on vehicle weight. That is, allowable carbon monoxide is 1.7% and hydrocarbon content is 305 parts per million for a car in the category of the R-100.

NSU went to work on a similar program. In the middle of June, 1969, an Ro-80 was launched on a 4,000 mile circuit through the Swabian lowlands. After running for two weeks, it completed its initial trials in West Germany. These were followed by final measurements on the NSU test beds and air freight to Ypsilanti, Michigan, for testing by the *Division of Motor Vehicle Pollution Control*. In the middle of July, 1969, another Ro-80 began a non-stop 50,000 mile test. Here, 50,000 miles are travelled on a course which has been tested and approved by the American authorities. Mountains and valleys, a certain amount of city driving, halting at traffic lights and crossroads for a minimum of 5 seconds, together with stretches at full acceleration—all have to be included. For this purpose, the NSU development department has its own special

Mazda R-100 passing the federal emissions tests.

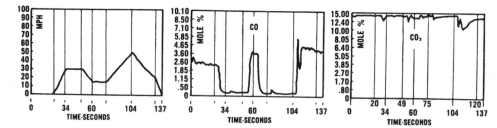

The Federal Test Cycle for Exhaust Emissions specifies starting with a cold engine and letting it idle for 20 seconds. Measurement of emissions begins at the moment of startup. After 20 seconds, the car speeds up to 30 mph in 14 seconds, and runs at that speed for 15 seconds before slowing down to 20 mph. The slowdown is performed with a closed, or nearly closed throttle, in a span of 11 seconds. After running at 20 mph for 15 seconds, the car accelerates to 50 mph in 29 seconds, decelerates to 25 mph in 16 seconds, and finally comes to a full stop in a 17-second span.

The second graph shows carbon monoxide (CO) emissions, as measured on a typical piston engine during this cycle. CO emissions are high while the engine is idling, and reach their peak during deceleration. They are lowest during acceleration and at steady speeds, with small peaks occurring as a result of sudden changes in throttle opening during acceleration. Complete closing of the throttle produced the sharp rises in the curve after approximately 50 and 105 seconds running.

In the third graph the level of carbon dioxide (CO_2) emissions, which are not considered polluting, is shown. However, mapping CO_2 volume during the test cycle gives researchers valuable data. The proportion of carbon dioxide remains fairly constant, regardless of vehicle speed and engine r.p.m., although a drop in CO_2 emissions occurs whenever the accelerator is suddenly depressed or released. Closing the throttle at high speed reduces CO_2 output, as more carbon monoxide is then formed.

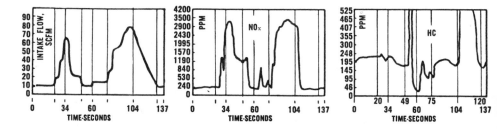

The first graph measures intake air flow in cubic feet per minute. This is an indication of the amount of oxygen available for combustion. Air flow is approximately proportional to vehicle speed and throttle opening. Lack of oxygen during deceleration is a primary factor in the formation of high volumes of carbon monoxide output, with a loss of carbon dioxide content in the exhaust gases.

In the second graph, the emission level of oxides of nitrogen (NO_x) are plotted (in parts per million) against time. NO_x occur as nitric oxide and nitrogen dioxide. They are toxic, and contribute to the formation of photochemical smog. NO_x emissions rise sharply under acceleration, but have little relation to vehicle speed or r.p.m. Lowest NO_x emissions are recorded during idling and during deceleration. Strangely, a rich mixture helps reduce NO_x content in the exhaust gases.

The third graph shows unburned hydrocarbon (HC) emissions (in parts per million) plotted against time. HC emissions are the direct result of incomplete combustion, and the pattern of HC emissions during the test cycle is closely related to engine r.p.m., load and road speed. A peak is reached during deceleration from high speed (up to 2,000 parts per million). In general, HC emissions are high when air consumption is low, and vice versa.

track which is about 25 miles in length. On this circuit, rally procedure is followed using a stopwatch and rigid schedule. It is necessary to adhere to an average circuit speed of around 30 m.p.h. in summer and winter and, after all, it takes more than an occasional thunderstorm or stretch of sheet ice to discourage the marathon drivers (naturally, in alternating shifts) from completing their total of 1,900 circuits. During this phase, the car returns at 4,000 mile intervals to the test bed where the "California test," a prescribed simulated sequence of specified driving conditions, is conducted (in ten stages from idling to partial load. with pushing operation and acceleration) over a total period of 137 seconds. This procedure is carried out *seven times in succession,* ac ompanied by a simultaneous check of exhaust gas constituents. According to some expert conclusions, which are not entirely unopposed, this represents, on the average, an accurate and virtually true-to-life operating situation. Runs one to four are regarded as cold start cycles. The fifth run is disregarded, because it is used only to warm up the engine to ideal temperature. Runs number six and seven then are used as the "hot" phases for conclusive checking of the actual exhaust constituents. Most of the noxious agents in exhaust gas are odorless and must be analyzed with the aid of complicated and expensive apparatus.

On December 10, 1969, the final decisive tests were carried out at the laboratory of the National Air Pollution Control Administration at Ypsilanti. On December 18, 1969, the official certificate of successful completion of all of the various tests was signed. This certificate issued by the United States Department of Health, Education and Welfare finally cleared the NSU Wankel engine in the Ro-80 series for travel in the rarified atmosphere of North America, and thus gave NSU every reason to expect success in the coming year. The Ro-80 vehicle gave 150 parts per million hydrocarbon emission and only 1% carbon monoxide. The dramatic reduction in emission levels was accomplished by installing a thermal reactor as part of the exhaust manifold.

In order to understand the emissions problem, one must have an understanding of gasoline as a motor fuel. Gasoline is a hydrocarbon, made up of about 15% hydrogen and 85% carbon. Air is a mixture of 21% oxygen, 78% nitrogen, and 1% other gases. But only the oxygen combines with the gasoline. Gasoline will not burn in a liquid state; it must be changed to vapor and supplied with an adequate volume of air to allow combustion. The air/fuel mixture must be adjusted to speed, load conditions, and temperature, and the total volume of the mixture also must vary according to these conditions. For cold starting, the air/fuel ratio must be as rich as 3:1 or 4:1; a ratio of 10:1 is best for a warm engine at idle. For part throttle, light load operation, a lean ratio of 15 or 16.5:1 will suffice, but for full throttle operation, the ratio should rise progressively to about 12:1.

In general, a rich mixture will produce more toxic emissions than a lean mixture. The exact emission quantity and composition depends mainly on throttle opening and engine r.p.m. With a rich and/or poor mixture there is the risk of incomplete combustion in the Wankel engine. The problem is most serious under light load conditions, when exhaust dilution is significant. The Wankel engine seems to have an advantage over the piston engine in this area because tests have shown it to operate well on a lean mixture. In addition, the Wankel has intense combustion chamber turbulence, which tends to promote mixture homogeneity and complete combustion.

Gasoline must be in vapor form at low temperature to ensure easy starting. It must vaporize at an increasing rate as carburetor and manifold temperatures rise to allow fast warm-up, smooth acceleration and even fuel distribution among the cylinders. The vaporizing characteristics must be in keeping with the climate and altitude to prevent vapor lock and fuel boiling inside carburetors, fuel pumps and lines. Gasoline should contain few extremely high-boiling hydrocarbons to ensure good fuel distribution and freedom from crankcase deposits and dilution. A high *anti-knock* quality (octane number) throughout its boiling range is needed to give freedom from knock at all engine speeds and loads. Gum content must be low to prevent valve sticking, carburetor difficulties and deposits inside the engine and intake manifolds. Gasoline also must have good stability against oxidation to prevent deterioration and gum formation in storage.

The anti-knock property of a gasoline is indicated by its octane number. The octane scale was created by giving the number 0 to heptane (C_7H_{16}) and the number 100 to iso-octane (C_8H_{18}). Numbers between 0 and 100 indicate the proportion of each if the two are mixed. The octane number of a gasoline is determined by a test comparing it with a mixture of heptane and iso-octane. For example, if the gasoline shows the same tendency to knock as a mixture containing 6% heptane and 94% iso-octane, its octane number is 94. There are two methods of establishing a fuel's octane number. The *research method* requires that the test engine be run under closely controlled conditions of speed, air intake temperature, and ignition timing. The *motor method* requires that the test engine be run with variations in speed, air intake temperature and ignition timing. The difference between *motor* and *research* method ratings for the same fuel is called *sensitivity*. It is impossible to obtain more than 100% iso-octane in a reference fuel blend, but some fuels have ratings above 100 octane. This means that the reference fuel becomes iso-octane plus a certain amount of tetra-ethyl lead. The knock value of such fuels is defined as milliliters of tetra-ethyl lead per gallon of iso-octane.

Anti-knock compounds usually are lead-based. The active anti-knock agents are alkyl-lead compounds consisting of ethyl and/or methyl groups attached to an atom of lead—tetra-ethyl lead is the most common of these. Because lead is a known poison, and petroleum industry spokesmen generally admit that about 50% of the lead content in gasoline escapes into the atmosphere via the car's tailpipe, the automobile industry, led by General Motors, has pressured the oil companies into providing non-leaded or low-lead content gasolines of 91 octane. All American piston engines now in production will run on this 91 octane fuel. How will this affect the Wankel engine? As part of its research on the use of non-leaded gasoline and low-grade gasoline in rotary engines, Toyo Kogyo carried out tests on the fuel octane value required. They wanted to see to what extent the octane value of the gasoline could be lowered without adversely affecting performance. The tests were carried out at the Miyoshi Proving Ground beginning on May 29, 1970, in collaboration with an oil company. Two Mazda R-100 coupes were used for the test, and various measurements of both fuels and running performance were made. The results were more favorable than predicted:

Octane value required	67 (full throttle at 2,000 r.p.m.)
	66 (part throttle at 2,000 r.p.m.)
Maximum speed	113 m.p.h.
Acceleration	16.6 sec. standing-start ¼ mile

Piston engine cars of similar performance require a high-octane gasoline of 98 to 100, but the rotary cars are an exception. Mazda rotary vehicles normally run on regular gasoline with octane values of 87 to 91. The test proves that the Wankel-powered car fully retains its performance, without any drop in maximum speed and acceleration, even with far lower octane gasoline totally free of lead additives. This ability of the rotary engine to operate on low-octane gasoline attracted Toyo Kogyo's attention at earlier stages of development, and research has since continued on the relationship between non-leaded gasoline and combustion. Toyo Kogyo engineers explain that the rotary engine does not require leaded gasoline because of certain basic characteristics intrinsic to its combustion process.

Generally, knocking (abnormal combustion) occurs due to slow flame propagation or high temperature at some corner in the combustion chamber. In the combustion chamber of the rotary engine, the mixture moves at high speed in the direction of rotor rotation, thus causing rapid flame propagation in the same direction. Also, the temperature at the trailing end in the combustion chamber is low due to the cooling effect of the large surface area at this portion of the chamber. Dr. Froede, of NSU, remarked at one time that displacing the rotor face recess in the leading direction increased the tangential gas velocity considerably and was an effective way of speeding up flame propagation. NSU also points out that, in standard Ro-80 form, their KKM-612 engine produces less than one-half the unburned hydrocarbons emitted by the Mazda engine despite the apparent disadvantage of considerable port overlap. NSU also confirms that the KKM-612 has low-octane, non-leaded fuel capability.

The various pollutants from automobile engines were listed earlier, but perhaps a little more should be said about the nature of these pollutants. More than 95% of the exhaust is composed of innocuous substances: water vapor, carbon dioxide, and nitrogen. The main air pollutants emitted by the internal combustion engine are hydrocarbons (HC), carbon monoxide (CO), and oxides of nitrogen (NO_x). Carbon monoxide and oxides of nitrogen are formed primarily in the bulk gas; hydrocarbons are formed in the quench area. Automobile exhaust gas is a minor source of particulates and sulfur dioxide (SO_2) content is insignificant.

Emissions parallel power output. Complete utilization of the fuel gives 100% power and no undesirable exhaust emissions; it's the unburned portion of the fuel that produces the problem. Carbon monoxide emission is caused mainly by poor combustion, in combination with rich air/fuel ratios and partial combustion in the quench areas. Most of the conditions that result in reduced CO emissions also result in reduced HC emissions. Just as is the case with the reciprocating piston engine, carbon monoxide emissions in Wankel engines depend on the air/fuel ratio of the mixture. The leanest mixture gives the lowest emission levels, but the ratio must be kept high enough to maintain stable combustion. Atomization of the fuel (pre-heating of the mixture and its flow speed in the intake manifold), distribution of the mixture and the turbulence of the mixture also are factors that influence exhaust emissions. With relation to these three factors, the Wankel engine has definite advantage over the reciprocating piston engine.

NO_x emissions probably are created in the high-temperature flame front. Because peak combustion temperatures in the Wankel engine

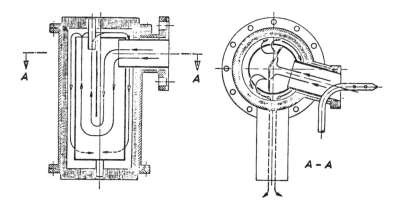

Elevation and cross-section of the afterburner used for the NSU Ro-80.

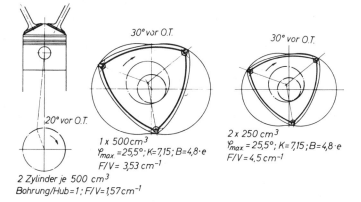

30° vor O.T.

30° vor O.T.

20° vor O.T.

1 x 500cm³
$Y_{max.}=25,5°; K=7,15; B=4,8·e$
$F/V= 3,53 cm^{-1}$

2 Zylinder je 500 cm³
Bohrung/Hub=1; F/V=1,57cm⁻¹

2 x 250 cm³
$Y_{max}=25,5°; K=7,15; B=4,8·e$
$F/V=4,5 cm^{-1}$

Comparison of surface to volume ratios in a two-cylinder piston engine with individual cylinder displacement of 500 cc. to that of a single-rotor 500 cc. Wankel engine, and a twin-rotor Wankel engine with 250 cc. chamber volume. F/V = Surface to volume ratio. Y_{max} = Leaning angle. K = K-factor, or R/e ratio. B = Bore (in piston engine). B = Rotor width (in Wankel engine). e = Eccentricity.

are slightly lower than in piston engines, NO_x emissions are less of a problem. One way that has been suggested to prevent the formation of NO_x is the addition of an inert gas to the mixture to reduce the peak cycle temperature without increasing the oxygen concentration. Experiments along these lines have been run with conventional piston engines, using exhaust gas recirculation to provide the inert gas, and one source gathered convincing data showing that there was a reduction in the amount of NO_x formed as the amount of exhaust gases recirculated was increased. There also was good correlation between the predicted reduction and the observed reduction.

When the exhaust emissions of the Wankel engine are compared to those of the reciprocating piston engine, the hydrocarbon emission level of the Wankel engine is higher and the oxides of nitrogen level is lower. For this reason, most research and development of rotary engine emission control systems is being directed to the reduction of hydrocarbons. The factors that determine the hydrocarbon concentration in the exhaust gas are not easily isolated, but the main cause of HC emission is considered to be the flame quenching action which takes place near the wall of the combustion chamber, plus the misfiring and leakage of fuel into the exhaust port. This flame quenching phenomenon is thought to be due mainly to the very poor surface-to-volume ratio at the trailing end of the combustion chamber. A large quench zone is formed in this section, and it becomes difficult for the flame front to reach the end gas. The following example may help illustrate the severity of the problem. If a handkerchief is dipped in gasoline and held in the open air, setting a match to it will cause it to burn very quickly. If another handkerchief, soaked in gasoline, is wrapped tightly around a flat iron and a match held to the gasoline-wet fabric on the iron, it will not burn. The conclusion is that a thin layer of explosive mixture on a relatively cold surface will not ignite. In Wankel engines, it means that fresh air/fuel mixture in the far portions of the combustion chamber may stay unburned throughout the combustion process and then escape as unburned hydrocarbons, partially burned or cracked fuel, carbon monoxide, and other less harmful products.

The possibility of wall quenching in any one engine is a function of that engine's combustion space surface-to-volume ratio. A piston engine with a big bore and a pancake-shaped combustion chamber has a chamber with small volume but large surface area. This combustion chamber is more prone to wall quenching than the compact space formed by a small bore and a cylinder head cavity shaped like half a pear. The small bore engine tends to have a low surface-to-volume ratio. In the Wankel engine, the combustion chamber is inherently flat (except for the cavity in the rotor) and elongated. Many experts have maintained that the Wankel engine, with its long, flat combustion chamber, is inherently at a disadvantage in respect to surface-to-volume ratio. While the combustion chamber passes through the minor axis, quenching occurs in the space on the trailing side, and, therefore, high concentrations of hydrocarbons occur in this trailing area. Quench may be partly compensated for, however, by higher wall temperatures due to the localized combustion and the frequency of combustion in one specific area of the housing.

The results of recent tests prove that the shape of the combustion chamber cavity formed on the rotor face, the position of the spark plugs, and the ignition advance are closely related to hydrocarbon concentration. Going to a lower compression ratio aids the surface-to-volume ratio by increasing the volume of the combustion chamber without adding much surface. The exact surface-to-volume ratio is determined by the engine's basic geometry. For engines that are otherwise identical, it is fixed by the angle of obliquity.

The angle of obliquity is half of the largest possible angle contained by the rotor's radial centerline (passing through the apex seal), and a line normal to the tangent of the epitrochoidal surface where the apex seal contacts the surface. With radially disposed apex seals it corresponds to the leaning angle. The higher the leaning angle, the more favorable the surface-to-volume ratio. For a given total engine displacement, the single-rotor Wankel offers an advantage in surface-to-volume ratio. A twin-rotor unit, with 250 cc. chamber volume, has a less favorable surface-to-volume ratio than a single-rotor 500 cc. Wankel engine.

In addition to the quench zone problem, the Wankel engine also suffers from a "crevice effect," the inability of gases to burn when the surfaces enclosing them move very close together. This crevice effect results in unburned gases being trapped along the edges of the rotor sides, in a small band limited by the side seals. The corresponding phenomenon in reciprocating engines is a thin layer of unburned gas below the piston crown around the piston top land, limited by the top compression ring.

A Wankel engine's thermal efficiency is raised when the compression ratio is increased. This means more power and reduced emissions. Similar improvements can be produced by allowing the engine to run at higher operating temperatures, simply because an engine that is running hot will provide the best mixing of the charge and the most complete combustion. In piston engines, there are a number of factors that limit operating temperature, the most critical of which is the exhaust valves. The Wankel engine promises reliability at far higher temperatures than are now possible with piston engines.

Fuel supply at high temperatures may be a problem, and this is the area where direct fuel injection inside the working chamber has much to offer over carburetors. The validity of this claim is proved by the performance of the Mercedes-Benz C-111. In 1967, Curtiss-Wright began tests with the Conelec electronic fuel injection system (under

Bendix patents). This system's characteristics are low-pressure, long-duration metering, and Curtiss-Wright was looking for improved fuel vaporization. However, because the Conelec system works on low pressure with port-mounted injectors, it did not give the predicted results.

Pre-heating of the intake air is part of the emission control systems on many cars powered by reciprocating piston engines. Toyo Kogyo so far stands alone in using such a device on Wankel engines. It performs a dual function on the Toyo Kogyo Wankel. The low-temperature area

Mazda engines responded well to secondary air injection. The exact position of the air hole or nozzle had strong influence on emissions. Position A (injection holes in the wall of the working chamber, near the exhaust ports), gave the lowest emissions. Position B (air nozzle inside the exhaust port) results were acceptable, and there was no risk of clogging the nozzles with combustion products. Position C (air nozzle downstream from the port) showed only a small improvement over running without secondary air injection.

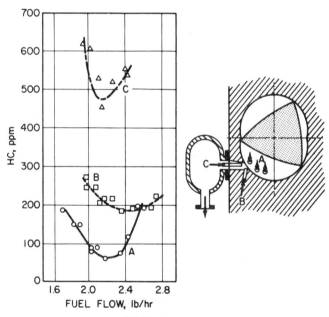

EMISSIONS, ppm

	SINGLE-CELL TYPE	DUAL-CELL TYPE
WARMUP	810	376
HOT	187	42
COMPOSITE	406	159

in one working chamber corresponds to bottom dead center after an intake stroke. Remember that operational events identified by *time* in the piston engine are defined as *points* on the working surface in the Wankel engine. The area in question was cooled by the intake mixture, and it was at one time thought that the extremely low temperature of this portion was detrimental to the atomization of the fuel mixture. This led to the design of a system for pre-heating the intake air by exhaust gas. The housing was redesigned to function as an exhaust heat exchanger. Toyo Kogyo's heating system used only a part of the exhaust gas for circulation in the low-temperature areas of both the rotor housing and the sidewalls. The system also served to equalize housing temperatures and thereby minimize the dual problems of thermal distortion in the housing and deterioration of the lubricant on the sliding surfaces. It proved impossible to completely eliminate the temperature differences and thermal distortion, and an amount of distortion ultimately was judged tolerable insofar as it did not noticeably hinder the operation of the moving parts in relation to the housing or cause malfunctions in overall engine operation in service.

Blowby gases in a piston engine are gases that force their way past the piston rings into the crankcase. Blowby is a significant source of pollution in these engines. In the case of the Wankel engine, blowby leaks past the apex seals in both directions. Leakage at the leading end of the combustion chamber means that raw mixture is vented into the exhaust port, while gas leakage at the trailing end is included in the next charge and is not lost. The gas sealing performance of the side seal also has an influence on blowby. Gas leakage here goes into the housing oil sump and reduces engine performance.

If both the oil seal and the side seal function equally well, the harm brought about on the oil seal by the blowby gas is far smaller in the case of the side port design than in the case of the peripheral port. This is because the side intake port is located between the side seal and the oil seal so that if there is a gas leak through the side seal, escaped gases will be drawn back into the intake port. This means that the gas pressure acting on the oil seal's periphery will be kept low.

The utilization of thermal reactors, made by Du Pont, allowed both the NSU Ro-80 and the Mazda Wankel to pass the current emission standards. Thermal reactors similar of these may play an important part in the acceptance of the Wankel engine in coming years. A thermal reactor is simply an "afterburner" that exists solely for the purpose of burning leftover combustion products to prevent them from escaping into the atmosphere. Afterburners do not contribute to the efficiency of the engine; in fact, they detract from engine performance. Tests with thermal reactor systems have indicated power losses up to 20% and increases in fuel consumption of about 10%.

The thermal reactor provides a high-temperature zone in which hydrocarbon and carbon monoxide emissions are burned almost completely before being passed on to the atmosphere. The reactors, consisting of a casing containing a tubular core, replace the conventional exhaust manifolds. Hot exhaust gases are mixed in a tubular core with air from a separate air injection system operated by an engine-driven pump. Oxidation takes place in the core, and the oxidized gases leave through a series of holes along the length of the core, pass around the core's outside to help keep it at a high temperature, then are expelled into the exhaust pipe.

Research and development conducted by Toyo Kogyo on exhaust emission control of the Mazda 0813 engine was carried out with this type of exhaust reactor. At first, tests were made to find which secondary air injection location was the most efficient. Injection holes were provided inside the exhaust chamber close to the exhaust port, by utilizing the unique structure of the rotary engine. In another test, an air nozzle was installed inside the exhaust port orthogonally to the exhaust gas flow, and, in a third test, an air nozzle was installed and aimed counter to the exhaust gas flow. At idling speed, both the first two in-

Gas flow in the thermal reactor used on the Curtiss-Wright engine.

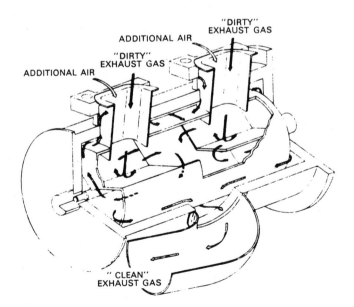

stallations yielded low-emission levels. Because it is feared that the injection holes can become clogged by combustion products during operation, tests are at present being performed with the orthogonal nozzle installation. However, if the injection hole configuration can be further improved, there is a great possibility that the test position having the greatest potential, the one nearest the port, can be restored.

Curtiss-Wright RC2-60 U5 automotive engine fitted with an afterburner.

Certain design criteria were established by the Toyo Kogyo engineers. The reactor had to allow sufficient time for the efficient mixing of exhaust gas and injected air, had to have efficient heat insulation, rapid warm-up time, durability, and had to be made of low-cost material. In order to hold the temperature of the reactor core below 1,800°F., a durability determining factor, the supply of secondary air had to be automatically cut off by a control device when engine revolutions exceed a certain limit.

The base line of hydrocarbon emission level on the 0813 engine was recorded as being 2,000–2,500 parts per million under the California cycle. With various engine modifications and the adoption of a thermal reactor with air injection, this was reduced to 120–200 parts per million, and the CO and NO_x emission levels were 0.6–1.0% and 400–600 parts per million, respectively. The hydrocarbon emission level in the cold cycle could not be considered as being satisfactory, but the level in the hot cycle was down to a very low figure.

The installation used by NSU for the Ro-80 is essentially the same, and Daimler-Benz has done a considerable amount of work with the Du Pont reactor. In 1969, Uhlenhaut felt that the C-111 would meet the California exhaust emission standards when hot, but would fail the cold-starting test. The problem was that very high temperatures are necessary for proper reaction, and the materials in use up to that time did not have enough durability. Exotic metals had to be avoided because of their high cost, which would prohibit mass production, then and in the future. However, in April, 1970, a Mercedes-Benz spokesman said, "Major progress in the area of exhaust emission control has been made with the use of afterburners." He added that a new thermal reactor system, developed specifically for the C-111 engine, is currently undergoing intensive tests at the company's research department in Stuttgart.

Curtiss-Wright had started in 1966 to investigate the exhaust emission problem. Bentele suggested that the exhaust system be equipped with devices which promote oxidation of the exhaust gases by prolonging their residence time, and through turbulence and mixing with air. A patent was granted for this system.

Curtiss-Wright and the University of Michigan have jointly developed an insulated cylindrical reactor manifold. A series of baffles within the manifold slow down the exhaust gases and keep the particles suspended. At the same time, air is pumped into the exhaust manifold by a vane-type Saginaw compressor. This air mixes with the exhaust gases and helps complete the combustion process at 2,000°F.

Professor David E. Cole of the University of Michigan, son of GM President Edward N. Cole, and Charles Jones of Curtiss-Wright delivered a joint paper to the annual SAE Congress in Detroit in January,

1970, concerning the emission control studies conducted at the University. The paper explains a full-scale investigation of the emission characteristics of the RC2-60 U5 engine and presents many interesting conclusions.

The exhaust emissions of the RC2-60 U5 were measured, with and without an exhaust reactor, both under steady-state conditions at the University of Michigan and in a vehicle operated on the simulated California cycle at an independent facility under contract with Curtiss-Wright. The big question was whether the Wankel engine had any inherent features that tended to produce high exhaust emissions. Next came the identification of the design parameters that affected exhaust emissions, and the determination of how they could be modified to reduce the emission levels. The study began with an investigation of how changes in air/fuel ratios and spark advance affected hydrocarbon emissions. The study was restricted to hydrocarbons because only limited equipment was available. In addition, it was believed that carbon monoxide emissions would follow a similar pattern, and consequently no separate study on CO was undertaken.

First, the engine was tested in its basic form to establish a base line for results obtained with various emission control systems. The test engine was run at 1,000, 2,000, and 3,000 r.p.m. under varying load conditions, connected to the dynamometer through an automatic transmission. Instead of merely adopting the federal test, which seems arbitrary although it includes a variety of common city and suburban driving conditions, Cole was more interested in studying the overall emission characteristics of the Wankel engine than in making modifications that would suit the present federal test method. He discovered that the emission problem was most severe at low engine speeds. Consequently, only low-speed dynamometer tests were made. At high speeds, the emission levels were surprisingly low, but the problem with high r.p.m. was that as breathing and emission levels improved, the rubbing speed of the seal tips became very high. At low r.p.m., the engine proved to have considerably higher emission levels than modern automobile piston engines. That was because the Curtiss-Wright RC2-60 U5 had been designed without any regard for emissions. Detroit's engine designers have fought with the emission problem for many years and have had time to incorporate their knowledge into all-new engine designs as well as to apply it to existing engines in the form of detail modifications.

The engine first was tested with a thermal reactor *without* secondary air injection, then *with* air injection. To ensure complete burning, temperatures in the area of 1,100°F. were considered to be the allowable minimum. Oxygen shortage is a constant impediment to proper burning, but this can be overcome by injecting air into the reactor. Also, the gas flow inside the reactor has to be turbulent to ensure complete burning, and the reactor must be designed to give adequate residence time for the gases to complete proper reaction. This meant a long and sinuous gas flow pattern was necessary, maintained by internal baffles in the reactor core. Such a baffle system was developed, and the baffles also ensured mixing of the secondary air with the exhaust gases. Hot "clean" gas was recirculated around the core to minimize heat loss from the core. The hot "clean" gas also circulated around the reactor neck to maximize exhaust gas temperature as it entered the reactor. Air injection was arranged by mounting a direct port nozzle upstream from the reactor, and an insulation blanket was attached to the inside of the casing to minimize heat loss by radiation. The high-temperature parts were made of a high-grade turbine alloy called Hastelloy X.

Dynamometer tests indicated considerable additional improvements in emission levels were obtainable through adjustments in spark timing, carburetor setting, choke control, heat riser flow, and air pump pressure. The total reactor volume was equal to the displacement of one chamber of each rotor; the volume of the inner core was equal to one chamber. Cole reported that he found indications that larger reactor volume could have further reduced emission levels by allowing longer residence time for the exhaust gases inside the reactor. The air injection nozzle was positioned at the earliest point behind the exhaust port considered feasible for maximum mixing of the secondary air and exhaust gas before it entered the reactor. No tests were made using alternate positions.

The engine responded very well to the addition of the thermal reactor. Hydrocarbon emissions at 1,000 r.p.m. were reduced from over 900 to less than 100 parts per million. Running at 2,000 r.p.m., with a constant manifold vacuum of 20 inches of mercury without the reactor, gave a hydrocarbon emission level of about 500–510 parts per million with 50 degrees spark advance. Retarding the spark to 30 degrees (before the minor axis) brought the level down to 340–350 parts per million. Parallel improvements were found possible with similar spark retardation under lower load conditions.

Air/fuel ratios were studied at 2,000 r.p.m. without the exhaust reactor, running on regular gasoline. Manifold vacuum was varied in the tests (15, 10 and 20 inches of mercury). Air/fuel ratios around 17.5:1 gave the lowest hydrocarbon concentration in the exhaust; except at the highest manifold vacuum, which gave the cleanest exhaust at 15.5:1 air/fuel ratio.

Next, the engine was fitted with a reactor on the test bench and put through a cold-starting test to see what effect temperature had on starting ability and emission levels. The engine was started and run at 1,000 r.p.m. Hydrocarbon emissions dropped drastically after 30 seconds, when core temperature reached 1,100°F. When core temperature leveled off at about 1,700°F., HC emissions also leveled off—to 150 parts per million. Within two minutes, the reactor had reached 90% of its full effectiveness. Test results with the reactor showed temperatures of 1,900°F. in the core and 700°F. on the outer shell surface.

It is characteristic of the Wankel engine to run with high exhaust gas temperatures and thereby contribute to the efficiency of the thermal reactor. Each exhaust port receives exhaust gas during one-half of the four-stroke cycle. This compares to between one-sixth and one-eighth in a piston engine. The higher frequency of exhaust pulses, directed against the same sector of the reactor, raises the temperature in that area. The Wankel engine was found to have a very high tolerance to retarded ignition and lean mixtures. In combination, these conditions produce late burning and lower flame propagation, which in turn means hotter exhaust gas. The Wankel engine, then, is well suited to withstand the high exhaust gas temperatures and backpressures created by the use of an afterburner. Because there is no exhaust valve, high exhaust gas temperature is less critical than in a piston engine to begin with, and high backpressure is no threat to valve life and operation. The lack of a valve guide and its support, and the simple shape of the port, makes design for adequate coolant flow around the port a simple task.

Hydrocarbon emission levels proved extremely sensitive to air/fuel ratios. With 17.5:1 air/fuel ratio, the RC2-60 *without* the reactor realized a 25% reduction in hydrocarbon emissions. With the same 17.5:1 air/fuel ratio, *plus* the thermal reactor *without air injection*, a reduction of 75 to 90% in hydrocarbon emissions resulted.

In a parallel program, another RC2-60 U5 engine was installed in a 1964 Ford Galaxie 500 sedan. This experimental vehicle was equipped with an air injection reactor system incorporating a Saginaw air pump with a capacity of 19.3 cubic inches, a 10 inches of mercury relief valve, a Delco back flow check valve, and Rochester diverter valves. The air pump was driven at 1.3 times mainshaft speed and vehicle weight was 4,366 pounds. The vehicle was tested on a Clayton chassis dynamometer by the Esso Research and Engineering Company, Linden, New Jersey. Exhaust emissions measured and recorded in this test included hydrocarbon, carbon monoxide, and oxides of nitrogen. In general, the vehicle tests confirmed the findings of the engine bench test results. Hydrocarbon levels were around 100 parts per million, carbon monoxide varied between 0.77 and 1.47% and oxides of nitrogen varied from 378 to 607 parts per million. The test vehicle, said Cole, illustrated that current legal limits could be met for hydrocarbon and carbon monoxide emissions. The NO_x emissions were above the anticipated restrictive level, but were believed controllable by relatively minor adjustment. The reactor principle was proved to be fully compatible with the Wankel engine, and the development of low-cost versions was recommended.

The last paragraph in David Cole's paper suggested exploring far higher speeds because the engine appears to have reduced emission levels in the upper r.p.m. ranges. "The RC engine's rapid decline of hydrocarbon emission levels as a function of increasing speed combined with the engine's propensity for higher r.p.m., by virtue of complete balance and no valves, may be exploitable. Raising the entire speed spectrum 50 to 100 percent is not an unreasonable development goal today."

Section III

APPLICATIONS

16

Introduction

THE DUTY OF PROVIDING MOTIVE power for a passenger car is only one of a multitude of applications—actual and potential—of the Wankel engine. It can be regarded as a prime mover with the capability of replacing the piston engine throughout its whole range of applications. It also has the same multi-fuel capabilities as the conventional piston engine.

The Wankel engine does not, in principle, suffer from scale effects. It can be built to any scale, in any size, and operate successfully. At the small end of the scale, internal friction could become too great in proportion to the power output, but Wankel engines have proved practical down to a displacement of 18.5 cubic inches (305 cc.) per working chamber. At the other end of the scale, problems of flame front travel may ultimately restrict rotor dimensions. However, Curtiss-Wright built and tested an experimental Wankel engine having 1,920 cubic inches chamber displacement. It ran well, and provided ample proof of the feasibility of up-scaling the engine to dimensions usually associated with marine navigation rather than with land transportation.

Even if there should be an optimum size for the Wankel engine rotor, there is not, necessarily, a limit to the power range of Wankel engines. There is no limit to the number of rotors that can be combined in one power unit, so Wankel engines can be compared with multi-cylinder reciprocating piston engines. Just as a V8 delivers torque more smoothly than a four-cylinder piston engine, the multi-rotor Wankel engine offers superior instantaneous torque and eliminates the negative torque periods of the single-rotor unit. The most obvious method of creating a multiple-rotor Wankel engine is to add rotors to

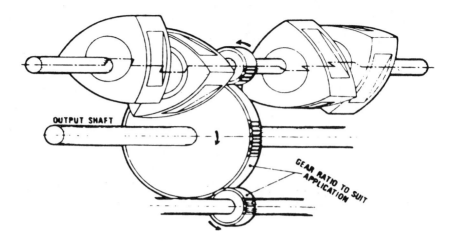

This is R. F. Ansdale's proposal for a modular multi-rotor Wankel engine. Three shafts, each with four rotors, geared to a common output shaft, produce a compact twelve-rotor unit.

the original mainshaft. The engine is so constructed that the designer has freedom to lengthen and add eccentrics to the mainshaft, fit rotors on them and build up a series of outer housings separated by partitions until all rotors are located inside working chambers.

The most promising layout of a multi-rotor Wankel engine was proposed in 1968 by the independent consulting engineer, R. F. Ansdale, in an article published in the magazine *Combustion Engine Progress,* entitled "Feasibility of High-Output Wankel RC Engines" (RC stands for Rotating Combustion). Ansdale proposed coupling a number of twin-rotor Wankel engine modules, suitably geared, to a common output shaft. Twin-rotor engines need only two main bearings and are easy to assemble as a unit. Using four or more rotors on a common shaft would require intermediate main bearings between the rotors, which would add complications to both production and maintenance. Placing two twin-rotor engines end-to-end, with a spur gear on the shaft that links them together, would result in a four-rotor, inline unit. If the spur gear was meshed with a larger gear "below" it, which was attached to a new output shaft, the mainshafts of the two combined engines would add torque to the new output shaft. This new shaft could be meshed with two additional spur gears, each of which would be connected to two modules. If the gears were properly arranged around a common output shaft, spaced 120 degrees around for example, a highly compact 12-rotor Wankel would result. Two such engines could be arranged in series to make a 24-rotor installation. According to Ansdale, it is conceivable that a 24-rotor engine might yield 6,000 horsepower, with extraordinarily low vibration levels, and only nominal torque fluctuations, while allowing full service accessibility to all modules.

The full spectrum of possible Wankel engine applications must include the following:

Branch	Application	Horsepower Range
Industrial engines	Portable welding generators	15–25
	Stand-by electric generators	5–50
	Portable battery chargers	5–25
	Portable power tools	1–10
	Portable lifting equipment	10–20
	Industrial utility vehicles	10–40
Engines for building & construction equipment	Portable building equipment	10–20
	Concrete mixers	5–20
	Portable conveyors	10–20
	Portable compressors	5–40
	Construction machinery	10–50
Engines for agricultural equipment	Lawn tractors	6–12
	Lawn mowers	3–5
	Soil & harvesting equipment	10–50
	Horticultural equipment	5–50
	Irrigation booster pumps	10–40
Aircraft, automotive & marine engines	Automobiles	40–400
	Trucks	100–600
	Motorcycles	10–80
	Scooters	3–5
	Outboard motorboats	20–200
	Auxiliary yacht engines	5–15
	Life boats	15–50
	Light aircraft	150–500
	Auxiliary glider engines	5–15
Engines for household & recreational equipment	Small lawn mowers	2–5
	Snowmobiles	10–60
	Golf carts	5–40
	Snowblowers	2–5
Engines for heavy transportation and stationary uses	Railroad locomotives	3,000–6,000
	Ships	3,000–6,000
	Pumping sets for atomic reactors	3,000–6,000
	Generator sets	3,000–6,000

Branch	*Application*	*Horsepower Range*
Miscellaneous	Fire fighting pumps	10–40
	Air-conditioning units (buses)	3–5
	Refrigeration units (trucks)	5–10
	Special military purposes (generator sets, gas turbine starter units, etc.)	100–500

The scope of this book is largely restricted to the application of Wankel engines to automobiles. Some of the results of applications in this area are found in the following chapters.

17

The NSU
Wankel Spider

UNDER THE TERMS of the standard Wankel engine license contract, all licensees are given the benefit of NSU's know-how, and they in turn pledge to share their experience and developments with NSU and each other. The "partners" hold regular technical conferences, and the open exchange of information is compulsory for all parties in the sense expressed in the agreement. This pooling of data and information relates to all aspects of design and manufacture of Wankel engines and their parts. This includes unlimited numbers of blueprints, samples and models, manufacturing drawings, specifications, parts lists, engineering data, test reports, design information, performance charts and similar engineering, manufacturing and technical data and information. In short, anything that is deemed necessary to the design, redesign, adaptation, operation, construction, manufacture, production, assembly, maintenance, service and repair of Wankel engines by each licensee, including equipment and tooling used in the manufacture of such engines.

As part of this program, NSU development engineer Hans Paschke was sent to Japan to observe Toyo Kogyo's work in 1962 and 1963. After his first ride in the Cosmo sports car, he sent a most enthusiastic report back to Neckarsulm. It was not received with great cheer at the NSU headquarters. NSU was not itself ready to start production of the Spider, and such a glowing report from Japan could only serve to stir up more troublesome questioning from minority stockholders. Then von Heydekampf, ready to make his epoch-making announcement of production plans for the Spider, received news on June 7, 1963 that Mazda (Toyo Kogyo) wanted to show a Wankel-powered car at the Frankfurt show that September!

The Toyo Kogyo announcement emanated from one of its vice presidents during a press conference, and the remark was picked up by the wire services and broadcast all over the world. The news hit Neckarsulm like an earthquake. Von Heydekampf discussed possible countermeasures with Dr. Henn, head of the legal department, and Dr. Henn sent Toyo Kogyo a brief letter from which the usual polite phrases were omitted, pointing out that Mazda's exploitation rights for the Wankel engine were confined to East Asia and that consequently the company had no right to exhibit its machines in Europe. This letter had the effect of a cold shower in Hiroshima. The Japanese coolly replied to the letter that the whole thing must have been a misunderstanding. Toyo Kogyo had no intention of exhibiting a Wankel-engined car in Frankfurt in September, but were arranging to show such a prototype at the Tokyo Automobile Show in October. Thus, the NSU Wankel Spider did not have to share the limelight at the Frankfurt auto show with a Mazda, and is recognized as being the world's first production car powered by a Wankel engine.

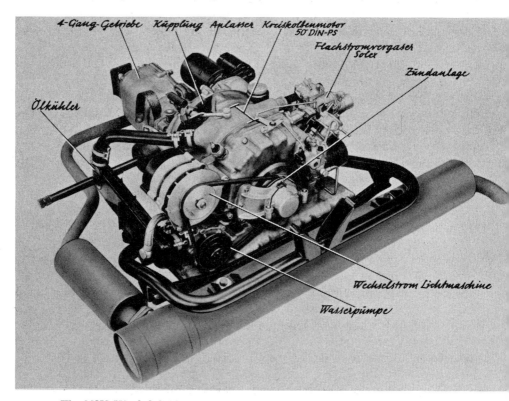

The NSU Wankel Spider engine and drive train, ready for installation.

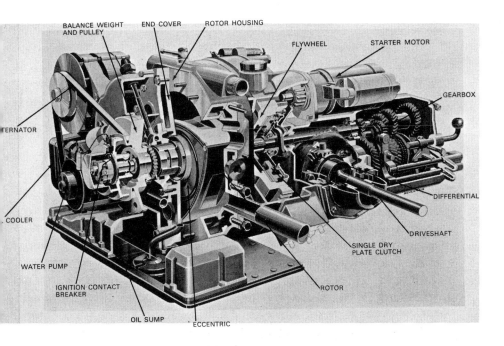

Cutaway drawing of the NSU Wankel Spider engine.

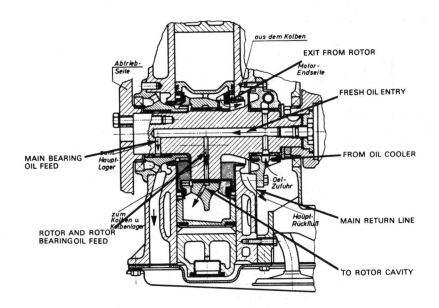

Lubrication and rotor cooling system in the KKM-502 engine.

The Spider body was a convertible version of the Sport Prinz, with some special features made possible by the use of a Wankel engine, such as a trunk of useful shape above the engine. The car was strictly a two-seater but also had generous luggage space behind the seats. It was well designed and the bodywork was cleverly thought out. Quality seemed up to or above Prinz standards. The convertible top was easy to put up or down, being a one-man operation taking about half a minute. It was stored under a metal panel rather than under one of those stretch-type vinyl covers usually found on American convertibles, and was completely out of the way. The chassis components for the Spider were taken directly from the Sport Prinz. The architectural layout remained unchanged, with the single-rotor Wankel engine positioned behind the rear wheel centerline, just as with the stock piston engine. Steering, front and rear suspension, and transmission remained essentially the same.

My first experience driving the NSU Spider came in 1965 when I borrowed one from the importers and distributors in New York, Transcontinental Motors, Inc., and took it to Bridgehampton race circuit on Long Island for a test. There were many conventional aspects to driving the NSU Spider, just as was the case with the Ro-80 and the Mazda R-100. You could actually perform all the normal driving functions without knowing the car had a Wankel engine, and find that it would respond by doing more or less what you would expect. It started up in

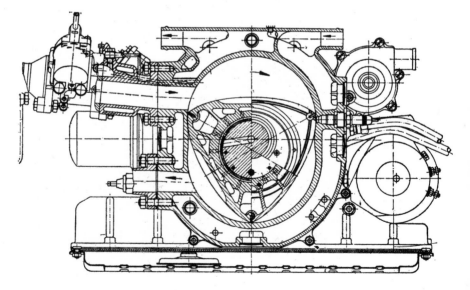

Cross-section of the KKM-502.

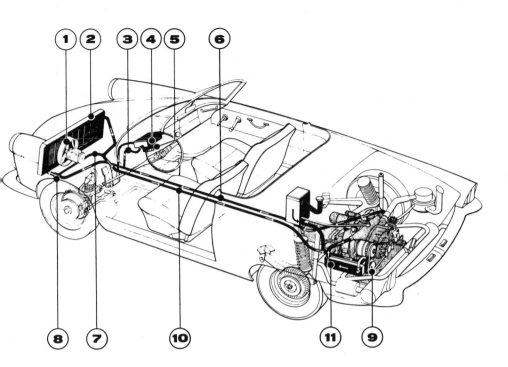

NSU Wankel Spider cooling system. 1 = Cooling fan. 2 = Radiator. 3 = Branch to heater. 4 = Heater element. 5 = Heater fan. 6 = Hot line. 7 = Return line from thermostat. 8 = Thermostat. 9 = Water pump. 10 = Cool water line. 11 = Heat exchanger (oil/water).

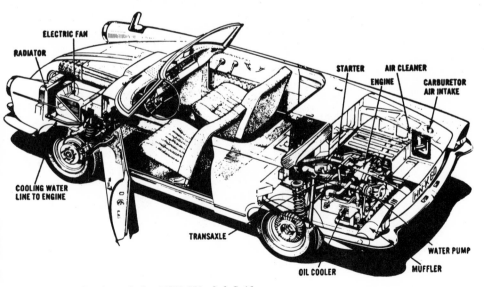

Cutaway drawing of the NSU Wankel Spider.

The NSU Wankel Spider showed such reliability, that German tourists showed no hesitation about taking them deep behind the Iron Curtain, far from parts and service facilities.

The NSU Wankel Spider.

the normal way, by turning a key, but as soon as it started running you knew it was not a conventional piston engine. Its silence, lack of vibration, and the 7,000 r.p.m. redline on the tachometer immediately told you that this was something different. More than the redline itself, it was the striking ease and rapidity with which the needle would climb there that impressed me.

The driving position was unusually correct for such a small car, in that the front wheel housing did not rob the driver or front seat passenger of too much foot room, and the steering column had a very slight tilt. I expected the steering to be both light and quick, and I was not disappointed. It was also uncommonly precise in its action. The car could be placed within an inch of its intended path at any speed.

The brakes were powerful and stable. But the most remarkable thing about driving this car on an extremely fast road circuit was the handling. I have driven a lot of rear-engined cars with swing-axle rear suspension under the same conditions, and expected a certain amount of oversteer with the NSU Spider. But it was perfectly balanced! There was no breaking loose, front end or rear. The whole car stuck to the chosen

It was in international road rallies that the NSU Wankel Spider showed its mettle. Its reliability record is even more impressive than the number of victories, with nearly all starters finishing.

The racing career of the NSU Wankel Spider began in 1966, when private owners combined with NSU engineers acting independently, to get more power and speed from the single-rotor engine.

line around each curve, in an attitude that can best be described as neutral with an understeering tendency. With a front roll center located 1.18 inches above ground level and the very high rear roll center on the Spider (11.37 inches above ground level), body roll was practically non-existent. Front wheel bump travel was more than adequate, at 5.31 inches, and rear wheel bump travel was an even more generous 5.56 inches. As a result, ride comfort was excellent for a sports car, but still would be deemed harsh by family car standards. Because the car was so light, it bounced a bit on big bumps.

Directional stability was strong for such a light car, mainly due to the very high front wheel caster angles (12 degrees) combined with a swivel axis inclination of 7°30′. Still, steering effort, with its 16.25:1 overall ratio, was quite light. Static front wheel camber was 2 degrees positive. The rear suspension was designed for zero static wheel camber, but tended to go positive under transient conditions. The tires were small, but they were radial-ply Michelin X, and I could not have felt more confident.

Acceleration, too, was a surprise. The first thing that had to be learned about this car was to watch the tachometer, simply because the engine gave no hint of audible protest when it reached its rev limit. On the contrary, it seemed happier the faster it spun, and no doubt would have

gone on to 9,000 r.p.m. or more if the throttle had been kept open
a little too late beyond the redline. But over-revving was not encouraged,
since tests had shown that engine wear and oil consumption increased
at an alarming rate when shaft r.p.m. exceeded 6,000. There was no
torque below 3,000 r.p.m., so that the shift had to be used frequently.
For any unexpected opening in traffic, a downshift was necessary to
give proper acceleration. This was not objectionable to the clientele
that was attracted to the NSU Spider, since they were performance-
oriented enthusiasts and were used to small piston engines with similar
torque characteristics.

The long linkage between the floor-mounted shift lever and the
actual transmission made some lack of precision in the gate inevitable,
but shifts were fast because the synchromesh was unbeatable, and the
shift movements were very, very light. In turnpike driving, the car was
more than able to hold its own. Acceleration was more than adequate as
long as the engine could be kept above 4,000 r.p.m. Engine noise levels
were extremely low at all steady speeds, and cruising at 70 m.p.h. in the
Spider was as pleasant as in some far more powerful sports cars.

SPECIFICATIONS

MAKE	NSU
MODEL	SPIDER
Year introduced	1964
Year discontinued	1967
Price	$2,979
Type of body	Open two-seat roadster
Type of construction	Unit body
Driving wheels	Rear
Power unit position	Rear
Curb weight	1,543 pounds
Weight distribution front/rear	44/56%
Power/weight ratio	24.2 pounds per horsepower
Fuel tank capacity	9.2 gallons
Fuel tank position	Front
Power unit	Wankel
Number of rotors	1
Chamber displacement	30.51 cu. in. (498 cc.)
Equivalent total displacement	61.02 cu. in. (996 cc.)
Compression ratio	8.6:1
Power output	64 horsepower
at r.p.m.	6,000
Torque	52 foot pounds

SPECIFICATIONS

MAKE	NSU
MODEL	SPIDER
at r.p.m.	2,500
Carburetion system	One Solex 18/32 HHD
Ignition system	Coil and battery
Cooling system	Water and oil
Clutch	Single dry plate
Transmission	Four-speed all-synchromesh
Gear ratios 1	3.08:1
2	1.77:1
3	1.17:1
4	0.85:1
5	—
R	3.43:1
Final drive ratio	4.43:1
Front suspension	Upper and lower A-frame control arms with stabilizer bar
Front springs	Coil springs
Rear suspension	Swing axles and semi-trailing arms
Rear springs	Coil springs
Steering system	Rack and pinion
Turning diameter	28.8 feet
Overall steering ratio	16.25:1
Turns, lock to lock	2.8
Brake system	ATE-Dunlop discs F; drums R
Disc diameter F	8.94 inches
Disc diameter R	—
Drum diameter F	—
Drum diameter R	—
Lining area	—
Swept area	—
Parking brake	—
Tires	Michelin X
Tire size	135 × 12
Wheelbase	79.53 inches
Front track	49 inches
Rear track	48.3 inches
Overall length	141 inches
Overall width	59.84 inches
Overall height	49.61 inches

SPECIFICATIONS

MAKE	NSU
MODEL	SPIDER

Test Results

Acceleration times

0–30 m.p.h.	3.8 sec.
0–40 m.p.h.	6.3 sec.
0–50 m.p.h.	9.8 sec.
0–60 m.p.h.	13.8 sec.
0–70 m.p.h.	20.1 sec.
0–80 m.p.h.	28.2 sec.
0–90 m.p.h.	44.0 sec.
0–100 m.p.h.	—
Top speed 1	27 m.p.h.
2	47 m.p.h
3	71 m.p.h.
4	95 m.p.h.
5	—
Average fuel consumption	28 m.p.g.

18

The NSU
Ro-80

NECKARSULM IS A SMALL INDUSTRIAL
town in the valley of the river Neckar, about an hour's drive north of
Stuttgart in southern West Germany. I went there in June, 1966, to look
at the twin-rotor Wankel engine that NSU was experimenting with. A
young liaison engineer named von Manteuffel let me see the new engine
undergoing dynamometer tests, showed me the dismantled test engines,
and explained the design features. Then he took me into the garage and
pointed to a Jeep-like vehicle. It was a DKW Munga, a four-wheel-drive,
off-road car built by Auto Union from 1954 to 1965. This was *the* test
car for the twin-rotor NSU Wankel engine.

At von Manteuffel's invitation, I clambered in behind the wheel. The
engine started easily but seemed noisier than the single-rotor unit that
powered the NSU Spider. That was mainly because the engine was installed in the front of the chassis, without any kind of sound deadening
materials around it. It had excellent power and gave the Munga a higher
top speed than its chassis was ever intended for. It pulled smoothly from
quite low revs, but had little torque below 4,000 r.p.m. It took a lot of
shifting to drive it on the hilly country roads in the Neckar valley.

In town traffic, we experienced what von Manteuffel described as the
biggest drawback to this engine. It was a snatch in the drive train that
occurred at low speeds on a closed throttle, such as when easing towards
a street corner. The result was a stop-and-go bucking motion in the
whole car, which disappeared when the clutch was disengaged. This had
to be cured before the engine could be judged suitable for a high-grade

passenger car such as NSU was proposing to build. The Wankel Spider was a small sports car without the luxury and refinement of the Ro-80 and, while it had the same problem, it was considered tolerable in this type of car.

Nobody at NSU made any secret of the fact that a completely new car was being designed around this engine, but no details were available on the project that was to become the Ro-80 car. It was officially introduced a little over a year later, at the Frankfurt auto show in September, 1967. Production began in October of that year. I saw a few of them on European highways from time to time, but it was not until 1969 that I actually got to drive an Ro-80. By then, NSU had produced over 14,000.

The Ro-80 is not a sports car, but an extremely well-equipped and well-finished four-door, four-passenger sedan. The first thing you notice when sitting down behind the wheel is the nearly flat, tunnel-free floor,

Ro-80 instrumentation is clear and simple, the entire interior package having been designed with full regard for safety. Spacious and comfortable, the car is also very well appointed and finished.

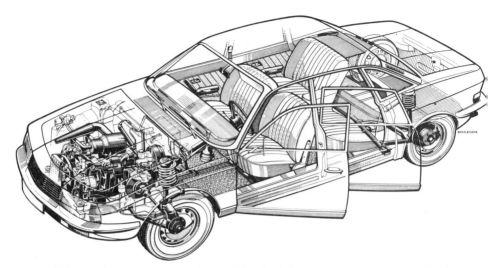

All drive train components are located ahead of the passenger compartment, leaving all space free for occupants and luggage rear of the cowl structure.

sunk six inches below the sills. Wide, individual seats are body-contoured to cradle you on corners, and backrest angle is fully adjustable. In addition to more-than-generous fore-and-aft movement, the driver's seat also has height adjustment. This is done by ingenious double runners, with the lower pair sliding up ramps to raise the seat. The broad seat back and the firm cushion make for tireless driving over long distances. The steering wheel is large and has a padded "safety" center. The instrument panel is well laid out, with big dials grouped in front of the driver. A small padded hood above them prevents reflections on the windshield.

The vital switches are carried on stalks branching off from the steering column for finger-tip control. A four-way switch on the right operates the three-tone horn, two-speed wipers and electric washers. The left-hand stalk dims the headlights and handles the turn signals. Small controls on the instrument panel include a pull-out master light switch and a switch for four way emergency flashers. An easy-to-read 130 m.p.h. speedometer is matched with an 8,000 r.p.m. tachometer.

The engine sounds smooth, even when idling, but has little torque below 3,500 r.p.m. Acceleration at lower engine speeds is definitely sluggish. On hills, there is enough power to maintain high speeds, but little power reserve for acceleration. Top gear acceleration is poor in relation to the claimed horsepower. To get performance, the gears must be used frequently. Top speed is remarkably high, and the Ro-80 will cruise in uncanny silence at 100 m.p.h. At top speed, the engine is still quiet, and there are no signs of stress or vibration. Full-throttle acceleration from a standstill is good, but not startling in view of the engine's 136 horsepower rating. For brisk driving at speeds under 50 m.p.h., you have to watch the tachometer and use the gears.

The Saxomat transmission used on the Ro-80 is a "selective automatic" made by Fichtel & Sachs to NSU specifications. It has a hydraulic torque converter with an electro-pneumatic clutch and a manual-shift, three-speed gearbox. A rocking switch in the gearshift knob disengages the clutch instantly when you grab it to shift. There's no need for a clutch pedal. This semi-automatic transmission was the solution to the "snatch" problem that occurred in the standard-transmission Spider and the experimental Munga. By interposing a hydraulic torque converter between the road wheels and the engine, the drive train obtained the necessary cushioning. Naturally, the engine does not give much braking effect on a closed throttle, but that is not considered significant. The absence of a clutch pedal does not mean that the Ro-80 can be driven as a car with automatic transmission, the driver still has to shift for himself. The car does not, however, possess the sporting character that a purely mechanical four- or five-speed floorshift would give.

The car will start off in second, or even third, gear, but only at a crawl. It is possible to stay in the original gear as long as desired without ever shifting. The lazy method is O.K. for leisurely driving, but for best results all the gears have to be used; the fine synchromesh gives quick and smooth shifts. But, you have to remember not to rest your hand on the gear knob. The moment you touch it, the clutch disengages.

The long, floor-mounted stick shift works the three-speed gearbox with secure and precise movements. Four speeds instead of three would have improved performance, although to some extent the torque converter makes up for the lack of low-range torque. The Saxomat is cheaper to make than a fully automatic transmission and it absorbs less power. Because a full automatic requires a large-capacity hydraulic pump and an expensive control unit to actuate the bands and clutches, which can take as much as 15 horsepower to operate even at idling speed, the Saxomat is a good compromise for small-displacement engines. NSU sees this low-cost arrangement as the transmission for future cars with small power reserves. It is already available on the Saab, Fiat 850, Simca 1000, and Volkswagen 1500.

In highway driving, the Ro-80 is incredibly quiet and deceptively fast. There is no wind noise at all. The body is so beautifully wind-cheating that it just skips through the air without causing any audible disturbance. The engine and drive train are quiet, too. There seems to be a complete absence of stress and vibration, even when driving at very high speeds. Directional stability at highway speeds is excellent, prob-

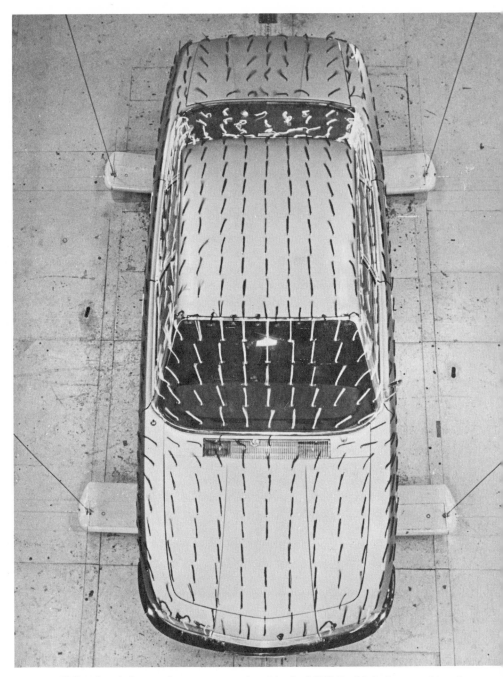

Full-scale wind tunnel tests were made with the NSU Ro-80 body, to achieve low aerodynamic drag and high stability in crosswinds. The wool tufts—several hundreds of them per car—show the direction of air flow at a multitude of points on the surface of the body. The Ro-80 has both low drag and low lift.

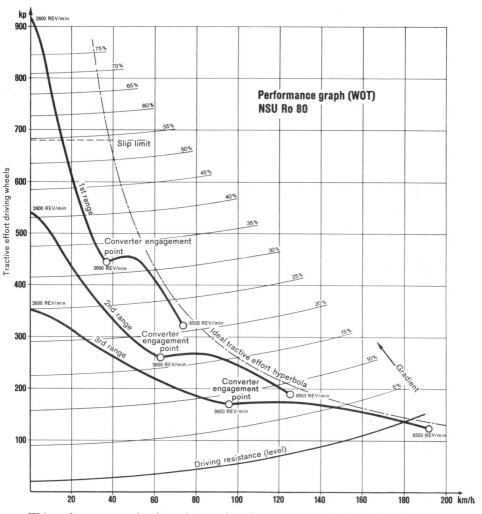

Performance graph (WOT)
NSU Ro 80

This performance graph relates the actual performance potential of the Ro-80 to the ideal tractive effort hyperbola at wide open throttle (WOT). Under 45 k.p.h. (28 mph) its potential is uncomfortably distant from the ideal curve, but at all normal driving speeds, its potential is only a few points from the ideal curve.

ably mostly as a result of its front wheel drive. All earlier NSU cars were of the rear-engine type. Was there a risk of unacceptable weight distribution with the twin-rotor engine mounted in the back? Probably not.

NSU chose to position the Wankel engine ahead of the front wheel axis, thus freeing a maximum amount of interior space for the passengers and driver. The front mounting also enabled the NSU engineers to simplify the cooling system and radiator installation. NSU also felt that the driving wheels should carry most of the weight; therefore once the

Compact on the outside, yet full-size inside, is how some critics have summed up the Ro-80. The body has extremely wide doors, yet the rear door does not cut into the fender well. Glass area is generous, and visibility is excellent in all directions.

forward location of the engine had been decided upon, front wheel drive followed as a matter of course. NSU never considered rear wheel drive for their front engined car. Doing away with the long driveshaft meant a lower floor, and a possible reduction in frontal area and aerodynamic drag. It also meant a lower center of gravity, with the consequent promise of improved ride and handling. A low center of gravity gives a good ride because carrying the load of occupants and luggage closer to the ground means smaller side movements in the vehicle when crossing asymmetrically applied bumps. It also permits a lower roll center without excessive risk of body roll on curves. Eliminating the driveshaft also cut production costs.

With a rear engine, the engine and transaxle form a unit. This unit is mounted in the tail of the chassis, and connected to the driving wheels via a transaxle and short driveshafts. The same situation exists with front wheel drive, with the extra complication that the wheels also are steered. Front wheel drive offers better traction under most conditions. The need for directional stability in NSU's new high-speed car also favored the choice of front wheel drive. Front wheel drive usually brings improved directional stability, while many rear-engine cars suffer from a

NSU Ro-80 shows high cornering ability, with moderate roll, on this narrow country lane. Superior roadholding means extra safety in everyday conditions, and can be an important difference in an emergency situation.

lack of it. The reason is that putting the drive through the front wheels produces an inherent self-centering tendency that is effective at all speeds. It also means that front wheel drive cars understeer.

The kind of understeer that comes with front wheel drive makes for excellent safety. It moderates the curve path if the driver makes a sudden steering wheel movement, or accelerates or brakes too hard for the curve indicated by the steering angle. Such features may not give the greatest maneuverability, but do give most drivers a feeling of being at ease when driving such a car. Front wheel breakaway, at the limit of adhesion, is a stable motion with a self-diminishing tendency. Getting off the throttle will bring the car back on a shorter turn radius, the car will slow down and the driver will retain control.

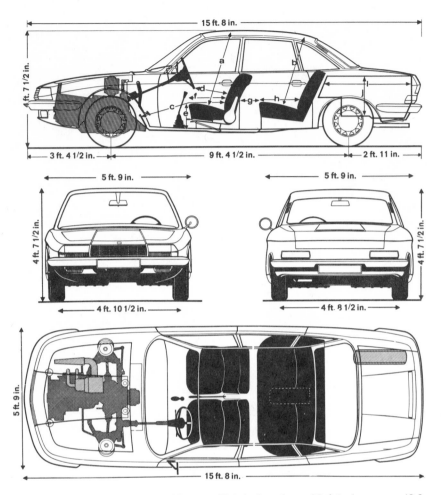

Dimensional sketches of the Ro-80. a = 38.8 inches. b = 38.6 inches. c = 43.3 inches. d = 16.0 inches. e = 11.8 inches. f = 19.3 inches. g = 11.4 inches. h = 18.1 inches. i = 47.2 inches. j = 20.5 inches.

The need for power steering with the front wheel drive Ro-80 stems from the fact that a large percentage of the car's weight is resting on the front wheels, the same feature that gives superior traction. The drive wheels on a front wheel drive car always point in the same direction the car is being steered. Inertia in the vehicle itself thus causes understeer, and a difference between steering angle and direction of travel, but traction is always applied in the direction the front wheels are pointed. Because the car is never in any doubt about the direction it is headed, its structure can be lighter, and the need for suspension reinforcements is smaller than on rear wheel drive cars.

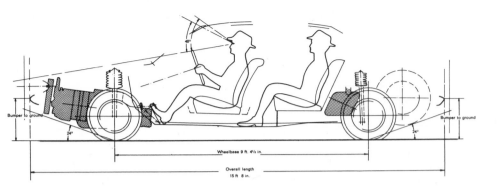

The Ro-80 was designed around a comfortably dimensioned passenger compartment. The next considerations were engine bulk, drive system, cooling requirements, weight distribution, and target price. The overall dimensions are then fixed on the basis of all other design parameters.

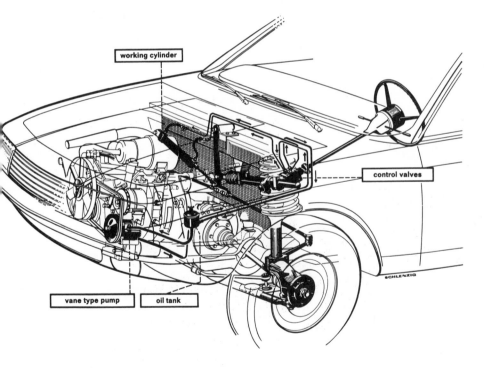

The hydraulic power cylinder for the steering is driven by a vane-type pump, which is driven by the engine. The power cylinder is not integral with the steering gear, but delivers hydraulic pressure to the steering valve at the head of the steering gear.

The Ro-80 is a well-balanced car with mild understeer. When pressed to the limit, it will not "plow" like a Toronado but drift on all four like a Lancia. It is not very particular about taking a corner with power-on or power-off. Most front wheel drive cars understeer with power-on, and oversteer with power-off. With its low center of gravity and wide track, there is little roll when taking curves at highway speeds, but on sharp, slow turns, the car leans over more than expected. With Michelin XAs tires, which give an excellent grip and unfailing side bite in all kinds of weather, the Ro-80 can be taken around curves at amazing speeds, in unusual stability, comfort and safety. The steering is power-assisted and is geared to give 3.75 turns from extreme left to extreme right. It is possibly lighter than some drivers like, but is extremely accurate. It works with unusual smoothness, without loss of precision in its response, or any loss of feedback.

The steering wheel is completely free of drive train vibrations and road shock. Return is firm and pleasant, making for relaxed cornering even under high centrifugal force conditions. The steering gear is mounted on the cowl structure at the rear of the engine compartment, far back from the front of the car. A rack-and-pinion gear, pivoted at its

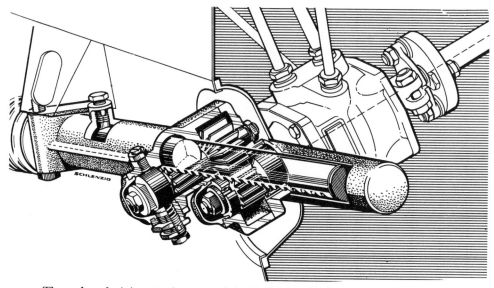

The rack and pinion steering gear of the Ro-80 is self-adjusting and offers resilient and phased application of steering force. High stresses in the steering gear occurs only when there is no hydraulic servo assist (i.e., when the engine is not running). In normal driving, it is the driver's steering effort which controls the valves that admit hydraulic pressure to the servo cylinders, which are pin-jointed to the drop arm.

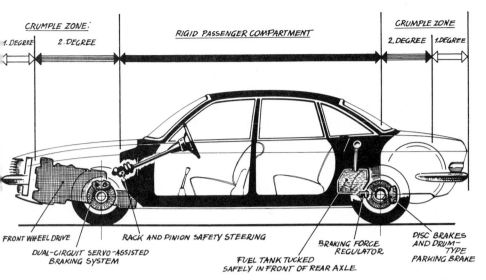

FRONT WHEEL DRIVE

DUAL-CIRCUIT SERVO-ASSISTED BRAKING SYSTEM

RACK AND PINION SAFETY STEERING

FUEL TANK TUCKED SAFELY IN FRONT OF REAR AXLE

BRAKING FORCE REGULATOR

DISC BRAKES AND DRUM-TYPE PARKING BRAKE

This sketch shows the distribution of the principal masses in the Ro-80, its general construction, and the provision of front and rear crush zones for occupant protection in front and rear impact situations. The hood and trunk absorb the impact energy, while the passenger compartment remains relatively intact.

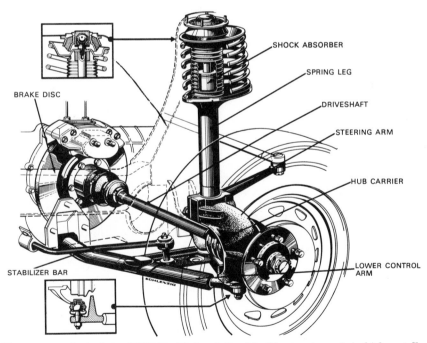

BRAKE DISC

STABILIZER BAR

SHOCK ABSORBER

SPRING LEG

DRIVESHAFT

STEERING ARM

HUB CARRIER

LOWER CONTROL ARM

Front suspension of the NSU Ro-80 is of the MacPherson type, in which a tall spring leg gives the wheel hub its upper locating point, and no upper control arm (A-frame) is needed.

outer end, is coupled to a long drop-arm. The hydraulic power cylinder is placed on the other side of the arm, which in turn is linked to the two tie-rods that operate the steering arms on each wheel.

The suspension system that helps endow the Ro-80 with such excellent road manners is quite advanced, without belonging in the "breakthrough" category. The front suspension is of the MacPherson type, which consists of a lower A-frame control arm and an almost vertical spring leg. The control arm is anchored to a pivot shaft on the chassis and supports the wheel hub. The spring leg is mounted in a socket on the spindle support arm and its top end is lodged in a spring tower abutment that forms part of the inner fender structure. Between them, the A-frame and the spring leg effectively locate the wheel. The spring is a coil wound around the upper part of the leg, which also contains a concentrically mounted shock absorber. The roll center is located about 2½ inches above ground level. For added roll stiffness, a stabilizer bar is used. This stabilizer consists of a special torsion bar, which is attached to the left and right A-frames and whose center piece is reversible. It turns elastically and resists the twisting motion when one wheel is raised and the other lowered in relation to the body—as always happens when taking a curve. When the springs of both wheels are in action, it revolves as a whole around its center piece without adding anything to the spring rates.

The rear suspension is designed on the same principles, but, because this is a front wheel drive car, the rear wheels have no duties other than to carry the load—they are just along for the ride. Rather than using a dead I-beam axle and leaf springs like a trailer (and favored by some constructors of front wheel drive cars), the NSU engineers decided on independent rear suspension. The wheels are held in place by semi-trailing arms and almost vertical spring legs. The arms are like A-frame control arms turned around so that the pivot axis runs across the car instead of lengthwise. The spring legs are almost exact duplicates of the front ones.

The suspension systems on the Ro-80 ensure precise guidance of the wheels—all four of them—when hitting bumps or potholes as well as in cornering situations, with minute changes in track, camber and caster. All four wheels are linked to the car so that they provide high lateral resistance to centrifugal force.

If handling is an important element of vehicle safety, brakes are even more important. The Ro-80 has big disc brakes on all four wheels. The front discs, which do two-thirds of the braking, are mounted inboard, flanking the transaxle, where they're exposed to a flow of cooling air. Because they are away from the wheels, they lower the unsprung

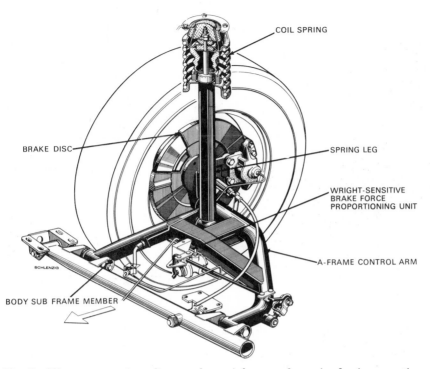

COIL SPRING

BRAKE DISC

SPRING LEG

WRIGHT-SENSITIVE
BRAKE FORCE
PROPORTIONING UNIT

A-FRAME CONTROL ARM

BODY SUB FRAME MEMBER

SCHLENZIG

The Ro-80's rear suspension relies on a lower A-frame and a spring leg incorporating a hydraulic shock absorber. The lower A-frame is linked to a weight transfer sensor, which regulates brake force distribution to the rear wheels. This reduces the risk of premature rear wheel lockup during hard braking. This suspension system keeps the rear wheels parallel to the car at all times, which means that single-wheel bumps never produce any undesirable camber changes. On the other hand, the camber angle changes with body roll, which adds a certain roll steer effect.

weight to allow better adhesion and traction. The rear discs are carried on the wheel hubs. For parking, there are cable-operated drums inside the rear discs. The brakes are power-assisted and are as sensitive as those on an American car. They sometimes tend to come on too strong for the pedal pressure, but are easy enough to modulate so that the car decelerates smoothly at an even rate and comes to a complete stop at the intended point. Brake action is always prompt, almost sudden, and effective. Brake force distribution between front and rear is flawless. There is little nose dive, and no threat of premature rear wheel locking on hard stops due to a load-sensing valve that limits pressure to the rear brakes according to weight. The car stops from 60 m.p.h. in under 150 feet, without any threat of side pull or locking. From 90 m.p.h. it stops in just over 300 feet. Such brake performance is considered well above average.

In a land like Germany, with its high-speed Autobahnen, normally unrestricted by speed limits, the ability to cruise near maximum speed is demanded of even high-powered cars. To make high-speed cruising

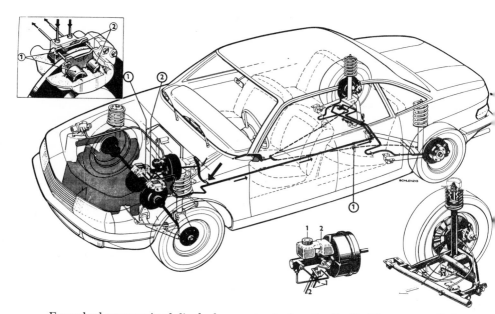

Four-wheel power-assisted disc brakes are standard on the Ro-80. The master cylinder operates dual circuits, but the system differs from the minimum prescribed by U.S. Federal law. Rather than merely separating front and rear wheel brake circuits, the Ro-80's basic circuit operates on all four wheels. The secondary circuit operates individual wheel cylinders in the front disc brake calipers. A leak anywhere in the primary system does not affect the secondary system, which gives about two-thirds of maximum braking effectiveness.

comfortable, quiet, and economical, NSU decided when the design goals for the Ro-80 were set that a "streamlined" body was going to be of extreme importance. The basic shape of the car was designed in 1961 by a graduate of the Schule Für Formgestaltung in Ulm, who is now employed by NSU. Then began four years of wind tunnel testing, in collaboration with the Stuttgart Polytechnic Institute. Wind tunnel work began with a one-fifth scale model, and full-scale tests started in August of 1963. As was the case with Daimler-Benz and the C-111, these tests were made not only for drag, but also for lift and stability in crosswinds. After a number of minor modifications, the final car had a drag coefficient of 0.355, which compares favorably with many popular shapes and approaches the best ever realized in the way of a four-door sedan:

1930	Ford Model A	.832
1960	Volkswagen	.50
1960	Chevrolet Corvair	.43
1960	Pontiac	.53
1966	Citroën DS 21	.326

The general design goals for this vehicle were the production of a car with the same seating capacity, luxury and performance as the most popular Mercedes-Benz and BMW sedans, while taking the fullest possible advantage of the lightness and compactness of the Wankel engine. NSU saw the possibility of making a lighter and lower vehicle, with a better power-to-weight ratio, than was available from the competition. NSU had no experience with cars of this caliber, and they were strangers to the high-priced market, too. But there was one thing more—NSU had no existing parts that they might be tempted to include in the new design for reasons of production economy and, as a result, they were able to start with a truly clean sheet. The entire Ro-80 package was designed around the Wankel engine, in contrast with the earlier NSU Wankel Spider. The Spider was basically a convertible-

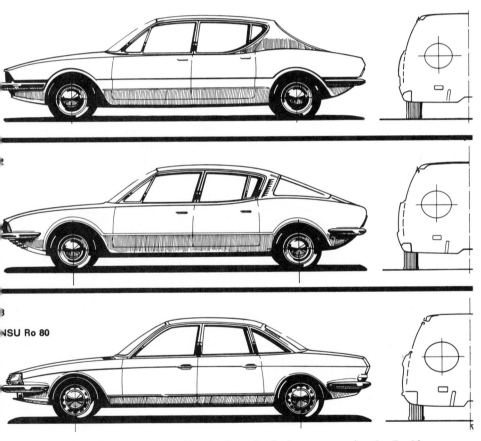

These three designs were considered when the final prototypes for the Ro-80 were to be built. The first and second designs were rejected for lack of practicality. Number three was the choice.

bodied Sport Prinz chassis, with a single-rotor Wankel engine merely taking the place of the stock Prinz piston engine.

The engine used in the Ro-80 is the KKM-612. It weighs 280 pounds and develops 136 horsepower at 5,500 r.p.m. The unit is only 18 inches long, 16 inches high and 17 inches wide. It was small and light enough to be installed in any conveniently available space, and for reasons we have already discussed, NSU decided to place the power unit in the nose of the car. Because of the small bulk of the KKM-612, the hood is short and has a low profile. Because of its low weight, positioning the engine overhanging the front wheel centerline posed no problems in terms of weight distribution. It raises the polar moment of inertia, which in turn strengthens the inherent understeering characteristic of the vehicle. Understeer remains, however, well within acceptable limits.

Now that the Ro-80 has been thoroughly tested, NSU has decided to increase daily production in 1971 to some 75 units per day.

With the Ro-80 in the background, the K-70 shows a clear family relationship. Both have front wheel drive. The prototype K-70 had a four-cylinder water-cooled engine, but the engine compartment had room for the KKM-612.

SPECIFICATIONS

MAKE	NSU
MODEL	Ro-80
Year introduced	1967
Year discontinued	—
Price	$3,732 in the home market
Type of body	4-door 4/5 seater sedan
Type of construction	Unit body
Driving wheels	Front
Power unit position	Front
Curb weight	3,035 pounds
Weight distribution front/rear	63/37%
Power/weight ratio	22.3 pounds per horsepower
Fuel tank capacity	21.9 gallons
Fuel tank position	Rear
Power unit	Wankel
Number of rotors	2
Chamber displacement	30.3 cu. in. (497.5 cc.)
Equivalent total displacement	121.2 cu. in. (1.9861 liters)
Compression ratio	9.0:1
Power output	136 horsepower
at r.p.m.	5,500
Torque	117.2 foot pounds
at r.p.m.	4,500
Carburetion system	One 2-stage Solex 18/32 HHD
Ignition system	—
Cooling system	Water and oil
Clutch	Single dry plate
Transmission	Torque converter with manual three-speed gearbox
Gear ratios 1	2.056:1
2	1.208:1
3	0.788:1
4	—
5	—
R	2.105:1
Final drive ratio	4.857:1
Front suspension	MacPherson spring legs and A-frame lower control arms with stabilizer bar
Front springs	Coil spring and telescopic shock absorber
Rear suspension	Semi-trailing A-frame control arms and spring legs
Rear springs	Coil spring and telescopic shock absorber
Steering system	Power-assisted rack and pinion
Turning diameter	38 feet 8½ inches

SPECIFICATIONS

MAKE	NSU
MODEL	Ro-80
Overall steering ratio	18.3:1
Turns, lock to lock	3.8
Brake system	ATE-Dunlop 4-wheel disc brakes
Disc diameter F	11.18 inches
Disc diameter R	10.71 inches
Drum diameter F	—
Drum diameter R	6.5 inches (parking brake only)
Lining area	43.07 square inches
Swept area	—
Parking brake	Mechanical duo-servo on rear wheels
Tires	Michelin XAs
Tire size	175 HR 14
Wheelbase	112.6 inches
Front track	58.27 inches
Rear track	56.46 inches
Overall length	188.19 inches
Overall width	69.29 inches
Overall height	55.51 inches

Test Results

Acceleration times	
0–30 m.p.h.	4.3 sec.
0–40 m.p.h.	6.0 sec.
0–50 m.p.h.	8.8 sec.
0–60 m.p.h.	12.8 sec.
0–70 m.p.h.	16.0 sec.
0–80 m.p.h.	22.4 sec.
0–90 m.p.h.	32.3 sec.
0–100 m.p.h.	52.0 sec.
Top speed 1	46.6 m.p.h.
2	80.2 m.p.h.
3	111.8 m.p.h.
4	—
5	—
Average fuel consumption	21 m.p.g.

19

The
Mazda Cosmo

BY THE END OF 1962, the design and basic layout of the Cosmo Sport 110-S were determined, and early in the following year, detailed design work commenced full scale. In October of 1963, Toyo Kogyo displayed two types of rotary engines—a 24 cubic inch (400 cc.), single-rotor engine and a 48 cubic inch (800 cc.) twin-rotor engine. In 1964, a twin-rotor engine with a single chamber displacement of 400 cc. and a four-rotor engine of the same chamber size were displayed at the 11th Tokyo Motor Show.

The production Cosmo engine is a water-cooled tandem twin-rotor unit, not unlike the NSU KKM-612 in basic configuration, having 491 cc. chamber displacement. Here are its vital statistics:

Radius	105 mm.
Eccentricity (e)	15 mm.
Rotor width	60 mm.
R/e ratio	7:1
R + e	120 mm.

The Cosmo engine puts out 110 horsepower at 7,000 r.p.m. and weighs only 225 pounds. It is also small enough to fit comfortably in existing Mazda cars—length is 20.2 inches, width 23.4 inches and height 21.4 inches. The engine housing is built up as a sandwich structure composed of two rotor housings and three side housings, which are clamped together in one body by 19 tension bolts. The rotor housings are made of molded aluminum alloy castings for higher radiation efficiency. Their strength at high temperature has been improved by heat

The Mazda Cosmo 110-S was first shown to the public at the Tokyo Auto Show in 1966.

treatment. Hard chromium plating, which has a high resistance to wear and possesses outstanding friction characteristics, is applied on the trochoidal surface against which the apex seals slide, and the surface is finished by profile grinding.

Both front and rear side housings and the intermediate housing are made of a special high rigidity cast-iron. The castings are made by the shell-molding process, which has high productivity and produces uniform thin-wall castings. The sidewalls are induction-hardened in a radial stripe pattern. By this method, a superb resistance to wear is obtained and at the same time the distortion caused by the hardening process can be held to a minimum.

The two rotors are placed in line on the eccentric shaft with a displacement phase 180 degrees apart. The rotors are made of ductile cast-iron for durability and their interiors are partitioned into several chambers by ribs so that they can be efficiently cooled by lubricating oil. The carbon steel rotor gear that controls the trochoidal movement of the rotor is compactly fitted on the rotor with six double spring pins. These pins also serve as buffers against oscillating loads acting on the rotor gear.

By late 1963, the first layouts for the Cosmo sports car had been completed, though they differed in some details from the final design. The fuel tank was shown under the trunk floor instead of above the rear axles, where it later was located, and drum brakes originally were specified for the front wheels. Using that first twin-rotor engine, the prototype Cosmo was complete and running by July, 1964. It then weighed 1,980 pounds; the engine weighed 198 pounds without the transmission. Field tests on the Mazda 110-S vehicle were performed in parallel with engine bench tests. Early in 1965, 60,000 test miles had been accumulated.

During the next 12 months, Toyo Kogyo switched from peripheral ports to side intake ports to improve the engine's torque and low-speed performance. The revised engine idled at 500 r.p.m. and pulled smoothly in any gear from 800 r.p.m. Peak torque occurred between 3,000 and 4,000 r.p.m.; however, peak power and overall performance were sacrificed.

As a countermeasure, a new engine was designed in 1965. This engine had 120 cubic inches displacement instead of the original 98, and was

Powerplant for the Cosmo, complete with gearbox and all accessories.

equipped with dual ignition. The larger engine also was given side intake ports, two per "cylinder," the inner ports being fed by the primary throat of a progressive carburetor and the outer ports being supplied by the two larger secondaries.

With this power unit, Toyo Kogyo produced a small series of some

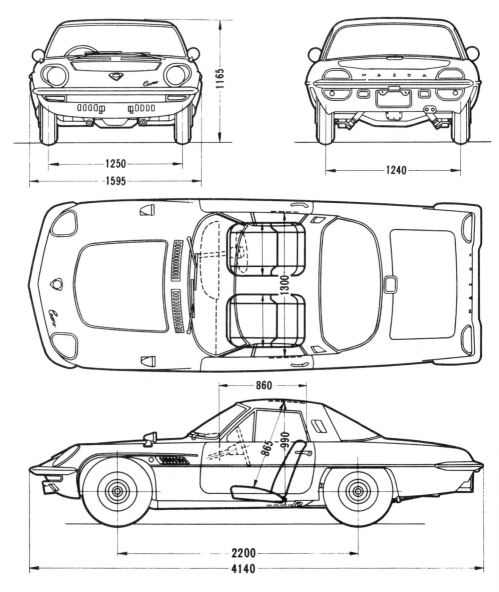

Dimensional sketches for the 110-S Cosmo.

80 Cosmos. Of these, 60 were loaned to the firm's key suppliers and to their 79 factory-owned main dealers for field evaluation, not unlike that undertaken by Chrysler with its turbine cars. It was a six-month test program, initiated in April, 1966. At the end of its self-imposed field test program, Toyo Kogyo knew it had a reliable engine with seals good for 80,000 miles. Its users reported good acceleration and fuel mileage, returning 24 m.p.g. at a steady 70 m.p.h., but the car's high-speed performance, with a maximum of only 98 m.p.h., was below par for a two-liter vehicle. At this point, the decision was made to keep the inner side intake ports for low-speed operation, and to add peripheral intake ports to fill out the high-speed part of the curve. The change, effected by late 1966, extended the power range from 6,000 to more than 7,000 r.p.m. Early in 1967, final design changes were made to improve distributor drives and general accessibility.

With the production engine, the leading plug fires 2 degrees after top dead center and the trailing one at 7 degrees after top dead center. Due to higher temperature in the combustion chamber (combustion is always occurring at one side of the rotor casing), high temperature plugs are specified (NGK B-8EPD for normal use and B-9EPD for high-speed runs).

With correctly metered breathing and efficient flame propagation, the twin-rotor engine has attained a remarkable degree of flexibility through a wide r.p.m. range. The engine is slightly quieter than piston engines with similar power and high r.p.m. Mostly it offers a different sound, particularly at high speed. Instead of the clattering roar of a Honda, for example, the Cosmo gives off a humming growl at 8,000 r.p.m. At idle, and up to 2,000 r.p.m., this Wankel engine is distinctly quieter than a piston engine. Acceleration, particularly in the lower

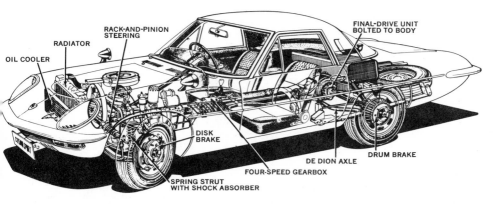

Cutaway drawing of the Mazda Cosmo.

speeds, is sensational. A four-speed manual transmission takes it from 0–60 m.p.h. in 8.9 seconds and from 25–70 m.p.h. in 9 seconds flat. Up-shifting is recommended at 5,000 r.p.m. and the engine seems to grab hold of each gear as if it were waiting for it. A torque curve that is re-markably flat on top provides quick response to all demands. From 65 foot pounds at 1,000 r.p.m., torque climbs to 80 at 1,500 and 85 at 2,000. At 3,000 r.p.m., 92 foot pounds is produced, 98 at 4,000 and 99 at 5,000. At 6,000 r.p.m., torque falls off slightly to 96 foot pounds.

The Cosmo is capable of loafing along at 1,000 r.p.m. in top gear, without lugging, at a steady 12 m.p.h. When the accelerator is floored, it moves out quickly and without complaint. The rotors also reduce vibration to the point where it is only barely perceptible. This is as true at idle, just over 800 r.p.m., as it is at 8,000 r.p.m.

The cooling system is sealed, and has an expansion tank which col-lects coolant that would otherwise be lost through the overflow pipe. This system also keeps the radiator permanently filled with liquid in-stead of air, so that coolant foaming is prevented and cooling efficiency is kept high at all times.

The car handles with assurance. The steering ratio is 17.3:1, giving three quick turns of the wheel from full right turn to full left turn (lock-to-lock). Very firm suspension and a low profile give the car good cornering characteristics. It is stable, even in sharp turns on rough roads. Front suspension is a double wishbone arrangement, with un-equal length fabricated upper and pressed lower A-arms, coil spring/

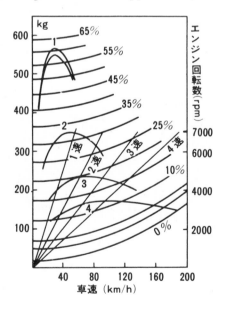

Tractive effort and hill-climbing ability in all gears, for the Cosmo 110-S.

damper units and stabilizer bar. Front spring rate (i.e., wheel rate in bounce) is 112 pounds per inch; bump travel 2.74 inches and static wheel deflection is 3.55 inches. The front end of the car has a ratio of sprung-to-unsprung weight of 5.3:1, which is excellent. The stabilizer bar is strong—168 pounds per inch. Total roll stiffness is 494 foot pounds per degree; enough to keep roll angles down even at extreme lateral accelerations.

The Cosmo cockpit.

The Cosmo front end has some built-in roll understeer, amounting to about 0.55%. Static camber on the front wheels is 1 degree positive and the roll center is located 2.5 inches above ground level. Swivel axis inclination is 8 degrees and the caster angle 3°30′. On the rear, the Cosmo has a de Dion axle, with semi-elliptic leaf springs and trailing radius arms. The de Dion tube which connects the two rear wheel hubs is solid, and the halfshafts incorporate ball sliding splines. The driveshafts are open, and each has two universal joints. The shafts, in other words, are relieved of all locating duties and have only to transmit the drive to the wheels. The de Dion tube is free to move up and down with the wheels, but is anchored to the chassis by radius rods which keep it in proper fore-and-aft location under all conditions of load and thrust. The final drive unit is bolted to the unit-construction body. The de Dion axle is not clamped directly to the laminated springs, but is attached to them via pivots. The de Dion tube runs between the frame and the springs. Radius rods on each side help take up the driving thrust and brake torque reaction forces; the leading portions of the leaf springs take the rest.

Rear roll center is about 14 inches above ground level and total roll stiffness is 142 foot pounds per degree. Static wheel camber is zero, wheel rate in jounce is 140 pounds per inch, static wheel deflection is 3.55 inches, and wheel travel in jounce is 2.56 inches. Rear sprung-to-unsprung weight ratio is 6.9:1, even with the de Dion tube. Braking, with discs up front and drums in the rear, is straight and sure. The original drum brakes were replaced by discs on the front wheels in 1966, prior to the beginning of production. The rear drums have one leading and one trailing shoe each. There is no power brake option.

On May 30, 1967, the Mazda Cosmo 110-S, with the world's first production model twin-rotor Wankel engine, was placed on sale. This car was built mainly for the purpose of determining the market acceptance of the rotary engine; therefore it was produced only on a limited scale. It was followed on July 13, 1968, by the introduction of the R-100 Coupe on the domestic market. That date marks the arrival of the first volume-production, Wankel-powered car.

SPECIFICATIONS

MAKE	MAZDA
MODEL	COSMO SPORT 110-S
Year introduced	1967
Year discontinued	1969
Price	$5,812 in the home market
Type of body	Two-seater coupe
Type of construction	Unit-construction body
Driving wheels	Rear
Power unit position	Front
Curb weight	2,223 pounds
Weight distribution front/rear	47.9/52.1%
Power/weight ratio	20.2 pounds per horsepower
Fuel tank capacity	15 gallons
Fuel tank position	Rear end of chassis
Power unit	Wankel
Number of rotors	2
Chamber displacement	29.96 cubic inches (491 cc.)
Equivalent total displacement	119.7 cubic inches (1,994 cc.)
Compression ratio	9.4:1
Power output	110 (DIN)
at r.p.m.	7,000
Torque	102 foot pounds
at r.p.m.	3,500
Carburetion system	One Hitachi KCA-306-1 triple-throat carb.
Ignition system	Coil and battery
Cooling system	Water and oil
Clutch	Single dry plate
Transmission	Four-speed all-synchromesh
Gear ratios 1	3.379:1
2	2.077:1
3	1.390:1
4	1.0:1
5	—
R	3.389:1
Final drive ratio	4.111:1
Front suspension	Coil springs and telescopic shock absorbers with stabilizer bar
Rear suspension	De Dion tube and radius arms
Rear springs	Semi-elliptic leaf springs and telescopic shock absorbers
Steering system	Rack and pinion
Turning diameter	32.2 feet
Overall steering ratio	17.3:1

SPECIFICATIONS

MAKE	MAZDA
MODEL	COSMO SPORT 110-S
Turns, lock to lock	3
Brake system	Disc brakes front—drums rear
Disc diameter F	—
Disc diameter R	—
Drum diameter F	—
Drum diameter R	—
Lining area	52.6 square inches
Swept area	—
Parking brake	—
Tires	—
Tire size	165 HR 14
Wheelbase	86.6 inches
Front track	49.2 inches
Rear track	48.8 inches
Overall length	163 inches
Overall width	62.8 inches
Overall height	45.85 inches

Test Results

Acceleration times	
0–30 m.p.h.	3.1 sec.
0–40 m.p.h.	4.3 sec.
0–50 m.p.h.	5.9 sec.
0–60 m.p.h.	8.9 sec.
0–70 m.p.h.	11.0 sec.
0–80 m.p.h.	14.0 sec.
0–90 m.p.h.	17.8 sec.
0–100 m.p.h.	22.0 sec.
Top speed 1	37 m.p.h.
2	61 m.p.h.
3	91 m.p.h.
4	115 m.p.h.
5	—
Average fuel consumption	—

20

The Mazda
R-100

A DEMONSTRATION of the high-speed reliability of the Wankel engine was given in July of 1969, when the Japanese firm of Toyo Kogyo entered three Mazda R-100 coupes in a 24-hour endurance race at the Francorchamps road course in Belgium. The Wankel-powered cars were matched against the fastest production touring cars in the world and finished an impressive fifth and sixth overall after one car crashed during practice. The success of the Mazda R-100 in this race was but a small indication of the basic soundness of the standard production model.

Unlike the C-111 and the Ro-80, the Mazda R-100 is not a car created around the Wankel engine, but rather the installation of the Wankel in an existing car. The R-100 body and chassis are borrowed from the 1200 Coupe, normally powered by a 71.33 cubic inch four-cylinder sohc piston engine developing 75 horsepower at 6,000 r.p.m. and having a top speed of 93.2 m.p.h. The twin-rotor Wankel engine fits easily inside the same package and gives far superior performance. The Wankel is rated at 110 horsepower at 7,000 r.p.m. and gives a top speed of 110 m.p.h. The chassis layout is conventional, with a front engine and transmission and a one-piece open driveshaft to the I-beam rear axle. The only modification made to the 1200 to prepare it for the additional power of the Wankel engine was to install a completely new brake system, including front discs. The body structure, suspension members, and drive train components were considered to be adequate and were not changed.

The Wankel engine is an outstandingly smooth and quiet engine in the Mazda, having a flexibility that is found in very few drive train com-

The Mazda R-100 stays in front of two Porsche coupes on a curve during the endurance race.

binations. The engine will pull smoothly from 10 m.p.h. in fourth gear (about 600 r.p.m.) all the way to the 7,000 r.p.m. limit. At 7,000 r.p.m., an electrical device shuts the secondary chokes of the twin tandem-choke carburetor, and immediately reduces the engine speed to the factory-prescribed limits. The Wankel also produces a surprising braking effect when the throttle is closed. Closing the throttle closes both main throats in the carburetor and the engine immediately tends to revert to idling speed. This is not a new innovation—the same thing is accomplished by the same procedure with a piston-type engine—but the Wankel will slow the car to such an extent that often the brakes need not be used except for a complete stop. A "long" final drive ratio gives restful gearing, which allows the car to reach 64 m.p.h. in second gear and 85 m.p.h. in third. At 5,250 r.p.m. in top gear, the speedometer reads 100 m.p.h. and the engine is still smooth and totally unstrained.

The steering lacks precision, but is unusually light. The linkage is

In racing trim, the Mazda handled outstandingly well. On this hard turn, there is hardly any body roll, and the inside front tire is not lifted from the ground.

In January, 1970 Toyo Kogyo entered a team of Mazda R-100 coupes in the Monte Carlo Rally. Here one is shown on an Alpine road.

completely free of kickback caused by ruts, holes and bumps in the road. The steering gear is a low-friction recirculating ball design, having a variable ratio that causes a certain vagueness near the straight-ahead position. On very slight curves, it has a strangely "dead" feeling.

The chassis may be more than adequate for the 1200 piston engine but, no matter what the manufacturer says, it can hardly cope with the additional speed capacity and torque of the Wankel engine. Its short-comings exist mainly in the rear suspension system. The handling characteristics are not exactly poor, but they detract from the dis-tinctive excellence of the power unit. On smooth surfaces the car handles very well, with slight understeer, and it can be taken around curves at surprisingly high speeds. Body roll is slight, and the ride is pleasing without being soft. On rough roads, however, it soon becomes clear that the shock absorbers allow the car to float badly on wavy surfaces. They are too weak for their task. But the really weak point is the rear suspension. The axle bounces at the slightest provocation, resulting in lost traction, and directional corrections become necessary. On old country roads, driver fatigue sets in earlier than in other light cars of comparable power and speed.

In certain cornering situations, a change in vehicle attitude takes

The Mazda R-100 coupe is a two-door four-seat sedan of subcompact dimensions, but surprising performance. Based on a production model using a four-cylinder piston engine, the R-100 design has not taken advantage of the small size of the Wankel engine to make gains in useable space.

The R-100 cockpit has a sporty appearance. Instrumentation is ample, and the controls are convenient and functional.

place. This probably can be traced to roll oversteer. The basic characteristic is understeer, which can be countered by power induced oversteer due to the ease with which the inside rear wheel will lift and spin. The Bridgestone "Superspeed Radial" tires made under Goodyear license provide less adhesion than good European radials. They squeal even on relatively easy turns and break away early.

The axle is carried by two almost straight leaf springs placed very close to the wheel hubs so as to give maximum stability. There are no other links or arms locating the axle in relation to the body. This axle suspension system is called Hotchkiss drive, and is the least expensive way known to hang a back axle on a car. The hydraulic, telescopic shock absorbers are positioned "sea-leg" fashion, with axle mountings about halfway between the differential and the hubs. The body mountings are

The R-100 utilizes MacPherson-type front suspension, with an almost vertical spring leg and a transverse lower control arm. The stabilizer bar is not shown in this sectional drawing.

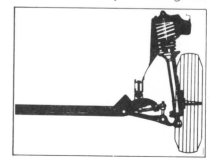

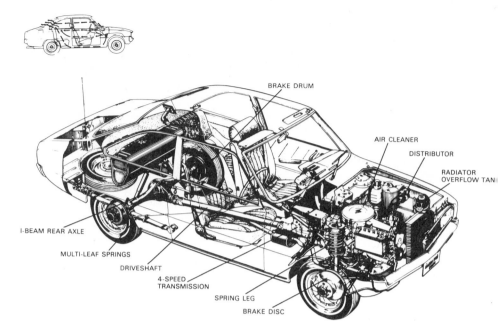

BRAKE DRUM

AIR CLEANER

DISTRIBUTOR

RADIATOR
OVERFLOW TAN

I-BEAM REAR AXLE

MULTI-LEAF SPRINGS

DRIVESHAFT

4-SPEED
TRANSMISSION

SPRING LEG

BRAKE DISC

The Wankel engine is practically invisible beneath the air cleaner, and partly hidden by the battery. The ignition distributor sticks up on the left of the engine. Chassis engineering is simple and straightforward, the overall concept being similar to that of the Ford Escort.

much closer together. This has no effect whatsoever in reducing body roll, but helps somewhat in restricting lateral movement of the axle in relation to the body on bad roads.

The problem with Hotchkiss drive is that in addition to the normal springing and load-carrying functions, the springs are called upon to carry out the axle locating duties. They have to take the full driving thrust and whatever braking loads the rear wheels can produce. Using springs as suspension links also has the drawback that they tend to flex, not only in the intended plane but also in other planes. Above all, the springs give the axle the freedom it needs to move up and down; one wheel up and the other down, or both up and down together. They permit some fore-and-aft motion in the axle—which causes a rear wheel steering tendency when one wheel is twisted forward and the other backward. The leaf springs cannot positively prevent sideways movement of the axle when subjected to side force, and they do not eliminate axle rotation around the driveshaft axis. The shortcomings of springs when used as control arms are not so obvious in low-powered cars, but in high-powered cars the springs must be stiffened or supplemented by traction bars.

Why does the axle want to make all these undesirable movements? They are mainly caused by torque reactions, which are the result of putting drive torque into the axle. This goes for axles just as for all other bodies when any force exerted in any direction is met by an equal and opposite force. As the driveshaft turns the pinion in the axle, the whole axle housing tends to revolve in the direction opposite to the pinion's rotation. The ring gear in turn drives the ring, which sets up another reaction which tends to force the axle housing to revolve in the direction of the ring gear's rotation. This torque reaction to the ring gear's rotation causes a lift in the forward end of the pinion shaft on acceleration and bends the springs into an S-shape. This is called *spring wind-up*, and usually affects the left spring more than the right. Uneven spring wind-up twists the axle diagonally in the chassis and causes rear wheel steering. The torque reaction to the pinion's rotation tends to lift the right wheel on acceleration while pressing the left one harder to the road surface. This causes *axle tramp*, which usually manifests itself as a periodic jumping up and down of the right wheel and leads to loss of traction and wheelspin. To regain traction, the driver has to let up on the accelerator. Because of this same torque reaction, the left wheel has extra traction until the moment the right wheel starts to spin, and this is what makes the left spring more susceptible to wind-up than the right one. Wind-up ends, of course, when wheelspin sets in.

The front suspension is MacPherson type, with near-vertical spring legs and a lower A-frame control arm. This is identical in principle to the front suspension design of the NSU Ro-80, minus the complication of the front wheel drive components. The brake pedal is very hard, but it is easy to modulate brake force once the high pedal effort is overcome. The front discs are strongly fade-resistant and perfectly stable in operation. The front to rear balance in the brake system is remarkably well chosen, and the brake force seems to be distributed just right whether the car is driven with one occupant or four, on a wet or dry road.

The R-100 obviously was not put into production for a long-term run. It must be regarded as an interim model, representing a hybrid solution relying partly on obsolete chassis engineering and partly on innovative and thoroughly refined power unit design. The Mazda range now includes two newer, and far more advanced, Wankel-powered cars that have not yet become available for road testing. Both are described in detail in the next chapter.

SPECIFICATIONS

MAKE	MAZDA
MODEL	R-100
Year introduced	1969
Year discontinued	—
Price	$3,045 in the home market
Type of body	Two-door 4-passenger sedan
Type of construction	Unit-construction body
Driving wheels	Rear
Power unit position	Front
Curb weight	1,775 pounds
Weight distribution front/rear	56.5/43.5%
Power/weight ratio	16.15 pounds per horsepower
Fuel tank capacity	16.2 gallons
Fuel tank position	Rear end of chassis
Power unit	Wankel
Number of rotors	2
Chamber displacement	29.96 cubic inches (491 cc.)
Equivalent total displacement	119.7 cubic inches (1,994 cc.)
Compression ratio	9.4:1
Power output	110 horsepower
at r.p.m.	7,000
Torque	100 foot pounds
at r.p.m.	4,000
Carburetion system	One Hitachi-Stromberg 4-barrel KCB 306
Ignition system	Coil and battery (dual)
Cooling system	Water and oil
Clutch	Single dry plate
Transmission	Four-speed all-synchromesh
Gear ratios 1	3.737:1
2	2.202:1
3	1.435:1
4	1.0:1
5	—
R	4.024:1
Final drive ratio	3.70:1
Front suspension	MacPherson spring leg and lower A-frame control arm with stabilizer bar
Front springs	Coil springs and telescopic shock absorbers
Rear suspension	I-beam axle and Hotchkiss drive
Rear springs	Semi-elliptic leaf springs and telescopic shock absorbers
Steering system	Ball and nut; recirculating ball
Turning diameter	30.0 feet
Overall steering ratio	(variable)

SPECIFICATIONS

MAKE	MAZDA
MODEL	R-100
Turns, lock to lock	3.7
Brake system	Disc brakes front; drums rear
Disc diameter F	—
Disc diameter R	—
Drum diameter F	—
Drum diameter R	—
Lining area	57.05 square inches
Swept area	—
Parking brake	—
Tires	Bridgestone Superspeed Radial
Tire size	145 SR 14
Wheelbase	88.98 inches
Front track	47.24 inches
Rear track	46.85 inches
Overall length	150.79 inches
Overall width	58.27 inches
Overall height	52.95 inches

Test Results

Acceleration times	
0–30 m.p.h.	3.6 sec.
0–40 m.p.h.	5.7 sec.
0–50 m.p.h.	7.8 sec.
0–60 m.p.h.	10.7 sec.
0–70 m.p.h.	15.1 sec.
0–80 m.p.h.	19.6 sec.
0–90 m.p.h.	28.2 sec.
0–100 m.p.h.	44.0 sec.
Top speed 1	34 m.p.h.
2	64 m.p.h.
3	85 m.p.h.
4	110 m.p.h.
5	—
Average fuel consumption	20.2 m.p.g.

21

The New Mazda
Rotary Cars

PERHAPS IT WAS THE INSPIRATION of the Ro-80, or perhaps it was a desire to obtain some first-hand experience with front wheel drive. Whatever the reasons, Toyo Kogyo's engineering staff went to work combining an enlarged version of the 0820 engine with a front wheel drive sports coupe. The result, the RX 87, was shown to the public at the 1968 Tokyo Auto Show. In October of 1969, a production version was announced under the name R-130. Production is limited, however, and the R-130 is sold only on the Japanese home market.

On May 13, 1970, the Capella Rotary series and the Capella 1600 series were introduced on the domestic market. The Capella series is a new line of passenger cars, available with either a sohc piston engine or a Wankel engine developed from the 0820. The piston-powered Capella replaces the Mazda Luce, and will be marketed as the Mazda 616; the rotary engine version is designated RX-2.

The RX-2 fills the gap between the Mazda R-100 and R-130, and is powered by a high-performance twin-rotor Wankel engine (573 cc. x 2) developed especially for this vehicle. The R612 engine is basically the same as the R-100 engine (491 cc. x 2), but its rotor width has been increased to raise engine volume and its exhaust port shape has been changed to minimize exhaust noise. By modifying the exhaust port, the inherent quietness of the rotary engine has been further improved. Like the R-100 engine, this engine is highly flexible over its wide speed range and runs practically without vibration. Intermediate-range and high-speed performance have been further improved, adding to the performance of the RX-2 as compared with the R-100.

In announcing the Capella series, a Toyo Kogyo official said: "The RX-2 is an ideal car for the Seventies, combining high performance, luxury, safety and economy. We introduce it on the market in the belief that the rotary engined cars will take over the main current of vehicles in 1970."

Demonstrating the performance of the Wankel engine through participation in racing continues, but Toyo Kogyo has not shown any inclination towards building racing prototypes. They have found a simpler method, by which they merely develop a special 200 horsepower racing version of the 0823 engine for installation in a ready-made racing prototype. The car chosen by Toyo Kogyo is British-built, carrying the little-known name of Chevron. The Chevron car is the creation of Derek Bennett, who built himself a dirt track midget car at the age of 18. He raced it for two years all over England and won 16 events, often setting new lap records.

The first Chevron was a Ford-powered 1,172 cc. sports car. Bennett stopped racing the Chevron in 1962 and bought a Lotus Elite; then, in 1964, he switched to a Formula 3 Brabham. The following year he built another car in his Bolton, Lancashire workshop, a club-formula sports car with the Chevron name. Then orders began to come in, and he started production. The next step was a Group 6 GT car. For 1967, Chevron concentrated on the two-liter class and used BMW engines. Later that year, Bennett decided to make a series of Formula 3 cars, followed by a Formula 2 model. Late in 1968, he began to build

The Mazda RX-2 coupe.

The Mazda RX-2 cockpit.

The Mazda-Chevron at Le Mans, 1970.

Mazda engine installation in the Chevron car.

The Mazda-Chevron at speed (Spa, 1,000 km.).

The Mazda R-130 coupe.

During 1969, the Familia sedan was made available with the 0820 engine and marketed as the Familia Rotary SS.

Formula B cars with 1,600 cc. twin-cam engines for export to America.

During 1969 and 1970, the B8 Chevron proved almost invincible in the two-liter class of international sports car racing. For the Mazda engine, Bennett designed a new model, the B-16. The B-16 is a sports-racing prototype with the engine placed midships in a tubular steel chassis. The whole car weighs only 1,298 pounds. The Chevron-Mazda campaigned in Europe during the 1970 season. It finished second in its class in the 1,000-kilometer endurance race at the Nürburgring in Germany in May, but retired in the opening stages of the 24-hour race at Le Mans due to unspecified engine trouble.

Strangely, the 200 horsepower Wankel engine is one of the noisiest power units on the scene. It has proved to have an impressive speed capability, but it may be overstressed, a point which is supported by its fragility. As an engineering research exercise, the racing program may bring valuable lessons, but results in terms of publicity or sales promotion are elusive.

If Toyo Kogyo decides to get serious about racing, the company will have to field a team. They are equipped to run every race on computers and make every conceivable test in the laboratory before their future racing cars even have to come face to face with competition. But at this writing, no firm decision has been made.

The RX-85 prototype was the final pre-production version of the R-100.

SPECIFICATIONS

MAKE	MAZDA
MODEL	RX-2
Year introduced	1970
Year discontinued	—
Price	—
Type of body	Two-door coupe
Type of construction	Unit body with front sub-frame
Driving wheels	Rear
Power unit position	Front
Curb weight	2,116 pounds
Weight distribution front/rear	52.4/47.6%
Power/weight ratio	17.6 pounds per horsepower
Fuel tank capacity	18 gallons
Fuel tank position	Rear
Power unit	Wankel
Number of rotors	2
Chamber displacement	34.9 cubic inches (573 cc.)
Equivalent total displacement	139.6 cubic inches (2,292 cc.)
Compression ratio	9.4:1
Power output	120 horsepower
at r.p.m.	6,500
Torque	108 foot pounds
at r.p.m.	3,500
Carburetion system	Stromberg 4-barrel
Ignition system	Coil and battery
Cooling system	Water and oil
Clutch	Single dry plate, diaphragm-type
Transmission	Four-speed
Gear ratios 1	3.683:1
2	2.263:1
3	1.397:1
4	1.00:1
5	—
R	3.692:1
Final drive ratio	3.70:1
Front suspension	MacPherson spring leg and lower control arm
Front springs	Coil springs with concentric hydraulic shock absorbers
Rear suspension	I-beam axle and four-link location system
Rear springs	Coil springs and hydraulic shock absorbers
Steering system	Ball and nut
Turning diameter	31 feet

SPECIFICATIONS

MAKE	MAZDA
MODEL	RX-2
Overall steering ratio	17–19:1 variable
Turns, lock to lock	—
Brake system	Disc F, drums R
Disc diameter F	—
Disc diameter R	—
Drum diameter F	—
Drum diameter R	—
Lining area	—
Swept area	—
Parking brake	—
Tires	—
Tire size	155 SR 13
Wheelbase	97 inches
Front track	51 inches
Rear track	50 inches
Overall length	163 inches
Overall width	62 inches
Overall height	45 inches

Test Results

Acceleration times	
0–30 m.p.h.	—
0–40 m.p.h.	—
0–50 m.p.h.	—
0–60 m.p.h.	—
0–70 m.p.h.	—
0–80 m.p.h.	—
0–90 m.p.h.	—
0–100 m.p.h.	—
Top speed 1	—
2	—
3	—
4	—
5	—
Average fuel consumption	—

22

The Citroën M-35

IN JANUARY OF 1970, Citroën announced a program to build 500 Wankel-powered cars, with the model designation M-35. The M-35 is not a high-speed car. It's not a family car either, as it has only two doors and back seat room is too cramped for two passengers on a long trip. But it's a great little economy car for commercial travelers or young married people. It's not a sports car, although it is not entirely devoid of some sportiness in its character.

Over the past several years, Citroën, acting within the framework of Societe Comotor, has studied the rotating piston engine in collaboration with NSU. When an advanced prototype was finally built, Citroën decided to invite the public to cooperate with its engineering office in a large-scale field test. They offered a new vehicle to enthusiasts of the new technology. The M-35 body was designed strictly as a functional structure for this research vehicle. It does not indicate the shape of a new production model or the possible evolutionary direction of other models currently in production; it is only a prototype and production will be limited to 500 cars. The M-35 is assembled in the La Janais plant, near Rennes, at a rate of two cars a day. The cars are to be in continuous operation until a certain mileage is achieved. The engine is guaranteed for two years and the rest of the vehicle for one year with no mileage limit.

No M-35 is to be sold outside Metropolitan France. But Citroën made sure that M-35 cars would be delivered in all provinces of France, so as to obtain field experience under a great variety of conditions such as mountain roads, motorways, and urban traffic. It is expected that the lessons of this field test will bring Citroën's engineering office a sum of knowledge unobtainable by any in-house test method. A full evaluation of the 500 prototypes is scheduled for 1974. Not until that time will Citroën reach a decision about mass producing a Wankel-powered car.

When you get into the driver's seat of the M-35, your reactions depend on whether or not you have had previous experience with Citroën cars. If you have, things will be as expected. If you have not, you will be surprised. In typical Citroën fashion, the M-35 has a one-spoke steering wheel and a gearshift lever that is actually a bent rod sticking out of the instrument panel, with a knob at the end, reminiscent of some of the less practical umbrella handles I have seen. Some of the instruments are built into a little cluster attached to the steering post. The speedometer is a horizontal design, placed in the panel itself, but the tachometer is mounted in an "added-on" position to the left of the other instruments.

By their appearance, the seats give no hint of their real superior quality and comfort. Not excessively spongy or soft, they still give way

The M-35 is a four seat business coupe, powered by a single-rotor Wankel engine which has been built by NSU and partly designed by Citroën. As with all other current Citroën models, it utilizes front wheel drive.

Sharing a number of body and chassis parts with the Ami-8, the Citroën M-35 is unique only with regard to its engine. It is not proposed as a future production model, but strictly as a large-scale field test experiment.

but also give support exactly where needed most—including lateral support. The seats have reclining backrests. They are not hinged at the bottom, as is the case with most other types, but higher up, in the small of the back above the hip joint. The seat can be set to almost perfect position for any driver and the driving position is comfortable, although the almost-flat steering wheel position looks unusual at first. The driver soon gets used to this, and in fact, it gives a better grip for a hard pull. The unassisted steering sometimes calls for the use of force, as the combination of front wheel drive and the forward engine location tend to keep the car traveling in a straight line. The rack and pinion steering is faultless as far as precision is concerned, but gets awfully heavy at low speeds.

The funny gearshift lever in the dash is hard to reach, but it moves with reasonable precision. The single-rotor engine is perhaps a bit slow in starting, but responds instantly to the throttle. Engine noise is well muffled and insulated. The exhaust tone is pleasant, and the mechanical noise from the power unit and drive train is muted by a layer of sound-deadening material on the underside of the hood. There is only a small amount of gear noise that filters through from the transmission. The engine pulls strongly in top gear from 25 m.p.h. (2,000 r.p.m.)

without snatch or vibration. The ignition distributor is a new type of SEV unit with two-stage advance to overcome snatch on the overrun without loss of performance during acceleration.

As was noted with all other Wankel-powered cars, the engine seems to thrive on high rotational speeds, and never gives any hint that it is spinning at or near its permitted peak. There is an audible warning to prevent overrevving; when engine speed exceeds 7,500 r.p.m., a loud buzzer goes off. In my opinion, it would be preferable to just cut the ignition, as Mercedes does on the C-111. The engine runs satisfactorily on regular-grade French gasoline. Like other Wankel engines, it may lose a fraction of its power potential when running on low-octane fuel, but abnormal combustion phenomena do not occur.

Citroën claims a top speed just short of 90 m.p.h., but the M-35 actually will go faster on superhighways when given full throttle for longer periods. However, the engine does not have the power to maintain speed on uphill gradients. Low-range torque is poor, acceleration is mediocre, and top-gear torque is decidedly lacking. Fortunately, the gear ratios are well spaced, and third gear can be used to accelerate to 80 m.p.h. The shift is easy to operate, once you get used to it, and the synchromesh is flawless.

The technical details of the M-35 cannot be profitably discussed until you know what the basic package is. First of all, it has front wheel drive. As is the case with the NSU Ro-80, the engine is located right up in the nose of the chassis. This is the arrangement that gives the best space utilization within a short wheelbase.

The M-35 however, was not created as a design study in the manner of the Mercedes-Benz C-111 or the NSU Ro-80, but utilizes a large proportion of off-the-shelf parts. It has a platform frame, borrowed directly from Citroën's two economy cars, the 2CV Dyane and the Ami-8. Both the Dyane and the Ami-8 have two-cylinder, horizontally opposed, air-cooled engines, positioned ahead of the front wheel centerline, in addition to front wheel drive. Some of the body panels are the same as those used on the Ami-8, but the Ami-8 is a family car and the M-35 had to have a coupe body. An adaptation was quickly concocted by the Citroën styling department. The same hood, fenders and doors have been adopted unchanged. The rear quarters resemble those used on the Ami-8 station wagon. The front sheet metal has been extended forward to accommodate the radiator needed for the water-cooled NSU engine, which takes up 2½ inches more room than the air cooling fan used on the Ami-8. The M-35 body is built by Carrosserie Heuliez and shipped to Citroën for assembly. Technically speaking, however, the M-35 is an Ami-8 with an engine swap. Although the engine is manu-

The cockpit of the M-35 reveals its 2-CV ancestry, with its add-on tachometer and dash mounted gearshift lever. The seats are extremely comfortable, and the car has been designed with full regard for safety.

factured by NSU, it carries many signs of development work carried out by Citroën.

When the Wankel engine went into the Ami-8 chassis, the transmission was replaced too. The M-35 transmission is an entirely new design that has no parts in common with the gearboxes used on the Dyane or Ami-8. It's a simple countershaft type, using only two gears to obtain the correct output ratio. This is possible because of the front wheel drive, and the rotation can be reversed without complications. Other Citroëns with similar chassis and piston engines do not reverse the rotation but pass the power flow through two gearsets. The other gearboxes (Ami-8 and Dyane) have three separate shafts. Unfortunately, the M-35 shift pattern is "backwards" in relation to conventional cars. When upshifting from first to second, it's easy to go straight into top gear by mistake.

Citroën's other light vehicles use suspension systems with horizontally interconnected coil springs to provide equalizing suspension, but the M-35 has a new version of the oleo-pneumatic suspension system developed for the D series, the long-wheelbase "standard-size" cars in Citroën's model line. Citroën has been a pioneer of self-equalizing suspension systems, and the M-35 combines certain elements of the first with certain elements of the newest.

Citroën's, and the world's first, mass-produced car with an equalizing suspension system was the 2CV. Design work began in 1936, testing and development went on throughout World War II (the Citroën management apparently convinced the German occupation forces that they were working on a revolutionary type of military vehicle somewhat less than Jeep size), and the car was presented to the public in 1948. Front and rear wheels were carried by bellcrank levers, leading at the front and trailing at the rear, pivoted in such a fashion that they moved in vertical planes. A single spring assembly on each side of the car was connected to both front and rear suspension levers. The spring unit consisted of two coil springs enclosed in a common cylinder, one facing forward, the other backward.

For the initial production run, the springs were loaded in compression, but in 1955 the layout was modified so that the main springs worked in tension (the pushrods from the suspension arms simply be-

The Citroën M-35.

came tie-rods). Volute springs inserted between the cylinder and the tie-rod guides ensured the requisite stability. The suspension linkages provided a roll axis at ground level, giving the car very low roll stiffness, and even the pitch resistance was disappointingly low (which is understandable with a wheelbase of only 95 inches). Neither did it provide automatic leveling to compensate for load variations, and it could not be considered suitable for high-performance cars.

Citroën's oleo-pneumatic suspension has been used on the series since 1955. Its immediate forerunner was an oleo-pneumatic, self-leveling rear suspension used on six-cylinder Citroëns since 1953. The DS-19 was introduced at the Paris Auto Show in October, 1955, and caused a genuine sensation, although its existence had been rumored for a long time. The car had a curb weight of 2,775 pounds, and 66% of the weight was carried on the front wheels. The wheelbase was extremely long, 123 inches, while front track was 59.25 inches, considerably wider than the 51.5 inch rear track. The front wheels were also larger than the rear ones on the initial production run, but the rear wheels were later brought up to front wheel size. All the wheels had independent suspension. The front wheels were located by upper and lower transverse control arms and the rear wheels by trailing arms. The front arms were not triangular but elbow-shaped, tapering from the chassis attachment points to the spindle holes for the ball joints in the wheel hubs. A splined extension of the upper arm was coupled to a rocker arm which transmitted suspension movements to the air-and-oil spring unit. A similar rocker arm system was coupled to the rear trailing arms.

Each rocker arm was connected to a spherical container in which a flexible rubber diaphragm separated the gas element of the suspension from the oil-based element. The weight of the car and the suspension movements were sustained by nitrogen gas under pressure in the cylinder on which the spherical container was mounted. The effective length of the suspension leg was dictated by the amount of oil in the cylinder. The oil volume also controlled the height of the car above the ground and its automatic leveling, and the oil also worked as a shock absorber for all suspension movements. The damping action was obtained by fitting a two-way restrictor valve between the cylinder and the spherical container. Lamination of the fluid reduced the flow rate, with the result that movements were continuous and progressive, and always proportional to the speed of fluid displacement. When passengers entered the car, the gas in the cylinders was compressed and the car sank closer to the ground. A central hydraulic brain then restored the car to its normal height by pumping more oil to the cylinders in which the gas was compressed. Conversely, passengers getting out caused the car to rise somewhat before the leveling valves went into action and released oil (which then returned to the central reservoir).

The Citroën system is noted for its extremely low ride-frequencies (i.e., the motion frequency of the sprung mass in free oscillation). Modern passenger cars, touring cars, and sports cars have frequencies in the range of 70–80 cycles per minute. The Citroën design allows frequencies as low as 40 cycles per minute, and, as a direct result of such frequencies, it offers a unique degree of ride comfort. The ride rate is automatically variable according to load and deflection—a near-perfect combination. However, the DS-19 suspension geometry offers no roll stiffness at all, and heavy stabilizer bars are fitted both front and rear. This makes the car sensitive to changes in road camber as a penalty for its remarkably upright cornering style. The Citroën oleo-pneumatic suspension has proved its practical value in international rallies over a 15 year period, and now has been adapted to the M-35.

On the M-35, the oleo-pneumatic spring units are placed horizontally

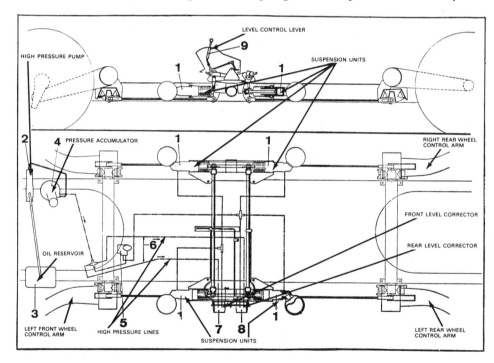

Oleo-pneumatic suspension system for the M-35 has four suspension units, centrally located, two on each side, with hydraulic connections to the wheels' control arms. Each wheel is located by a single control arm (leading in front and trailing in the rear).

near the center of the car, filling the same space normally occupied by the coil spring units. They are attached to the same levers and control arms that are used on the 2CV Dyane and Ami-8. In addition to providing automatic constant level control, this system gives the M-35 adjustable ground clearance, just as on the D series. For superhighway driving, the car is let down to the lowest suspension level. This has the effect of reducing suspension travel, and any sudden bumps then will be felt. Normal driving is done with the suspension system at medium height. For rough roads, the high position is used, which, incidentally, also gives the softest ride.

Aside from the engine, the greatest advantage the M-35 has over the Dyane and Ami-8 is the suspension system. The irritating pitching motions in the small cars has been practically eliminated in the M-35. The ride is luxurious, beautifully cushioned, far in advance of the average for cars of its size. It is very stable, and while roll is restricted by stabilizer bars, it has not been eliminated. The car leans sideways on curves, just like the coil-spring models using the same basic chassis. The M-35 has a firm understeering characteristic at normal road speeds. It wants to go straight, and resists turning. Self-centering action is very strong. The car is stable and highly predictable, with a generous margin of controllable front end slip before the wheels go into a slide. The inherent stability of the car is never upset by poor road surfaces, potholes, ripples or washboard pavement. It keeps right on its intended course, and road shocks are fully absorbed in the suspension system. Because of the shorter wheelbase, the ride is not quite up to D series standards, but it's close.

The M-35 is very safe on all kinds of roads, and takes L-bends, hairpins and S-bends with equal aplomb. When front end slip (understeer) gets critical, merely backing off on the throttle brings the car very nicely back to a normal control situation. The rear end never breaks loose, the rear wheels merely track behind the front ones wherever they go. The car is not only well balanced, but also unusually obedient. Tires are ZX 135-15, using a new finer-mesh steel belt than the regular X tire, in order to reduce any low-speed harshness. These tires have excellent side bite, and run with moderate slip angles at very high lateral accelerations. Steering is somewhat heavy for such a light car, but its turning circle is tight, and parking and maneuvering are quick and easy. The steering also seems slow, but it is necessary to provide more leverage than on the Ami-8 in view of the heavier front end. The Citroën engineers decided that a slower steering ratio would be more acceptable than the extra cost of power steering. Contrary to general claims for the Wankel engine, there is no weight advantage in this case.

Power brakes are standard with disc brakes front and drum brakes rear. The handbrake works on the front wheels, and the front disc brakes are mounted inboard for better cooling. The brakes are fed cool air via ducts, therefore they do not overheat and remain stable in operation even after hard use. Nose dive under braking is less pronounced than on the Ami-8. Wind noise at high speeds is bothersome, but the mechanical noise level is very low, which tends to accentuate wind noise. Road noise, too, seems higher than in the other small Citroëns, despite the hydraulic suspension system. Probably the fault lies not with the M-35 but with the Dyane and Ami-8. With their noisy air-cooled piston engines, a lot of the road noise and chassis rumble is drowned out.

The M-35 is experimental in character, even though the experiment is large-scale. Citroën has made it clear that the M-35 will not under any circumstances become a production car. What, then, is its purpose? It is a test for the economy car formula of the future, combining advanced concepts such as a Wankel engine, hydraulic suspension, and front wheel drive in one vehicle. The fact that Citroën is making 500 of these cars is eloquent testimony of the company's belief in this formula.

SPECIFICATIONS

MAKE	CITROËN
MODEL	M-35
Year introduced	1970
Year discontinued	—
Price	$2,740 in the home market
Type of body	Two-door four-passenger coupe
Type of construction	Platform frame and steel body
Driving wheels	Front
Power unit position	Front
Curb weight	1,793 pounds
Weight distribution front/rear	—
Power/weight ratio	32.6 pounds per horsepower
Fuel tank capacity	11.32 gallons
Fuel tank position	Rear
Power unit	Wankel
Number of rotors	1
Chamber displacement	30.03 cubic inches (497.5 cc.)
Equivalent total displacement	60.06 cubic inches (995 cc.)
Compression ratio	9:1
Power output	55 horsepower
at r.p.m.	5,500
Torque	50.6 foot pounds
at r.p.m.	2,745

SPECIFICATIONS

MAKE	CITROËN
MODEL	M-35
Carburetion system	One two-barrel Solex 18/32 HHD
Ignition system	Coil and battery
Cooling system	Water
Clutch	Single dry plate
Transmission	Four-speed manual
Gear ratios 1	3.811:1
2	2.312:1
3	1.454:1
4	1.077:1
5	—
R	4.182:1
Final drive ratio	4.125:1
Front suspension	Single leading arm on each side with stabilizer bar
Front springs	Oleo-pneumatic struts with automatic leveling
Rear suspension	Single trailing arms on each side with stabilizer bar
Rear springs	Oleo-pneumatic struts with automatic leveling
Steering system	Rack and pinion
Turning diameter	37.3 feet
Overall steering ratio	18:1
Turns, lock to lock	2.25
Brake system	Power-assisted; discs front; drums rear
Disc diameter F	10.65 inches
Disc diameter R	—
Drum diameter F	—
Drum diameter R	7.12 inches
Lining area	46.05 square inches
Swept area	—
Parking brake	Mechanical, on rear wheels
Tires	Michelin X radial ply
Tire size	135-15 ZX
Wheelbase	94.5 inches
Front track	49.6 inches
Rear track	48.03 inches
Overall length	159 inches
Overall width	61.25 inches
Overall height	53.25 inches

SPECIFICATIONS

MAKE	CITROËN
MODEL	M-35

Test Results

Acceleration times

0–30 m.p.h.	4.8 sec.
0–40 m.p.h.	8.2 sec.
0–50 m.p.h.	12.0 sec.
0–60 m.p.h.	18.0 sec.
0–70 m.p.h.	25.5 sec.
0–80 m.p.h.	40.0 sec.
0–90 m.p.h.	—
0–100 m.p.h.	—
Top speed 1	27.0 m.p.h.
2	44.0 m.p.h.
3	68.5 m.p.h.
4	89.5 m.p.h.
5	—
Average fuel consumption	24 m.p.g.

23

The Mercedes-Benz C-111

LATE IN 1969, Mercedes-Benz announced the existence of a mid-engined prototype sports car powered by the revolutionary Wankel rotary engine. This prototype, labeled the C-111, was designed to develop the Wankel engine, examine new suspension systems, explore the use of plastic for body structures and to pursue the study of high-speed aerodynamics. In accordance with these goals, the C-111 project is probably the most radical design that Mercedes-Benz has ever undertaken.

The engine in the C-111 sits behind the driver but does not hang out over the rear wheel centerline, as in a Volkswagen. The engine is carried entirely within the wheelbase, which has advantages over a conventional front engine/rear drive system because of better weight distribution, less weight, lower cost because of the fewer number of drive train components needed, less heat transfer to the passenger compartment and a lower noise level in the passenger compartment.

The Wankel engine is extremely well suited for mid-engine installation in a sports car because it is so light and compact. The C-111 engine delivers 330 horsepower and weighs only 308 pounds. In comparison, a 335 horsepower Ford V8 weighs 630 pounds and the new Mercedes-Benz 230 horsepower V8 weighs 495 pounds. The C-111 three-rotor Wankel engine is only 24.1 inches long, 35.8 inches high and 33.3 inches wide, while the 230 horsepower V8 is 33.4 inches long, 26.5 inches wide, and 27.15 inches high.

The C-111 has a slight rearward bias in weight distribution and an extremely low center of gravity. Static front/rear weight distribution is 45/55%. But, the least praised of its design features is the low polar

moment of inertia. The polar moment of inertia is an indication of the
car's resistance to a change in direction of travel, and is determined not
only by the front/rear weight distribution but also by the exact disposi-
tion of the various weight concentrations in the chassis (such as engine,
transmission, fuel tank, radiator, battery, spare wheel and occupants).
Having them all within the wheelbase gives a low polar moment of
inertia, which makes it easier to steer the car from one path to another.
Having the engine in the front of the car with a big fuel tank and a

In the early spring of 1970, Mercedes-Benz displayed a new version of the C-111,
the Mark II. It is shown in the foreground, in front of the model driven by the author
in September 1969, with the original prototype in the upper left-hand corner. The
Mark II has a four-rotor Wankel engine.

Parked in the middle of Mercedes-Strasse in Stuttgart, the C-111 shines in the morning sunlight. Mercedes-Strasse is the main approach road to the Daimler-Benz headquarters.

heavy spare wheel in the rear gives a high polar moment of inertia and the car resists any change in direction.

Because cornering is such an important part of racing, it is understandable why the mid-engine racing car has become universal in the past decade. Daimler-Benz has been aware of this fact for a very long time. Benz built a mid-engine racing car back in 1922, and there was a Mercedes-Benz 150 H mid-engine roadster in 1934. But the C-111 is the company's first attempt in modern times to investigate the concept of the mid-engine installation as applied to a high-performance Grand Touring car.

The architectural pattern for the modern mid-engine sports car was laid down by Porsche as early as 1954. Earlier cars from Porsche carried the engine in the tail, giving a high polar moment of inertia and intro-

The redesigned rear end of the C-111 Mark II offers considerably improved rear visibility in addition to increased luggage capacity. This test car is fully equipped for road use.

ducing an unpredictable element in the handling characteristics by the extreme rearward bias in weight distribution. For the Carrera model (Type 718), Porsche turned the entire power train around, placing the engine in the middle. This school of design quickly dominated the smaller categories of sports and racing cars. It was popularized by Cooper in Formula 1 Grand Prix racing in the late Fifties, and is now universal in that field, and has been regarded as a prerequisite for Le Mans proto-types of all sizes since the early Sixties. Beginning with the De Tomaso Vallelunga in 1963, a few mid-engine production cars have appeared. The most widely publicized have been the 1966 Lamborghini Miura, 1967 Lotus Europa, 1967 Dino (by Ferrari), 1968 Matra 530, 1970 De Tomaso Pantera, 1970 Monteverdi Hai and the 1970 Porsche 914.

Good weight distribution, a low center of gravity, and suitably low polar moment of inertia do not guarantee good roadholding and great handling precision—the suspension and steering systems' still must do their jobs. The C-111 has a unique suspension system. Wheels must be free to travel up and down when they meet unevenness in the roadway. Some horizontal compliance in such situations is also desirable. While they travel up or down, the wheels should have a minimum of freedom to move in other planes, such as in camber and toe-in. The extent to which such changes are allowed to occur is dictated by the suspension linkages—the same control arms that also determine the roll center height. In the C-111, the front roll center is located about 2 inches above ground level and the rear roll center about 5 inches above ground level. Coupled with a center of gravity positioned about 15 inches above ground level, this results in a naturally low tendency to roll. Roll, in this context, means lean or sway.

The Palatal body was built at Wagonfabrik Rastatt and shipped to Mercedes-Benz, where the complete vehicle was assembled. (*Photo: Weitmann*)

The rear suspension on the C-111 has tall but narrow-diameter coil springs, with concentric shock absorbers. Two steel bars, running forward from the hub carrier, are used as radius rods. The tubes in front of the spring are exhaust pipes, and the transmission can be seen behind the brake caliper. (*Photo: Weitmann*)

The front suspension has many novel features. The lower control arms are attached to the wheel at hub level, not below as is common on production cars. The lower control arms are not of an A-frame construction, but each is an I-bar located in the fore-and-aft plane by a semi-leading diagonal drag strut. The drag strut provides the necessary triangulation required for handling the weight transfer loads that occur during braking and hard cornering, while giving some measure of horizontal compliance. The upper control arms are constructed on the same principle, although the I-bars are shorter and the drag struts are almost parallel with the centerline of the car. The decision was made to give the car true center-point steering. That means letting the steering axis coincide with the center of the tire footprint. This has the effect of reducing self-centering action (or "returnability," as it is often called), but it does provide plenty of self-aligning torque, which is a measure of cornering force. The lower control arms are attached to the wheels at hub level, because of space considerations. The desired geometry could not be obtained with a mounting below the enormous brake disc,

The front suspension on the C-111 uses a tall coil spring leg with a built-in shock absorber. The disc brake is ventilated for maximum cooling. (*Photo: Weitmann*)

but it seemed quite feasible to place the mounting in the center of the disc and adjust the upper control arm accordingly. To understand this, you have to consider the slip angle of the tires.

The slip angle is the angle formed between the line where the tire

The three-rotor C-111 engine fits snugly into the space between the passenger compartment and the rear wheels. Behind it, there is room for a small trunk. The prototype engine makes use of modified, standard Mercedes-Benz components wherever possible, in order to minimize development time.

is pointing and the line it actually will follow. This slip angle is useful, for it creates side bite or cornering force. The cornering force opposes the side force that produced the slip angle to begin with. It enables the car to follow a curved path. As the car enters a curve, the tire is deformed and the footprint is distorted. At the front, the rubber has just been placed in contact with the road and is taking little or no load. At the rear, there is maximum side strain and force, while the vertical loads are centered on the footprint's geometrical center. The distance between the center of the vertical loads and the center of the side forces is called the "pneumatic trail." The amount of cornering force

The four-rotor engine is slightly longer than the three-rotor unit, and remains the same height and width. Engine accessibility for service and maintenance is very good.

generated in the tire, multiplied by the pneumatic trail, gives the amount of self-aligning torque. This assists caster action and tends to keep the wheels on a straight course, and it also gives the driver an indication of the force needed to steer the car.

The simplest way for the Daimler-Benz engineers to build returnability and directional stability into this suspension system was to incorporate an extreme caster angle in the steering system. Caster is set at 9 degrees, which makes the steering harder, but straightens the car after a curve and keeps it pointed straight.

The rear suspension design looks as if it came directly from a Grand Prix Formula 1 chassis. It has the same massive hub carriers, with hefty anchorage points for control arms. Fore-and-aft location is determined by four radius rods, two from each hub carrier, upper and lower. The lower ones are slightly longer and point towards the centerline between the two front wheels. The upper radius rods run almost straight forward. Lateral location of the hub carriers is taken care of by three control rods on each side, the two longer ones being attached at hub level in front and in back of the hub itself. The single short control rod is held by the top of the hub carrier. Rear stabilizer bars add roll stiffness with very little effect on spring rates during single-wheel deflections. Adding roll stiffness also adds to lateral weight transfer. For the road version, moderate weight transfer in the rear end was desired, to keep the rear wheels trailing the front at all times. For the track version, high weight transfer was judged desirable, so as to enable the skilled driver to provoke larger slip angles in the rear tires and thereby assist the vehicle around the curve. This should be thought of as adding oversteer rather than as reducing understeer, because the basic understeer in the front end was left intact and oversteer was added to the rear.

The present research program on plastic body construction at Daimler-Benz is directed towards the use of fiberglass-reinforced plastic as the material for a non-stressed body structure to be mounted on a metal frame. Other firms in Germany are experimenting with monocoque stressed-skin structures made of plastic materials, but the C-111 has a sheet steel frame, partly riveted and partly welded. The fiberglass-reinforced plastic body structure is bonded and riveted to the chassis frame.

Daimler-Benz found plastic an attractive enough material to include it in the specification for the C-111. There are overwhelming reasons why any automobile manufacturer should want practical experience with it. Fiberglass-reinforced plastic has several important advantages over steel for car bodies. The major one is lower weight. For a sports car, a weight saving of 40% could be realized. Fiberglass-reinforced plastic will not rust or corrode, which means longer body life in many parts of the world.

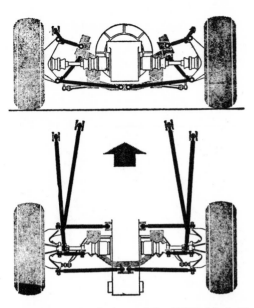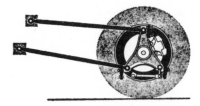

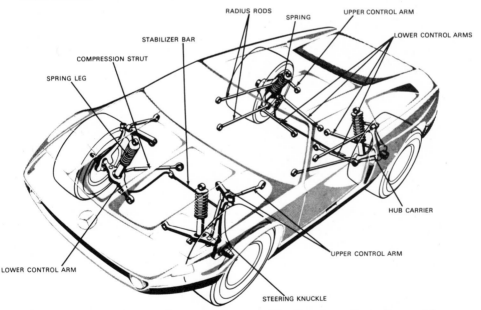

These sectional views of the C-111 Mark II rear suspension show the details of its geometry. Cross-section (above, left) makes it clear that positive camber cannot occur. Side elevation (above, right) shows how the converging extensions of the upper and lower radius rods meet ahead of, and higher up than the vehicle's center of gravity, to assure absence of acceleration squat. Plan view (below) shows how the absence of rear wheel steering phenomena is assured, by a double-triangulated system of rear wheel location.

RADIUS RODS
SPRING
UPPER CONTROL ARM
STABILIZER BAR
LOWER CONTROL ARMS
COMPRESSION STRUT
SPRING LEG
HUB CARRIER
LOWER CONTROL ARM
UPPER CONTROL ARM
STEERING KNUCKLE

C-111 Mark II chassis has minor suspension modifications from the earlier model. This sketch shows the overall configuration.

Trunk in the C-111 Mark II is small compared with a typical Detroit sedan, but quite roomy for a sports car of its type. Enlargement was made possible by rerouting the exhaust pipes and muffler.

Dents or damage from small blows tend to be localized, and therefore easier to repair. Fiberglass-reinforced plastic bodies do not respond to vibration the way sheet steel does, and the risk of stress failures is consequently reduced. It has acoustical properties that muffle road noise, and plastic claims to offer extra safety. There is some merit in this. Pound for pound, it is stronger than metals used for car bodies. Parts are made three times as thick as steel to obtain comparable structural stiffness, and this extra thickness gives the fiberglass-reinforced plastic twice the tensile strength for about half the weight of similar steel parts. Under severe impact, the plastic body absorbs a larger portion of the impact force and gives added occupant protection. The body structure itself is free of squeaks and rattles. It offers better heat insulation and gives improved noise insulation for an equal amount of sound-deadening material. These advantages have been amply demonstrated since Chevrolet began production of the Corvette back in 1953. But a large-scale replacement of steel by fiberglass-reinforced plastic has not occurred. This is due mainly to economic factors, and partly to a lack of public demand.

First of all, textile glass fibers are more expensive than steel. This is not because of the material itself, but because of the huge investment required for special equipment to produce them. Secondly, the conversion of fiberglass and resin into molded parts involves a higher labor content than is needed for metal stamping. Thirdly, finished body moldings have to be stored for drying, which means large storage spaces, a lot of time, and the risk of distortions due to improper storage.

From the desire to learn more about fiberglass-reinforced plastic, Daimler-Benz gained the extra advantage of being able to make frequent body changes on short notice. But Daimler-Benz did not actually make the C-111 body in their own plants, although it was designed by them. They farmed out the actual construction to a small firm that has been active in plastic body production for several years—the Wagenfabrik Rastatt.

The Mark II has an entirely revised interior. Instruments are regrouped and air vents relocated. The accelerator retains the auxiliary tab used on the original version. This tab is placed so that it is natural for the right heel to touch it under braking, to enable the driver to increase engine speed when downshifting. The tab works on a separate throttle linkage, with a different, slower ratio than the main accelerator, in order to prevent over-revving of the Wankel engine under zero-load conditions, such as when the clutch is disengaged.

The C-111 is not the first experimental Mercedes-Benz with a plastic body. The first experiments with fiberglass-reinforced plastic bodies were made in 1950 and 1951 on prototype chassis for the 300 SL sports car. At this time, no conclusions can be drawn about possible applications of this material to future production cars—it's just a research project. The body styling for the C-111 stems from a shape that Karl Wilfert, chief of body development at Daimler-Benz, designed in 1961. At that time he was toying with a project for a Le Mans prototype in the hope that the management would allow the engineering department to form a new official factory racing team, as they had done so successfully in 1954 and 1955. But, in November 1968, the engineering department got the orders to go ahead with another project—the C-101. C-101 was the original project designation for the C-111. The project number was changed to C-111 when it was announced to the public to avoid any conflict with existing or future Peugeot car model designations. All Peugeot numbers have one thing in common—the middle zero. Also, the directors of Daimler-Benz felt that three ones in a row symbolized the three-rotor Wankel engine much better than two ones separated by a zero.

Aerodynamic drag is an obstacle to economical high-speed driving, and one of the design objectives was a wind-cheating body shape. Wilfert and his men went to work on a GT coupe body based on his 1961 designs for a Le Mans prototype. Unfortunately, the cars with the lowest air drag are usually the ones with the least resistance to crosswinds. For the C-111, Wilfert had to find a shape that would not disturb the air very much when traveling through it, yet would remain relatively insensitive to sidewinds and changes in wind direction. These goals are incompatible because sidewinds act through a car's center of pressure in the same way that centrifugal force acts through a car's center of gravity. It makes no difference whether the pressure angle is straight from the side or just a little off the car's direction of travel—the center of pressure depends on shape alone. The lower the drag coefficient of a car, the farther forward its center of pressure. On most modern cars, the center of pressure is located in the area between the windshield corner posts. With more drastic streamlining, the center of pressure moves further forward. On speed-record cars with very smooth airflow, it can actually be ahead of the car itself. Sidewinds pushing on a center of pressure that is well ahead of the car's center of gravity will produce yawing, and the car will veer off course in the direction of the sidewind.

For many years, aerodynamicists and engineers have been exploring various ways of pulling the center of aerodynamic pressure further back, toward the car's center of gravity. For this purpose they have added

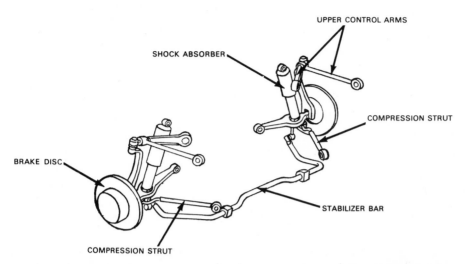

The front suspension system for the C-111 includes a thick stabilizer bar. The coil springs (not shown) are mounted over the shock absorbers. Strong anti-dive effect under braking is designed into the upper control arms. The lower control arms are located at hub level.

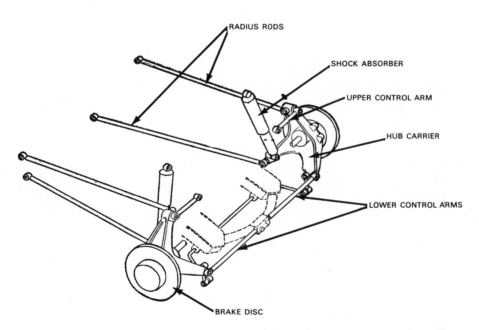

The C-111 rear suspension is based on modern Grand Prix racing car practice, with long radius rods and transverse control arms. This system permits minimal changes in camber angles, track and wheelbase during wheel deflections. It is a very lightweight construction.

tailfins and "spoilers." Spoilers are airfoils of various sizes and shapes, mounted in various positions on the car, some low down in front, others high up in the tail. The initial purpose of spoilers was to reduce aerodynamic lift forces, but it was soon found that they also added drag, which in turn improved directional stability. Tailfins present a greater windage area behind the center of gravity, which tends to move the center of pressure to the rear of the car. Daimler-Benz wanted to achieve its goals without recourse to fins and spoilers.

They started off with a thorough examination of the car's airflow needs—air for engine breathing, cooling air for radiator and brakes, and ventilation air for the occupants. By choosing the right size, shape and location for the scoops or slots to be used as air intakes, considerable gains were made in both drag and stability. The choice was not made,

The hastily cobbled-up C-111 test car began road testing in May, 1969. In this photo the air intake was positioned between the faired-in headlights, and the outlet was placed at the windshield base. Air flow for radiator cooling was considerably modified after wind tunnel tests, taking in air below the stagnation line, and letting it out in a low-pressure area located in the middle of the short hood.

of course, until thorough tests had been conducted. The tests were made in Daimler-Benz's own wind tunnels, some with scale models and others with full-size mock-ups. The wind tunnels provided wind from all quarters, and the company even rigged up a special machine for creating sudden sidewinds. It delivered gusts up to 50 m.p.h. right at the roadside.

At the end of five months of intensive work, a body shape of the desired stability, with low lift characteristics and a drag coefficient of only 0.33, had been created. The drag coefficient indicates the aerodynamic efficiency of a body. A body with low drag coefficient disturbs the air very little; a body with a high drag coefficient creates a substantial turbulence. The drag coefficient number was invented only for *comparing* body shapes—body size has no bearing on it. A drag coefficient of 1.00 is assigned to a particular shape—a cylinder seven times as high as it is wide. Drag coefficients are determined by wind tunnel testing. A square

Both a full-scale and a one-fifth scale model of the C-111 were constructed for wind tunnel test purposes. This was the stage of progress as late as February, 1969. (*Photo: Weitmann*)

plywood board has a drag coefficient of 1.11, while modern U.S. passenger cars have drag coefficients ranging from 0.42 to 0.50.

The main cause of lift is the high velocity airflow over the roof of a car—it creates a negative pressure area and aggravates the effects of crosswinds. A high pressure area under the car adds to the total lift force. A typical American sedan has a lift force of some 300 pounds at 100 m.p.h. High speed racing cars, like Ford's GT-40, have experienced situations where the lift was greater than the entire load on the wheels, and the car was actually flying. If the front wheels lift, the car cannot be steered at all, or braked very well; if the rear end lifts, traction is lost. At 150 m.p.h., the C-111 has a 33 pound aerodynamic lift force acting on the tail of the body, while the nose is pressed down by a 66 pound negative lift force.

Design work on the C-111 began in November of 1968. By the first of April, 1969, the car was completed. The first one was equipped with a provisional body and a three-rotor, 335 horsepower engine. Tests at the Daimler-Benz proving grounds in Stuttgart, on the open road and on racing circuits were satisfactory, and by July, the first of a small series of six cars equipped with a streamlined body were seen in various parts of Germany and in the neighboring Alps. Both the body and the chassis underwent many changes that summer. After some redesign, three new vehicles were built in August of 1969. The research division was authorized to build more C-111s, a limited number depending on the requirements of the research program, to enable the engineers to carry out research in several areas on different cars at the same time.

The C-111 was redesigned in the final months of 1969 to accommodate the four-rotor version of the Wankel engine. Adding one more rotor to the engine did not change its package size and weight very much. The four-rotor engine weighs 374 pounds complete—only 66 pounds more than the three-rotor version. The length of the four-rotor unit is only 29.4 inches. This insignificant growth did not necessitate a stretching of the wheelbase or redesign of the frame because the chassis of the first C-111 had, in fact, been designed with the ultimate installation of the four-rotor powerplant in mind. There is some vacant space ahead of the three-rotor unit, while the four-rotor engine fits snugly.

The body was completely redesigned to eliminate the most severe drawbacks of the first prototypes: poor visibility, lack of luggage capacity, and lack of elegance. The C-111 Mark II was first shown at the Geneva Auto Show in Switzerland in March of 1970, then it was flown to New York for its American debut. The original C-111 body was designed to have low air drag and to be functional only. Although it provided sufficient visibility for turnpike driving, improvement was necessary for city and country driving. Therefore, the Mark II body has an enlarged and reshaped windshield, and bigger side and rear windows. Rear vision in particular has been improved with a restyled engine cover. The roofline was completely changed. Instead of the rear quarter-panels, with louvres, the roofline is simply marked by a rib on each side, which improved rear visibility considerably. The panels were done away with altogether. In addition, the Mark II has 36% more glass area, and correspondingly improved visibility forward and to the sides. Both nose and tail were slightly lengthened. Although front spoilers and reverse airfoils at the

The C-111 instrument panel has large dials for the speedometer and tachometer (partly hidden by the steering wheel), plus small gauges for fuel tank, water temperature, oil pressure, and oil temperature. The gauge adjacent to the air vents is a clock.

rear were tested in the wind tunnel, high speed driving demonstrated that these aerodynamic aids were not necessary.

The Mark II also has a real trunk on top of the engine compartment. It has room for three suitcases, and even though it is located above the engine it is completely heat insulated. Uhlenhaut personally carried out the final "test" of its insulation. He put a pound of butter in the trunk, drove several hundred miles at high speeds, then looked into the trunk again. The butter hadn't melted.

The redesign brought about an 8% reduction in the drag coefficient, and the high speed lift characteristics are somewhat improved. The Mark II has a slight aerodynamic lift on the front end at speeds above 125 m.p.h. but the lift force never rises to the point where it is a problem, even at the amazing top speed of 186 m.p.h. No changes were made in basic chassis design or weight distribution. The front suspension was modified in detail, and the steering gear was changed to give lighter steering. Springs and shock absorbers are better matched and spring travel was increased for a more comfortable ride at high speeds.

Because of the higher output of the four-rotor engine, the cooling air intake in the front grille and the air outlets on the hood have been enlarged and redesigned for approximately 50% greater cooling capacity.

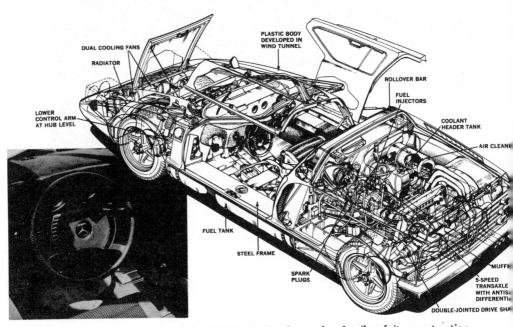

This cutaway drawing of the C-111 clearly shows the details of its construction. The entire drive train is positioned behind the seats, the front part of the car being taken up with controls and instruments. (*Drawing: Theo Page*)

The Mark II has a new interior, with a more civilized appearance. The new instrument panel alone makes it seem almost a different car from the one I had driven eight months earlier. The new interior has improved heating and ventilation and allows easy installation of a radio and an air conditioning system. The dual windshield wipers were replaced by a single wiper, which clears a larger area and is concealed when not in use. For better identification, the square tail lights were replaced by large round ones.

What about the future of the C-111? Will Mercedes-Benz go racing again, and try to win the 24-hour race at Le Mans with the C-111? Not very likely—the C-111 is not a racing car. The C-111 has many luxuries that are never found on racing cars, and much excess weight that would ruin its chances. The engine is not developed or prepared for racing, and is certainly less powerful than the piston engines in the cars that now dominate the sport. Because of new displacement limits coming into effect in 1972, the C-111 engine would not be permitted in the races that decide the World Manufacturers' Championship.

The car is strictly an experiment. Daimler-Benz decided to let the public see it because they are proud of it, and because they felt they needed to change their image somewhat. The bulk of Mercedes-Benz production cars are low-powered family sedans. The C-111 is the only car that Daimler-Benz has ever shown in public that was neither a production-ready vehicle nor a racing car. A spokesman for the company said, "We decided that although we had a long tradition of keeping the laboratory doors closed, there was no point in being bound by tradition." There are no plans to put the C-111 into production at this time, but it cannot be ruled out as a possible future rival to existing low-volume, high-priced Grand Touring cars, such as Aston Martin, Bizzarrini, Ferrari, Iso Grifo, Lamborghini, Maserati, and Monteverdi. This is a glamor market in which Mercedes-Benz has strong traditions, and the C-111 would be a fine representative. It is a car that stands out for its unequalled combination of technical novelty, high performance, first-class comfort, accurate steering, powerful brakes, and balanced road behavior.

SPECIFICATIONS

MAKE	MERCEDES-BENZ
MODEL	C-111 Mk I
Year introduced	1969
Year discontinued	Never in production
Price	Never listed
Type of body	Two-seater two-door coupe
Type of construction	Sheet steel platform frame with fiberglass body

SPECIFICATIONS

MAKE	MERCEDES-BENZ
MODEL	C-111 Mk I
Driving wheels	Rear
Power unit position	Midships
Curb weight	2,425 pounds
Weight distribution front/rear	45/55%
Power/weight ratio	7.2 pounds per horsepower
Fuel tank capacity	31.6 gallons
Fuel tank position	Midships in door sills
Power unit	Wankel
Number of rotors	3
Chamber displacement	36.6 cubic inches (600 cc.)
Equivalent total displacement	220 cubic inches (3.6 liters)
Compression ratio	9.3:1
Power output	330 horsepower
at r.p.m.	7,000
Torque	220 foot pounds
at r.p.m.	5,000–6,500
Carburetion system	Bosch direct injection with mechanical pump
Ignition system	Transistorized coil and battery
Cooling system	Water and oil
Clutch	Single dry plate
Transmission	Five-speed ZF (5DS-25/1)
Gear ratios 1	2.58:1
2	1.61:1
3	1.21:1
4	1.00:1
5	0.84:1
R	—
Final drive ratio	3.77:1
Front suspension	Triangulated control arms with non-parallel pivot axes. Stabilizer bar.
Front springs	Coil springs with concentric gas-pressurized shock absorbers.
Rear suspension	Trailing and transverse links with stabilizer bar.
Rear springs	Coil springs with concentric gas-pressurized shock absorbers.
Steering system	Recirculating ball.
Turning diameter	—
Overall steering ratio	—
Turns, lock to lock	—
Brake system	Ventilated disc F and R

SPECIFICATIONS

MAKE	MERCEDES-BENZ
MODEL	C-111 Mk I
Disc diameter F	10.8 inches
Disc diameter R	11.0 inches
Drum diameter F	—
Drum diameter R	—
Lining area	39.9 square inches
Swept area	—
Parking brake	Duo-servo drums on rear wheels
Tires	Michelin XVR
Tire size	195 VR 14
Wheelbase	103.2 inches
Front track	54.3 inches
Rear track	53.9 inches
Overall length	166.5 inches
Overall width	70.9 inches
Overall height	44.3 inches

Test Results

Acceleration times	
0–30 m.p.h.	1.9 sec.
0–40 m.p.h.	2.2 sec.
0–50 m.p.h.	4.0 sec.
0–60 m.p.h.	4.9 sec.
0–70 m.p.h.	7.2 sec.
0–80 m.p.h.	9.3 sec.
0–90 m.p.h.	11.9 sec.
0–100 m.p.h.	14.9 sec.
Top speed 1	55 m.p.h.
2	85 m.p.h.
3	112 m.p.h.
4	140 m.p.h.
5	162 m.p.h.
Average fuel consumption	—

SPECIFICATIONS

MAKE	MERCEDES-BENZ
MODEL	C-111 Mk II
Year introduced	1970
Year discontinued	Never in production
Price	Never listed
Type of body	Two-seater two-door coupe
Type of construction	Sheet steel platform frame with fiberglass body
Driving wheels	Rear

SPECIFICATIONS

MAKE	MERCEDES-BENZ
MODEL	C-111 Mk II
Power unit position	Midships
Curb weight	2,734 pounds
Weight distribution front/rear	45/55%
Power/weight ratio	6.8 pounds per horsepower
Fuel tank capacity	31.6 gallons
Fuel tank position	Midships in door sills
Power unit	Wankel
Number of rotors	4
Chamber displacement	36.6 cubic inches (600 cc.)
Equivalent total displacement	293 cubic inches (4.8 liters)
Compression ratio	9.3:1
Power output	400 horsepower
at r.p.m.	7,000
Torque	289 foot pounds
at r.p.m.	4,000–5,500
Carburetion system	Bosch direct-injection with mechanical pump
Ignition system	Transistorized coil and battery
Cooling system	Water and oil
Clutch	Dual dry plate
Transmission	Five-speed ZF (5DS-25/1)
Gear ratios 1	2.58:1
2	1.61:1
3	1.21:1
4	1.00:1
5	0.84:1
R	—
Final drive ratio	3.77:1
Front suspension	Triangulated control arms with non-parallel pivot axes. Stabilizer bar.
Front springs	Coil springs with concentric gas-pressurized shock absorbers.
Rear suspension	Trailing and transverse links with stabilizer bar
Rear springs	Coil springs with concentric gas-pressurized shock absorbers.
Steering system	Recirculating ball.
Turning diameter	—
Overall steering ratio	—
Turns, lock to lock	—
Brake system	Ventilated disc F and R.
Disc diameter F	10.8 inches

SPECIFICATIONS

MAKE	MERCEDES-BENZ
MODEL	C-111 Mk II
Disc diameter R	11.0 inches
Drum diameter F	—
Drum diameter R	—
Lining area	39.9 square inches
Swept area	—
Parking brake	Duo-servo drums on rear wheels
Tires	Dunlop
Tire size	F:4.50/11.60-15; R:5.50/13.60-15
Wheelbase	103.2 inches
Front track	56.9 inches
Rear track	54.7 inches
Overall length	174.8 inches
Overall width	71.8 inches
Overall height	44.1 inches

Test Results

Acceleration times	
0–30 m.p.h.	1.8 sec.
0–40 m.p.h.	2.0 sec.
0–50 m.p.h.	3.8 sec.
0–60 m.p.h.	4.7 sec.
0–70 m.p.h.	6.6 sec.
0–80 m.p.h.	8.8 sec.
0–90 m.p.h.	11.2 sec.
0–100 m.p.h.	14.0 sec.
Top speed 1	62 m.p.h.
2	96 m.p.h.
3	124 m.p.h.
4	156 m.p.h.
5	186 m.p.h.
Average fuel consumption	—

24

The Mustang
RC2-60

ON THE OUTSIDE, the Mustang in front of me was indistinguishable from one straight off the River Rouge assembly line. But instead of a 200 horsepower 289 cubic inch V8 under the hood, I found a much smaller unit. This was Curtiss-Wright's RC2-60 twin-rotor Wankel engine. When I turned the key, the engine came to life instantly. It had a steady idle at 800 r.p.m. and a dab at the accelerator sent the revs up to 2,000 instantly. A bigger dab sent the needle on the tachometer up to 4,000 or 5,000 almost as fast. Without load, the acceleration of the RC2-60 was as rapid as would be expected from a Formula 1 Grand Prix racing engine.

The sound was totally unlike that of the NSU Spider, with its single-rotor Wankel engine. It sounded almost like a six-cylinder piston engine at idle, and almost like a gas turbine at speed. It shared with the NSU engine the same lack of indication that there was any peak r.p.m. Curtiss-Wright indicated the redline to be 6,000 r.p.m., conservative for a Wankel engine, but I went beyond it many times. The sound did not change at higher engine speeds, and the engine gave the impression that it was capable of going right on accelerating forever. It was extremely smooth in terms of torque output, and was completely vibrationless. Despite the fact that the engine mounts had not been tuned for the RC2-60, the engine sat perfectly still even at idle speed. The RC2-60 was coupled to a standard Ford three-speed Cruise-O-Matic automatic transmission. The transmission was the same one that had been bolted to the original Ford V8, and the shift points remained as they had been for the piston engine. As a result, the transmission was poorly mated to the torque curve of the RC2-60. Upshifts occurred at a little over 4,000

r.p.m. For best performance, they should have been delayed until some-
where between 5,400 and 6,000 r.p.m.

The RC2-60 engine behaved normally during my entire time with it.
I drove it from New York to Indianapolis in 1966, then to Chicago,
Detroit and back to New York. It gave better fuel economy at high speed
than the original Ford V8. Low speed economy was sacrificed for tract-
ability and smoothness. Oil temperature rose slowly to 150°F., where it
stabilized, and water temperature hovered around the 180°F. mark all
the time. Oil pressure was a rock steady 45 psi. Under load, throttle
response was immediate. When coasting, the engine acted like a high-
compression piston engine. Engine braking power was probably less than
with a piston engine of comparable power, but the automatic transmission precluded any testing of this aspect of performance. Under full
throttle acceleration, the RC2-60 was noisy. The noise was partly a roar
from the air intake, partly exhaust noise. There was a distinct lack of
vibration and no mechanical noise from the power unit. At steady speed

Ford Mustang experimental vehicle equipped with the Curtiss-Wright RC2-60 U5
engine.

the engine was quiet, no matter whether it was running at 3,000 r.p.m. or 5,500 r.p.m. under light load. Whenever it was laboring, however, the noise level immediately rose. Acceleration times were as follows:

	Ford V8 289	RC2-60
0–30 m.p.h.	2.8 sec.	2.9 sec.
0–40 m.p.h.	5.0 sec.	5.2 sec.
0–50 m.p.h.	6.8 sec.	7.2 sec.
0–60 m.p.h.	10.0 sec.	10.6 sec.
0–70 m.p.h.	13.5 sec.	14.7 sec.
0–80 m.p.h.	18.4 sec.	20.0 sec.

The automotive version of the RC2-60 U5 engine.

The original V8 was rated at 200 horsepower, the RC2-60 at 185. With a manual transmission, there is no doubt that the Wankel would have beaten the reciprocating piston engine on acceleration.

The engine in the Mustang carried the designation RC2-60 U5. This version had been created by Curtiss-Wright specifically for passenger car propulsion. The RC2-60 U5 differed from its predecessors in several small ways, though the geometry was identical to the RC1-60.

Rotor radius was 5.75 inches and rotor width 3.0 inches. Rotor eccentricity was .87 inch, giving a K factor of 6.9 and a potential compression ratio in excess of 10.0:1. The rotor sealing configuration and basic rotor and housing cooling systems remained unchanged from the RC1-60, but the U5 modification included minor improvements in the rotor and rotor housing, the end housing, bearings, and mainshaft. Completely new were the intermediate wall, the center bearing support, gas and water manifolds, accessory drives and housing, oil pumps, power take-off adapters, starter mounting, and oil transfer bearings.

Closeup of the RC2-60 U5 installation in the Ford Mustang.

The dual side intake ports were combined with peripheral exhaust ports to give the following timing:

RC2-60 U5 PORT EVENTS
Port Timing

	RC CRANK ANGLE		EQUIVALENT FOUR-STROKE RECIPROCATING ENGINE CRANK ANGLES		
	Opens	Closes	Opens	Closes	Duration
Present porting					
Intake	583°	840°	29°ATC	20°ABC	174°
Exhaust	202°	598°	45°BBC	39°ATC	266°
Overlap	15°		10°		

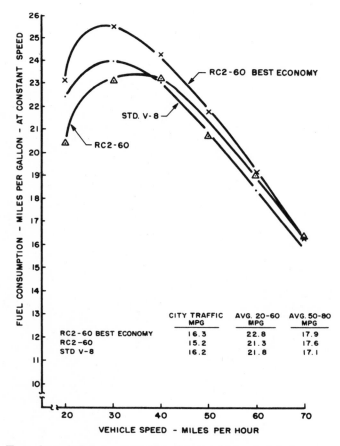

	CITY TRAFFIC MPG	AVG. 20-60 MPG	AVG. 50-80 MPG
RC2-60 BEST ECONOMY	16.3	22.8	17.9
RC2-60	15.2	21.3	17.6
STD V-8	16.2	21.8	17.1

Tests showed that the RC2-60 had better high-speed fuel economy than a production V8 of similar performance, but worse low-speed economy. If tuned for economy, the RC2-60 could surpass the economy of the V8 throughout the speed range, but at the cost of performance.

	Opens	*Closes*	*Opens*	*Closes*	*Duration*
Porting limits available in present castings					
Intake—max.	583°	855°	29°ATC	30°ABC	181°
Exhaust—max.	202°	611°	45°BBC	47°ATC	272°
Overlap—max.	28°		18°		

The spark plugs were mounted 25°, or 2.13 inches, before the minor axis and the distributor was set to give 40–45° spark advance at 5,000 r.p.m. All other accessories were standard automotive units. The two-barrel carburetor came from a Buick, and the alternator and ignition system was adapted from the system that came with the original 289 Ford V8. The RC2-60 U5 had dry sump lubrication and carried a separate oil tank and oil cooler, neatly installed under the hood of the Mustang.

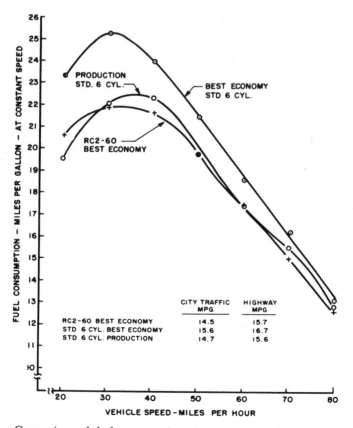

Comparison of fuel consumption at constant speed, with similar vehicles using three different power units—one RC2-60 U5 and two different production six-cylinder piston engines.

Cold-starting tests with the latest version of the U5 demonstrated consistent automatic starts at −20°F. with 5W-20 oil and a 90 A/H battery. The most dramatic improvement resulted from a drain in the exhaust manifold to prevent flooding. As a result of space limitations in the Mustang installation, the exhaust manifold was designed with an initial sweep upwards which would trap fuel against the rotor flank, making a start difficult unless it occurred on the first few turns. Other modifications were also made, including an automotive type starter. The automotive ignition system also helped, particularly when the spark plug electrodes were brought close to the trochoid surface.

SPECIFICATIONS

MAKE	FORD
MODEL	MUSTANG RC2-60
Year introduced	1965
Year discontinued	—
Price	Experimental only
Type of body	2-door coupe
Type of construction	Platform frame with steel body
Driving wheels	Rear
Power unit position	Front
Curb weight	2,575 pounds
Weight distribution front/rear	53/47%
Power/weight ratio	13.4 pounds per horsepower
Fuel tank capacity	16 gallons
Fuel tank position	Rear (under trunk floor).
Power unit	Wankel
Number of rotors	2
Chamber displacement	60 cubic inches (989 cc.)
Equivalent total displacement	240 cubic inches (3,992 cc.)
Compression ratio	8.5:1
Power output	185 horsepower
at r.p.m.	5,000
Torque	192 foot pounds
at r.p.m.	3,750
Carburetion system	One two-barrel Rochester
Ignition system	Coil and battery
Cooling system	Water and oil
Clutch	Hydraulic torque converter
Transmission	Cruise-O-Matic torque converter and three-speed planetary transmission
Gear ratios 1	2.46:1
2	1.46:1
3	1.00:1
4	—

SPECIFICATIONS

MAKE	FORD
MODEL	MUSTANG RC2-60
5	—
R	2.20:1
Final drive ratio	3.20:1
Front suspension	Upper A-frame and lower I-rod control arm with trailing drag strut and stabilizer bar.
Front springs	Coil springs
Rear suspension	I-beam axle
Rear springs	Multi-leaf semi-elliptic leaf springs
Steering system	Recirculating ball
Turning diameter	39.5 feet
Overall steering ratio	—
Turns, lock to lock	4.64
Brake system	Drums all around
Disc diameter F	—
Disc diameter R	—
Drum diameter F	—
Drum diameter R	—
Lining area	212 square inches
Swept area	—
Parking brake	—
Tires	Goodyear power cushion
Tire size	6.95 × 14
Wheelbase	108 inches
Front track	56 inches
Rear track	56 inches
Overall length	181.6 inches
Overall width	68 inches
Overall height	51.1 inches

25

Conclusions

DOES THE WANKEL ENGINE HAVE A future? We have seen that it has potential to match the reciprocating piston engine in terms of life between overhauls as well as emissions. What is more uncertain is whether both will be replaced by an unconventional powerplant, forced upon the world by strict regulations instituted to protect the environment and avoid imbalance in the ecology.

Arguments have been raised by our automobile manufacturers to the effect that it will be easier to meet future clean-air laws by developing the reciprocating piston engine than by investigating and experimenting with other power sources on the assumption (which may turn out to be unfounded) of their being pollution-free. Ten years ago it was impossible to make any realistic estimate of the status of the Wankel engine in world industry as of 1971, and today, it is just as senseless to try to set a timetable for the future adoption of the Wankel engine by the leading automakers. As is the case in all responsible reporting, it is best to try to look at the future of the Wankel engine by examining what is affecting its status today.

On November 2, 1970 GM issued the following press release: "General Motors has agreed to enter into a worldwide, nonexclusive paid-up license agreement with Audi-NSU, Wankel G.m.b.H., and Curtiss-Wright to facilitate its further intensive research and development studies of the Wankel rotary combustion engine to determine whether it is suitable for GM automotive applications. The tentative agreement among the companies, subject to the approval of the managements of the other companies on or before December 31, 1970, covers the manufacture and sale by General Motors of the Wankel rotary combustion

engine except as it applies to aircraft propulsion. The agreement is subject to termination by General Motors at the end of each contractual
year on one day's notice. The agreement provides that General Motors
will pay $5,000,000 by December 31, 1970 upon approval by the parties
concerned. Subject to GM's right to terminate the agreement General
Motors will pay $10,000,000 at the end of the fifth year. At the comple-
tion of these payments, General Motors will have the right to manufacture and sell the Wankel rotary combustion engine on a worldwide
basis without further payments."

The contract was signed formally on November 10, 1970. The $50
million will not be paid directly to Audi-NSU Auto-Union AG and
Wankel G.m.b.H. It is estimated that Curtiss-Wright will collect $23
million over the five-year period spelled out in the contract. Of the $27
million to be paid to the German firms, NSU will gain $16,200,000 and
Wankel G.m.b.H. $10,800,000. Because Audi-NSU Auto-Union AG is a
subsidiary of Volkswagenwerk, it is important to note that only 30% of
the $16,200,000 collected will be kept in the company's coffers. That
amounts to $4.86 million for NSU. The other $11,340,000 will be distributed to former NSU stockholders in proportion to their holdings of
Wankel vouchers.

Why does GM want the Wankel engine so much that it is willing to
pay $50 million for the rights to it? Obviously, even a company the size
of General Motors does not propose to invest the sum of $50 million in
a routine investigation of a possible future source of motive power. It is
a definite commitment, and one that commands other automobile
manufacturers throughout the world to take similar steps. It is clear that
GM is serious about building Wankel-powered cars at some future date.

The question has been asked, "Where Chevy goes, can Ford be far
behind?" Ford is planning to acquire a 35% interest in the Toyo Kogyo
Company—a step which would put Ford in on the ground floor as far
as Wankel engine research and production technology is concerned, but
which would bring the Dearborn giant no closer to the manufacturing
and marketing of Wankel-powered cars in the United States. That
could follow only after a negotiated contract—necessarily costly—with
Curtiss-Wright Corporation, Audi-NSU Auto-Union AG, and Wankel
G.m.b.H. When this step will be taken is mainly dependent on how
much and how fast GM can progress with the Wankel engine.

Because a great deal is known about the operations of General Motors,
the timetable for the Wankel engine can be set within certain limits. To
understand this, we must look at how technical innovations are developed
at GM, brought to maturity, and placed in production. Consider GM's
record of progress in applying gas turbines to automobiles. Last year,

General Motors' *Detroit Diesel Engine Division* announced its intention to be the first on the market with a commercial turbine engine for heavy duty vehicles. Detroit Diesel's engine is scheduled for initial production in mid-1971. Developing the heavy duty gas turbine engine from an interesting but impractical concept into a commercial reality was a process spanning more than 20 years. The GM Research Laboratories brought the engine from the concept stage to the point of technical feasibility, and in 1964 turned it over to Detroit Diesel. In the ensuing years, Detroit Diesel has met the challenge of making the turbine engine economically practicable.

The experimental vehicle series began with the Firebird I in 1953, the United States' first gas turbine powered automobile. A non-regenerative 370 horsepower engine propelled the sleek mobile laboratory—which today is a museum piece. Also in 1953, they built the Turbo-Cruiser I, a turbine-powered bus. The GM family of experimental turbine vehicles burgeoned with the arrival in 1955 of the Firebird II, a family car with a regenerative engine; the Turbo-Titan I, a heavy duty truck, in 1956; the Firebird III, with a more powerful GT-305 engine in 1958; and the Turbo-Titan II the following year.

The fifth generation GT-309 engine, boasting a rugged and simple design, was introduced at the New York World's Fair in 1964. It was field tested in several heavy duty vehicles, including Chevrolet's Turbo-Titan III truck and GMC's Turbo-Cruiser II bus.

General Motors has enormous investments in machine tools that produce parts for piston engines, and there is little more than normal inertia involved in getting a totally new kind of powerplant through its research and development phase and up to the point where it can be produced and marketed economically. Will GM put the Wankel engine through a similar program, and would a Wankel-powered Chevrolet be 20 years away? Convincing feasibility studies have been undertaken elsewhere. The work of NSU, Toyo Kogyo, Daimler-Benz and Citroën covers all the fundamentals, and furnishes eloquent testimony as to the advisability of (1) developing Wankel engines for passenger car applications and (2) designing new passenger cars around such power units so as to take full advantage of their small bulk and low weight.

If GM top management were to give the go-ahead order on Wankel engines today, it would nevertheless be years before you would be able to buy a Wankel-powered Chevrolet, Pontiac, Oldsmobile, Buick or Cadillac. It is naive to think that basic research has been completed. Research, testing, redesign, more testing, design alterations, and material development could take years. GM has unquestioned ability to undertake projects of this kind. The corporation has created the largest research organization of its type in the world. Dr. Laurence R. Hafstad was vice president of GM in charge of the research staff when the GM Technical Center at Warren, Michigan, was dedicated in 1956. In his opening address, he said: "It is in the tradition of science first to observe, then to understand, and finally to utilize the forces of nature. Man has been doing this since the dawn of history, but at one time discoveries were made in a 'hit or miss' manner by lone investigators poorly supplied with information and equipment. We have now learned to bring trained scientists and engineers together in such well-equipped laboratories as are at the Technical Center, to make discoveries and develop new ideas. This is an important new conception in our modern economy which has resulted in greatly accelerated technological progress."

"General Motors has supported a research organization for over 40 years. Research discoveries and developments have contributed to all the products of GM's manufacturing Divisions. Our automobiles, diesel engines, household appliances, locomotives, and jet engines have all depended upon a continuous research program for their constant improvement. The research staff is the one organization in General Motors that deals solely with fundamental, long-range research. Its scientists and engineers are concerned with projects that continuously explore the future. The program is divided into fundamental scientific research, long-range engineering research, and advanced engineering development. In the new Research Staff facilities at the Technical Center emphasis is placed on basic projects which, when successful, will result in technological improvement."

"The Research Staff has a two-fold responsibility to General Motors management and the manufacturing divisions. First and foremost, we discover and develop fundamental information which will become the basis for the new products of tomorrow. Second, our specialized personnel and facilities are available for use by the engineering departments of the various GM manufacturing units should they choose to use them."

The matter under study by the GM research staff is not simply to find a "Yes" or a "No" for the Wankel engine, but to specify technical directions for all products that fall within the corporation's sphere of interest. Any time the subject of a new type of powerplant comes up, the GM research engineers get down to the basics, which may not be involved specifically with rotary or reciprocating motion, but with the basic energy and materials.

Being the world's largest manufacturer of ground transport equipment, GM is extremely interested in future sources of energy. Liquid petroleum fuels are the mainstay of its products today. It is possible that in the future uranium and thorium may be just as commonly understood words as gasoline and fuel oil are today. Gasoline engines, diesel engines, gas turbines, steam engines, free-piston engines, and the fuels which they burn all provide promising projects for better powerplants in the years ahead. How to make the best use of atomic energy in industry and transportation of the future has become GM's newest long-range problem. A small example will illustrate the scope of nuclear power: one pound of uranium, U-235, has as much energy as 1,300 tons of coal, which is a ratio of 1 to 2,600,000—this is the incentive.

Another big problem is materials. At present it takes 18 tons of material per year to keep an American citizen at his present standard of living. This adds up to the astronomical figure of over $2\frac{1}{2}$ billion tons of material per year for the United States alone. Last year General Motors produced over ten million tons of automobiles. The GM research laboratories support a never-ending search for new materials, new processes, and new methods of fabrication. A part of this program is a group of engineering projects aimed at increasing the fatigue life of parts and utilizing materials more efficiently.

The research staff's projects are divided between two main groups of departments—applied science on one hand, and engineering research on the other. The applied science departments are concerned primarily with fundamental investigations into such fields as engine combustion, electronics, ultrasonics, paints and finishes, electroplating, instrumentation, and radioisotopes. The engineering research departments conduct basic studies in spark-ignition engines, gas turbines, vehicle suspension components, various types of bearings, and the fatigue life of automotive components. While these are only a few of the many areas in which research staff engineers and scientists work, they are typical and serve mainly to illustrate the general nature of the investigations continually in progress.

The *Automotive Engines Department* centers its activities around the development of more efficient piston-type automotive engines, engaging in research on basic engine design and on specific engine components. Studies of fuel economy, octane requirements, carburetion, and other performance factors are made on dynamometers and by road testing. Wankel engines have not yet been incorporated into this group. They are the responsibility of another section, the *Mechanical Development Department*, which is concerned with a variety of projects including unconventional powerplants, fatigue testing, diesel engines, free-piston engines, friction, and bearings.

When the research and development engineers have a design they feel is satisfactory, it is still a long way from mass production. Any item intended for production at the GM scale has to be redesigned for mass production, with due regard to materials handling, parts handling and assembly. These programs can be conducted side by side only to a certain extent, for there are distinct limits. Production engineers cannot get very far beyond preliminary studies until the design has been finalized; otherwise, much production engineering work could be suddenly rendered worthless by a basic design change originating in the research department. Only when the final production design is ready can tooling begin. It takes time to make tools, dies, jigs, and special machine adaptations. For an all-new engine line, it takes about 18 months from placing the first orders until the last deliveries are made. Then the purchasing department comes in. Parts not manufactured by the company itself have to be designed, developed, approved, and ordered from outside suppliers. Stocks have to be built up. All this takes time.

Even when production can begin, all is not cut and dried. No matter how thorough the methods, the engine line cannot run at full speed right from the start. Engines from the pilot line have to be tested to make sure they perform according to specifications and match the prototype. Minute design or material specification changes may be needed. And above all, tooling changes may be called for. A running-in period for any production line turning out an all-new engine is taken for granted. It is all the more important in the case of the Wankel engine.

On this basis, assuming a full-speed-ahead order, a blank-check budget, and no unforeseen setbacks, the Wankel-powered GM car is a minimum of five years away, more likely seven to ten years. That may seem a long time, but we are not talking about just another license agreement. We are talking about mass production by the largest auto company in the world. There can be no doubt that if GM goes to Wankel engines, Ford and Chrysler will have to follow. So will Volkswagenwerk (who may well beat GM's timetable, now that VW owns NSU), Fiat, British Leyland, and Japan's auto giants, Toyota and Nissan (Datsun).

It is no mystery why GM is so interested in the Wankel engine. GM takes an active interest in all types of automotive power units. The particular advantages of the Wankel engine have been set forth in an earlier chapter, but there is one area of unknown potential that has not been mentioned—automation.

Labor is a big part of the production cost, and labor costs are rising, while material costs have not increased proportionately. The industry, led by GM, is trying to cut costs by reducing the labor content in every product, along every step of the way. This is where the Wankel engine could revolutionize powerplant production. Because of the inherent simplicity of the Wankel engine and its low number of parts, it holds the promise of automated manufacture and assembly. Highly complex automatic transmissions are being assembled with a high degree of automation right now. The assembly processes of a Wankel engine seem considerably simpler. There are no basic tooling problems involved in manufacturing parts for Wankel engines, for even epitrochoidal chambers can be machined. All Wankel engine parts probably can be made on automatic transfer lines, the number of operations varying depending on design, finish and tolerances required.

When GM reaches the point where the research staff is ready to release the Wankel engine for production, the management has more decisions to make. GM could create a special Wankel engine division, similar to its *Detroit Diesel Engine Division*. The car and truck divisions then would be supplied with power units from the Wankel engine division to whatever extent the divisions choose, within the framework of their own autonomy. Detroit Diesel, by the way, has no monopoly on diesel engine production within General Motors. There are other possibilities. GM could assign Wankel engine production to an existing division, such as Detroit Diesel, Allison, or Electro-Motive. Or the corporation could give the Wankel engine to one of the car divisions, such as Oldsmobile which has a record of pioneering new concepts (automatic transmission, high-compression V8s, front wheel drive). The choice will be made depending on what scale the corporate management wants to build the Wankel engine. We have already seen that production would have to start gradually, which means that the Wankel engine could not suddenly become the standard power unit in a large-volume car line such as the Chevelle. Again, GM has a multiple choice. The Wankel engine could be made optional for one or more low-volume car lines, from one or more divisions. Or it could be made standard in one low-volume car line, from one or two divisions. The latter course is more likely; GM must be anxious to get the Wankel-powered cars out in the field, and that is best controlled by using it as standard equipment. Sales of options are less predictable.

It is conceivable that GM could assign Wankel engine production to one of its overseas divisions, such as Opel in Germany, Vauxhall in England, or Holden in Australia. However, if the corporation intends to install Wankel engines in a large number of American cars, it would make better economic sense to manufacture the engine within the U.S. When the decision is made, the license agreement allowing GM to exploit the Wankel engine will work in two ways. GM will get full access to test reports and research material compiled by other licensees, and the corporation will be required to share its own findings with the other companies engaged on Wankel engine development.

The licensing agreement with Audi-NSU Auto-Union AG, Wankel G.m.b.H., and Curtiss-Wright should allow GM to begin arming itself fully for a major assault on whatever problems still remain, and at the same time should advance the progress of the Wankel engine not only with strict reference to technological and industrial considerations but also in terms of public acceptance and support.

Appendix

OTHER ROTARY ENGINES

SOME OF THE MECHANICAL movements necessary to make rotary engines possible were invented long before the steam engine or the internal combustion engine had been thought of. These mechanisms were used mainly for water pumps. Since pumps share certain characteristics of variable displacement with heat energy machines, all types of pumps can be converted into heat engines of some kind. The reciprocating engine, for instance, corresponds to the kind of pump used to inflate bicycle tires. The bicycle pump has a piston and a cylinder. What the engine has in addition is fuel, and a spark plug. Naturally the valving differs in execution if not in principle. A centrifugal pump could be converted into a gas turbine. A vane type pump could become a rotary engine. Rotary engines in the strictest sense have unidirectional movement with uniform or variable velocity. They are not affected by alternating inertia forces due to changes in the direction of movement.

RAMELLI

The first evidence of a practical rotary pump appeared in 1588 in Genova, Italy, in a book published by a leading military engineer of the time, Captain Agostino Ramelli (1530–1590), entitled "Le Diverse e Artificiose Macchine del Capitane Agostino Ramelli." In it were illustrations of 195 inventions, ranging from pumps to gear transmissions. Ramelli's designs included over 100 different pumps, plus a variety of windmills, sawmills, screw jacks and derricks. They were designs—not realities. Some of the machines he illustrated and described could not have been made with the tools and materials then in existence.

The Ramelli rotary pump consisted of a cylindrical casing containing a drum able to rotate about an eccentric axis relative to that of the casing. The rotor was cylindrical and revolved eccentrically within a cylinder of greater diameter. Rotor movement was eccentric in relation to the outer cylinder, so that the rotor touched the casing at only one point on its orbit. The rotor had four radial grooves at right angles, with seal strips to separate the four working chambers. The vanes were loose in their slots, free to move in and out. The vane edges were maintained in continuous contact with the inner surface of the casing. However, the rotational speed of Ramelli's pump can hardly have been sufficient for centrifugal force to make the vanes

Ramelli's rotary pump, in an installation proposed by the inventor, driven by a water wheel.

provide adequate sealing. As no spring-loading was supplied, he seems to have relied upon gravity, placing the center of drum rotation in the upper part of the casing. It is possible that this reliance upon gravity accounts for the use of four vanes instead of two, which ought to have been sufficient for water-pump applications. Ramelli spoke of the vanes as "paddles" and described the pump as a "machine designed for the drawing of water from docks or foundations by the power of two men who turn the two cranks of an eccentric wheel within the covering cylinder. The cylinder is made of metal or other suitable material closed and well fastened by screws. It has but one aperture for the entrance of water, and is firm and immovable. When the water enters the cylinder it is forced around by the movement of the wheel and its four sliding pieces (which move easily back and forth as required) continually through a tube." The principle of Ramelli's pump is used to this day in certain types of compressors and pumps, and has inspired a host of inventors through the ages to try to improve upon the basic idea.

Schematic of the Ramelli pump, showing a version equipped with a single floating radial vane.

PAPPENHEIM

The gear-type oil pump most commonly used in modern automobile engines is based upon a rotary pump dating back to 1636. It is known as the Pappenheim pump, but the invention is variously ascribed to Grollier de Serviére or to Pappenheim. A grandson of the French inventor published a collection of his mechanical and other devices in 1719, which would indicate that it originated around 1640. But the pump was described by a number of German writers on record as early as 1636. The gear-type pump is in fact so old an invention that it remains uncertain whether Pappenheim was the name of its inventor or the village in which he lived!

The Pappenheim water pump used six-toothed gears of equal size, mounted side by side and meshing. The two shafts were parallel, but rotated in opposite directions. The teeth meshed in the center so as to assure positive displacement of the fluid. No provision for sealing was made beyond working with minimum clearances. Still, the Pappenheim pump is considered more advanced than Ramelli's because it did not have sliding, reciprocating vanes and no off-balance eccentric rotor. The Pappenheim pump was used for water fountains in many European cities: Rome, Prague, Regensburg and Salzburg. It was driven not by man-power but from an overshot water wheel set in a stream.

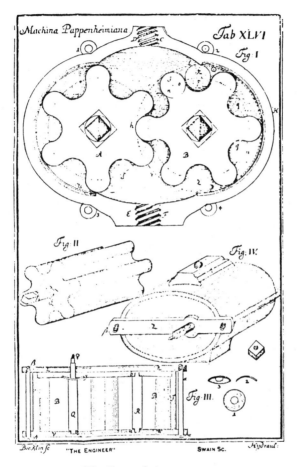

The Pappenheim pump.

WATT

The man most frequently credited with the invention of the piston-type steam engine, James Watt (1736–1819), also did much work on steam powered rotary engines. The son of a carpenter and shipping merchant, he spent much time in his father's shops and became a skilled artisan. At the age of 19, Watt went to London for training as an instrument maker. When he returned to Scotland several years later, he was appointed instrument maker at the University of Glasgow. The idea of a steam engine was suggested to him one day in 1759 by Professor John Robison of the University of Glasgow.

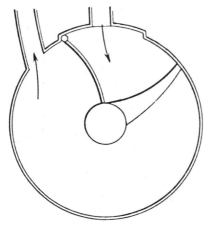

The rotary engine designed by James Watt (1782).

In 1763 a model of a Newcomen steam engine was submitted to Watt for study. He then began experimental work on steam power and soon developed many improvements on the Newcomen engine. Comprehensive steam engine patents were taken out by his company, Boulton & Watt, in 1769, 1776 and 1784, covering enough details of steam engine construction to assure a virtual monopoly for the production and marketing of steam engines for many years.

Watt's partner, Matthew Boulton (1728–1809) was an engineer from Birmingham. They joined forces about 1765, for the express purpose of manufacturing steam engines. In 1765 Watt invented a piston-type steam engine with a separate condenser. In 1784 he took out a patent which covered road vehicles and sea-going vessels propelled by steam engines. Watt never built a steam carriage himself, but contributed much theoretical work on the idea.

The first mention of Watt's rotary steam engine occurred in a letter to his friend, Dr. Roebuck, in February, 1766: "I have thought on a simpler circular steam engine than what I mentioned to you, and which I expect will be practicable." He worked on this engine for several months that year, but found that it was more difficult to make a model than he had expected. But the following year Watt suggested a new rotary machine consisting of a "right and left-hand bottle screw spiral involved in one another." This scheme could be for a turbine, but sounds more like two intermeshing right and left hand screws inside a common casing. The machine seems to have been built about 1771, but it was not till 1774 that it was actually tested.

Between 1772 and 1782 Watt designed a series of rotary-piston steam engines of different types. The first and rather primitive Watt rotary machine was made of a wing-shaped rotary blade inside a cylinder. The blade was not permitted to complete one full revolution. It opened an intake port and approached an exhaust port, then returned to its original position by reversing its direction of rotation. The intake and exhaust ports were separated by a curved radial wall anchored at the rotary blade axis and providing a seal against the housing.

Watt's enthusiasm for the rotary engine seems to have run hot and cold. Apparently he was ready to abandon all work on rotary machines as early as 1772, before the first one had even been tested. The first test report dates from 1776, when Boulton wrote that their last wheel had a power equal to three horses. The defects of the machine had then become apparent. There was too much friction and too much leakage.

In 1782 Watt patented another form of rotary engine which consisted of a rotor plate mounted on an axle and adapted to turn inside a cylindrical casing. An "abutment" was hinged to the casing and bore against the axle at the outer end. Steam was admitted on one side of this abutment and exhausted from the other. It is doubtful that this engine was ever built. However, there are distinct suggestions that the machine was patented with the object of discouraging one of the firm's workmen (Cameron) from spending further time on developing the idea.

The problems of inventor (employee) and industrialist (employer) relationships so common nowadays are not a phenomenon unique to the 20th century. There is reason to believe that Watt's 1782 rotary machine patent did in fact cover not one of his own ideas, but Cameron's. Cameron had approached Watt with an idea for making a rotary or perhaps a semi-rotary steam engine, but Watt said that he had himself thought of the same type of engine some years before and had even made a model of it. Cameron refused to be discouraged and got permission to work on his own idea, but all his efforts were in vain.

Shortly after Watt's patent was issued, Cameron left Boulton & Watt. In a patent application dated 1784, Cameron described a machine much like Watt's but having two reciprocating abutments instead of the single hinged flap, plus another interesting machine, incorporating a rotor which moved in a deep helix, right-hand for half a turn and left-hand for the rest, so that it completely encircled the shaft. This rotor was contained in a cylindrical casing of the same diameter as that covering the top of the helix blade. A sliding frame projected into the side of the working chamber and fitted over the helix, it was capable of moving up and down as the helix turned, acting as an abutment. The frame admitted the steam. This frame acted as a slide valve, and could be given a dual-directional action so as to give a continuous impulse towards turning the helix. Since the machine was dependent upon the use of a separate condenser, it infringed Watt's patent and was refused.

HORNBLOWER

Watt had many rivals, inside and outside his own firm. One of the outside rivals who gave him the most trouble over patent claims for rotary machines was Jonathan Hornblower, the inventor of the compound engine. His rotary, patented in 1781, is remarkable for its ingenuity, in terms of its conception as well as its construction. It consisted of two hollow vanes, mounted together like the hinges of a door inside a cylindrical casing concentric with their axles. The axis of one vane projected through the hollow center of the axis of the other vane, so that the vanes could move independently. The inside of the vane which was mounted on the longer shaft was supplied with steam through a passage in the shaft—the other vane was connected with the exhaust pipe in a similar manner. One vane had mushroom valves arranged in each of its faces and adapted to lift outwards. The other had similar valves but lifting inwards. The two valves on each vane were interconnected so that if one was open the other had to be closed. Despite its merits, the Hornblower engine could not be made to work on account of its friction and leakage problems.

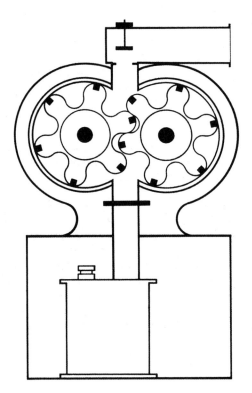

Murdoch's rotary steam engine of 1799 borrowed its geometry from the Pappenheim engine, but had apex seals in the form of wooden scrapers in the gear tooth tips.

MURDOCH

William Murdoch, another employee of Boulton & Watt, used the Pappen-heim gear-type pump to contrive a rotary steam engine in 1799. He fitted wooden apex seals in each gear tooth to improve gas sealing. Steam was admitted through a mushroom valve and was exhausted to a separate condenser mounted below the machine. The date of its appearance and the use of a separate condenser suggest that Murdoch may have delayed application for a patent until Watt's patent covering the condenser expired. The air pump for the condenser was driven by a crank from the main shaft. Each tooth of each gear carried in its head and around its flank a layer of packing in a slot.

Murdoch's engine had no external gears to keep the two rotors in correct angular relationship and to prevent wear of the contacting faces. Despite the rotor seals, the engine was still quite inefficient, as other sealing problems remained unsolved. It also suffered from excessive friction. An engine of this kind developing about ½ h.p. was set up at the Soho Foundry in 1802 to drive the machines in Murdoch's workshop. It proved not to be capable of meeting any useful purpose.

BRAMAH

Joseph Bramah, versatile engineer, large-scale inventor, and operator of a big factory in Birmingham, was also attracted to the idea of rotary machines. Bramah, of course, is famous for other inventions such as the hydraulic press and the flush toilet.

He patented his hydraulic press in 1796. He patented a screw propeller in 1785, the modern flush toilet in 1778. He also manufactured unpickable

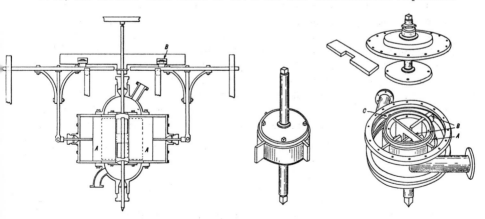

Various views of the 1890 Bramah rotary steam engine.

door locks and made many improvements to machine tools. In 1802 he built a rotary wood-planer for Woolwich Arsenal. But his first patent for a rotary engine dates back to 1785 and included a machine very much like that described in Watt's patent specification of 1782. The hinged abutment was to be closed by weights. Another had a sliding block on the cylinder actuated by a rack and sector wheel. Bramah thought of using rotary engines for marine installations. His patent specification of 1790 includes two different versions of a rotary engine.

The first type consisted of a piston mounted on a drum, and able to revolve inside a cylindrical casing. Two reciprocating abutments projected from the walls of the casing towards the drum, operated from an external guide ring mounted on the shaft. One had to be closed while the other was open to let the rotor pass. The steam and exhaust ports were positioned on opposite sides of the rotor, fed by passages in the shaft. The second design consisted of a hollow drum placed eccentrically inside a casing, having two blades projecting across it on diameters at right angles, and fitting against the walls of the containing cylinder. This was essentially nothing more than an elaboration on the Ramelli pump principle. Nothing more was heard of Bramah's rotary machines, but it is interesting that he was to play a significant part in the development of motorized highway travel in Great Britain. In 1821 Joseph Bramah had under construction in his Birmingham factory a steam stage-coach designed by Julius Griffiths of Brompton. In 1828 Bramah built the oft-depicted Chruch vehicle—a fancy-bodied steam-driven road vehicle built to ply between London and Birmingham on a regular schedule.

FLINT

The abutments used to control steam flow in the engines we have studied so far were mounted on simple hinges. The idea of using rotary abutments for rotary engines was not far off, however. And, of course, the rotary abutment is the ancestor of the rotary valve. The honor of having been the first to suggest a rotary engine with rotary abutments must go to Andrew Flint, who received a patent in 1805. In his machine there was to be a fixed outer cylinder carrying two nearly semicircular abutments operated by an external striker mechanism. These abutments did not rotate continuously but were to be swung aside to allow the rotor to pass. The single rotor was to have intake and exhaust ports arranged on each side. The engine would therefore, had it worked, have had no dead centers.

POOLE

Moses Poole patented a rotary engine in 1817. It had a cylindrical rotor inside a cylindrical casing. The rotor shaft provided no eccentricity, but gas sealing was supposed to be performed by hinged flaps carried on the rotor.

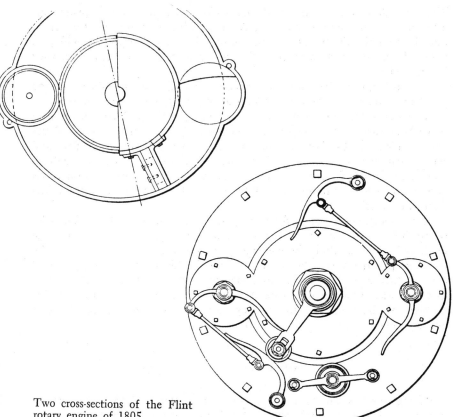

Two cross-sections of the Flint
rotary engine of 1805.

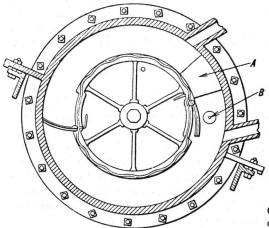

Cross-section of the Poole engine
of 1817.

These two flaps were positioned diametrically opposed. A control piston was integral with the stationary casing. A small wheel mounted on the casing forced down the abutments as they approached, so that they would pass under the piston, and an external "striker" gear re-erected them after they had passed.

The Poole patent description contained a great deal of thought on sealing problems. Poole did not approve of the sealing systems then common in steam systems. He showed a clear understanding of the problems involved in preventing waste and escape of steam between the sides of the rotor and the flat end covers of the hollow casing as well as between the rotor edges and the circular chamber. Other steam engines used a packing of hemp or cotton, lubricated with tallow. This form of sealing produced great friction losses particularly when compression was high enough to make the fittings perfectly steam tight.

The stuffing Poole specified was to consist of gaskets of plaited yarns inserted "in a groove cut in a serpentine or waving line, so near the outer edge of the wheel as to admit the packing at certain intervals to lie close to the steam stop and prevent the waste of steam." This statement is proof that even at this early date, the main difficulty in the construction of a rotary machine was recognized by capable engineers. It was an insurmountable problem at the time, due to the lack of precision in machining methods and equipment.

WRIGHT

Poole's patent seems to have led to nothing as far as Poole's own activity is concerned, but was probably the inspiration behind Wright's invention. Lemuel Wellman Wright patented a rotary engine in 1825 which appears to be based on Poole's. The rotor carried flaps for sealing. Instead of using a wheel to depress the flaps, the steam stop itself was formed like a ramp so as to force them gradually down. They were raised again by tail pieces mounted externally on their shafts striking against a wheel. But again—as in Poole's description—the specification is principally concerned with sealing problems.

Wright proposed to adapt for use in his rotary engine some of the same types of metallic packing which were then coming into use in reciprocating piston engines. Wright mentioned packing the ends of the rotor with plaited yarn in a serpentine slot. He also called attention to the packing of the flaps. The outer edge of each flap was to be made of a separate piece carried by means of slots on pins fixed to the main body of the flap. This loose piece was to be pressed outwards by a steel strip spring, as the spring working under the apex seal in a Wankel engine. Effective packing of the steam stop was to be obtained by a piece of brass, spring-loaded to keep it continuously pressed against the periphery. Packing at the ends of the rotor was effected by rings forced against the stationary part by helical springs.

MARRIOTT

In 1828 Henry Marriott and August Siebe, one an ironmonger and the other a machinist, patented a rotary pump which might have proved reasonably practical if it could have been made with sufficient accuracy. Sealing blades were to slide in the slots of a drum-type rotor, forced outwards by a stationary cam inside the rotor to maintain contact with the cylinder. To ensure proper sealing, a part of the cam was to be made as a curved arm pivoted at one end and forced outwards by a spring. The stop was to carry a sliding brass piece forced outwards into contact with the drum by springs and the ends of the blades were to be packed in a similar manner. The pump barrels were to be lined with copper or brass "by casting the iron over the copper or brass."

TROTTER

Trotter's engine was patented in 1805 and had a cylindrical housing with a concentrically mounted drum. It was the first rotary engine to use true eccentric rotation rather than cam-shaped rotors to obtain volume variations. But it was the outer rotor, not the inner one, that was eccentrically mounted. The inner rotor carried a blade which projected through an eccentric drum revolving about its own axis. On admitting steam, the blade revolved, causing the drum to turn about its own axis. In this engine, either shaft could be made to revolve at a steady unvarying rate, with the other turning at a varying speed.

Steam could be applied simultaneously on both sides of the inner rotor, forcing the blade around with very nearly even torque, the steam acting always on the whole of the blade face, except that part actually passing through the slot in the eccentric drum. As late as 1882 an example of an engine constructed to this design was shown at an exhibition at the Agricultural Hall in London.

GALLOWAY

Trotter's ideas were taken up by Galloway 35 years later. In 1846 Elijah Galloway patented the first rotary engine with an epicycloidal rotor and an outer envelope. The engine could be described as "paracyclic." The five-lobe rotor ran in a five-lobe housing, giving a 1:1 ratio of rotational movement. It had a circular rotary housing with a five-lobed inner surface. The inner rotor was a spider with five arms, each arm being captive within one lobe of the housing throughout the operational cycle. The inner rotor was eccentrically mounted on a crank. As the crank revolved, the center of the inner rotor described a circle around the center of the outer rotor. Correspondingly, the end of each arm traveled in a circle. It traveled along the curvature of the lobe from about 120 degrees before top dead center to 120 degrees after top dead center, then making a return, in circular motion, across the open gap from 60 degrees after bottom dead center, without being in contact with the housing surface at all. No other guidance than that provided by the crank position existed. During this process the spaces in the lobes were caused to expand or contract in a precisely timed fashion.

The geometry of the Galloway engine is simple in principle. It is based on a pentagon. If circles with a very short radius are drawn from each point of the pentagon, the ends of the five arms have been determined. Delete the inner segment of about 120 degrees. Next, determine the midpoints of the lines between the points of the polygon and make them centers of five circles of the same radius. On these circles, delete the outer 180 degrees. Now we go back to the points of the pentagon, this time with a longer radius. This long radius is equal to the half-distance between the points less the radius of the partial circles drawn before. Use this new radius to draw partial circles around the earlier ones, using all the centers used earlier, points and midpoints between them. This has the effect of linking all the segments together

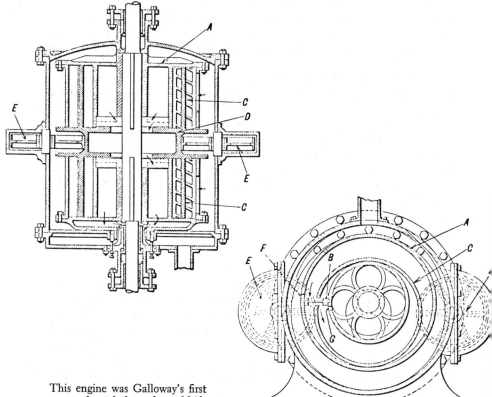

This engine was Galloway's first approach and dates from 1840.

into two paracyclic curves, one inside the other, with a common center. The outer one corresponds to the casing and the inner one to the rotor. All that is missing is the eccentricity. Move the rotor center away from the housing center in any direction until contact has been obtained. That fixes eccentricity at twice the short radius used for the initial sets of partial circles.

Galloway made the rotor a loose fit on the crank and spring-loaded the crankpin to assure permanent contact between rotor and casing at four points simultaneously. The spring is a circlip-type inserted in a slot on the crankpin. Four points maintain contact, because as soon as one arm leaves the working surface, another has just made renewed contact. This fact ensured some degree of sealing between the lobes, except the pair which communicated when the arm diametrically opposed to their dividing projection was at top dead center.

Galloway's 1846 engine showed considerable advancement.

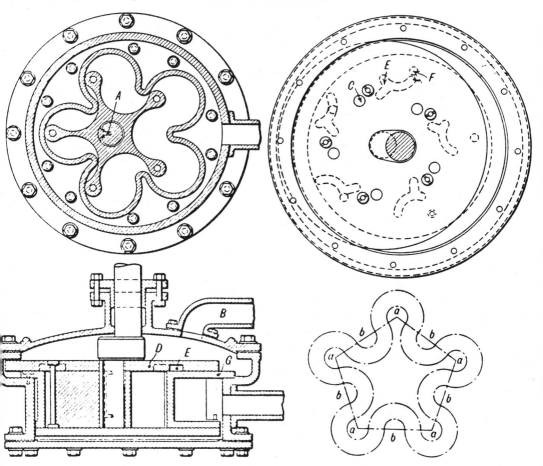

The Galloway engine ran on steam admitted through circular side ports in a cover plate which revolved with the rotor, into the several lobes. Each lobe had its own intake and exhaust ports. The exhaust ports branched out into spur-shaped recesses, curved to suit the motion of the arms. Due to its geometry, the Galloway engine produced torque over about 200 degrees of mainshaft rotation (40 degrees per rotor lobe). Although some packing was provided around the ports, Galloway did not apply rotor seals to his engine. Steam consumption was therefore rather high and efficiency quite low. It was used as a marine engine and developed about 16 h.p. at 400–480 r.p.m.

PARSONS

Sir Charles Parsons (1854–1931), father of the steam turbine, invented a rotary engine in 1882. It was not a true rotary engine, and was in no way a

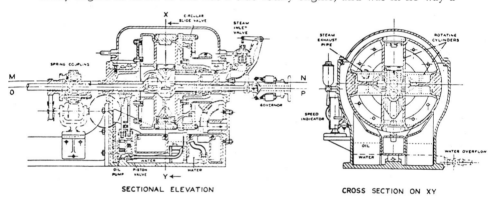

SECTIONAL ELEVATION   CROSS SECTION ON XY

Elevation and cross-section of the Parsons rotary steam engine.

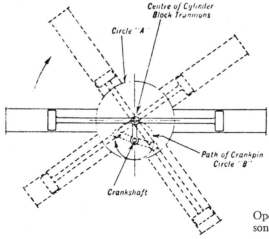

Operating principle of the Parsons rotary engine.

step on the way towards the Wankel engine. It had cylinders and pistons, but it was not a rotary in the same sense as the radial engines once popular in aircraft. The four cylinders were positioned on two diameters at 90 degrees, in "+" formation. The cylinders revolved on trunnions, and their rotation produced a relative movement between each cylinder and its piston. Several units were built and tested by Kitson's of Leeds. The Parsons engine was reported to have excellent balance, but it used up steam at the rate of 40 pounds per horsepower-hour.

ROOTS

The Roots pump would not make an efficient engine, chiefly because of low compression potential and cooling problems, but its mechanism is pure rotary, and has been extensively used for air compressors. The Roots-type compressor is a simplified version of the Pappenheim pump, with dual two-lobe rotors engaging to assure positive displacement. It was invented in 1860 by two Americans, Philander H. Roots and Francis M. Roots. Their father operated a wool mill, and for years their thoughts had mainly been occupied with problems of belt gearing and other machinery.

The two interlocking rotors revolve in an oval-section casing. A gear train keeps the rotors in phase with each other and maintains the inter-rotor clearance. Air is scooped in at the inlet opening on one side of the casing and discharged into the outlet duct on the opposite side. The only known rotary engine based on the Roots compressor was built in New York in 1867 by an engineer named Behrens. The rotors overlapped in the same way, but the lobes had a completely different shape.

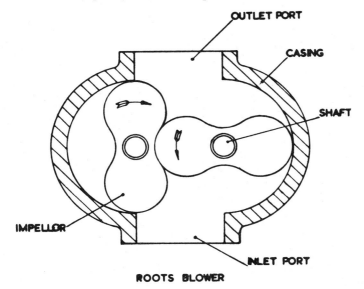

Roots-type blower, used for mine ventilation at Chilton Colliery.

The most important theoretician on the subject of rotary piston engines was Franz Reuleaux, Professor at the Technical Institute of Berlin, and author of many textbooks on machines that remain classics to this day. One book was "The Constructor" published in 1893. Earlier he had written: "Kinematics of Machinery" in 1876. But he was never able to define the workable rotary piston engine.

For the rotary engine inventors of the 20th century, there was a wealth of empirical knowledge to draw on, but no proper science, no clear directions. Still, three inventors are to be singled out for their contributions towards a

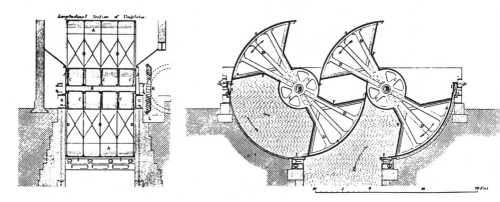

Simplified cross section of the Roots-type pump, as applied to a modern supercharger.

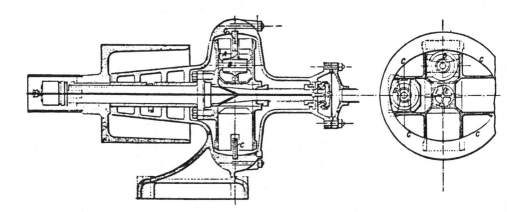

Swedish engineer Fredrik Ljungström invented this rotary steam engine in 1900. It has four radial cylinders cast into one block, mounted on a revolving shaft. The engine is crankless, and each piston has a roller pin instead of a wrist pin. The roller pins carry rollers which bear against an elliptical guide in a stationary housing. Centrifugal force kept the rollers in permanent contact with the guide. The engine is stated to have run at speeds up to 1,840 r.p.m.

workable rotary engine: John Francis Cooley, Dimitri Sensaud de Lavaud, and Bernard Maillard. Each was on the track of the geometry of the Wankel engine. But they lacked Felix Wankel's persistence and singleness of purpose. They gave up when their engines were still in the concept stage or in the early stages of development.

COOLEY

John F. Cooley from Allston, Suffolk County, Massachusetts, was a pupil of Reuleaux, and had been engaged to work on rotary piston engines in Berlin under his direction. After his return to the U.S.A. at the turn of the century, Cooley continued his research and experimental work on rotary engines, and formed a corporation to undertake this project: Cooley Epicycloidal Engine Development Company, Jersey City, New Jersey, and Boston, Massachusetts.

His invention was covered by U.S. patent number 748,348 dated December 29, 1903. The Cooley rotary engine was a two-lobe inner epitrochoid with a three-lobe outer envelope. Cooley was first to apply the inner epitrochoid shape to the rotor, with an outer hypotrochoid to form the working chamber. This is exactly the opposite relationship to the Wankel engine's geometry. Cooley has one more lobe in the chamber than on the rotor; Wankel has one more lobe on the rotor than in the chamber. He patented his engine for use both as a pump and as a steam engine. He described the operational cycle as allowing steam to enter through a port in a "spacer" into the pressure chamber, where it would force the rotor to turn. The spacer was actually the outer rotor, corresponding to the outer rotor in Wankel's original DKM-54.

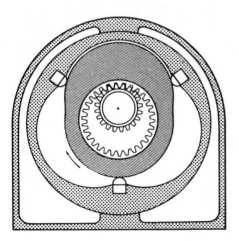

Simplified cross-section of the Cooley engine.

The inner rotor was eccentrically mounted, and this was the feature that assured rotation. The spacer surface was in continuous contact with the periphery of the rotor. It was this permanent contact which assured the volumetric variations and enabled the Cooley machine to operate as a steam engine. The inner rotor (piston) and outer rotor (spacer) were mounted on axes parallel to each other inside a cylinder (casing). Cooley described the cylinder as having radial partitions. These partitions were bearings for the spacer.

The rotor was mounted on a shaft eccentric in relation to the cylinder

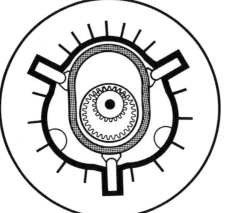

Simplified cross-section of the Umpleby conversion of the Cooley steam engine, to operation on the four-stroke internal combustion cycle.

Swedish inventors Wallinder and Skoog were issued a patent in 1923 for this true rotary engine, incorporating a hypocycloidal inner envelope, a five-lobe rotor, and a 5:6 rotation ratio, assured by central phasing gears. The patent covered both two-stroke and four-stroke operation.

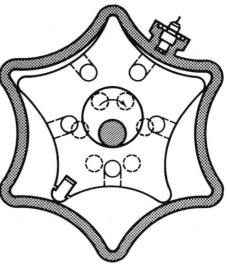

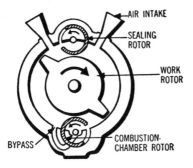

The Baylin engine, from Canada, is similar to Wankel's concentric engine of 1934, with its center rotor and dual rotary valves.

center. Bearings for the eccentric shaft were provided by the end covers of the casing, outside of the phasing gears. Rotor movement and phasing were directed by internal gearing. Both rotors revolved in the same direction, but at different speeds. The rotor gear was intermeshed with the spacer gear and assured the direction of rotation.

The spacer had entry and exit ports for steam between its points of contact with the rotor. The combined motion of rotor and spacer opened and closed the ports at the proper time. The sealing system was primitive, and steam consumption high. Apex seals were carried in the spacer, not in the inner rotor, as preferred by Wankel. Cooley was forced to discontinue development work on his engine. He made no attempt himself to convert it to operation on the four-stroke cycle as an internal combustion engine. But an English engineer named Umpleby did just that in 1908, only five years after Cooley had obtained his patent. However, Umpleby failed to develop the engine.

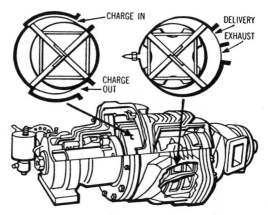

The Larsen engine, from Denmark, is based on Ramelli's principles, as applied to the four-stroke internal combustion cycle.

SENSAUD DE LAVAUD

Dimitri Sensaud de Lavaud was a Brazilian of French origin, who first gained prominence as the inventor of a novel method of casting pipe centrifugally, now used extensively throughout the world. He was a pioneer aviator, having built his own airplane and flown it in Brazil in 1909. He arrived in Paris in 1920 with the idea of going into retirement in the most civilized surroundings he could imagine. Instead, his inventive mind turned to ideas for perfecting the automotive transmission system and he invented a stepless drive system that would automatically select the right gear ratio for the car according to speed and load.

He took out patents for independent front wheel suspension and limited-slip differentials back in the twenties. In the thirties he began to study power plants and was attracted to the principles of the rotary engine. His work in this area resulted in the issue of French Patent 853 807 on December 16,

1938. The engine consisted of a five-lobe outer rotor with the shape of an inner hypocycloid, and an inner rotor with six lobes, running on a 5 to 6 reduction ratio.

The outer rotor was cylindrical in shape and was held inside a stationary casing. The inner rotor was eccentrically mounted inside the outer rotor.

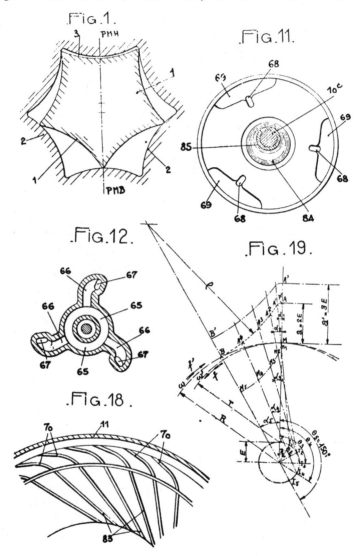

Illustrations from the D. Sensaud de Lavaud patent of 1938. The four-stroke cycle is completed in two revolutions of the inner rotor, and $\dfrac{4\pi \times 5}{6}$ rotations of the outer rotor, in the example shown.

Since the inner and outer rotors were kinematically interlinked, they revolved at speeds inversely proportional to the number of their lobes. The inner rotor had concave faces, matching an outer rotor with a working surface made up of six convex segments. The relationship between the inner and outer rotors provided a closed chamber between each two apices on the inner rotor. The volume of each chamber changed during rotation. Minimum volume was obtained when two inner rotor apices coincided most closely with two lobes of the working surface. Maximum volume was obtained when an inner rotor apex reached the peak of a convex segment—or exactly halfway between two lobes of the outer rotor. It had two coaxial shafts. The outer rotor completely enveloped the inner rotor and was fixed to the output shaft on one side and running in a ball bearing on a support shaft on the other side. This support shaft carried the inner rotor on an eccentric, ending before meeting the output shaft. Its eccentric end was carried in an eccentric bearing. The outer rotor had a cylindrical shape on the outside. The inner rotor was split down the middle—as in two discs. They were fixed to each other by springs loading them laterally to provide side sealing. The gap between them had strip-type gas seals.

If you follow the volume change during rotor rotation, it will soon be seen that one set of rotors only constitute a compressor—not an engine. The four-stroke cycle cannot be completed in one full revolution of the two rotors relative to each other. Starting from the point when the intake port opens, maximum compression will be reached when both rotors return to the same relative positions. For the Sensaud de Lavaud compressor to be converted into a four-stroke internal combustion engine, two sets of rotors are needed, with one chamber performing the duties of intake and compression, the other going through the expansion and exhaust phases. For this to be possible, they need a communicating port which admits compressed mixture into the combustion chamber at the right time and in the right volume to maintain its state of compression up to the moment of ignition.

Fresh mixture is admitted inside the casing through an axial pipe. An annular chamber around the output shaft feeds gas inside the outer rotor of the compression chamber. The gas passages have seals to avoid gas leaks between the housing and outer rotor surfaces close to output shaft bearing —to give low rubbing speeds. The sealing problem here is simple—because it is a depression area during the intake phase, pressure never goes higher than atmospheric. A side port in the outer rotor admits the compressed gas to the working chamber.

On the exhaust side, the outer rotor carries vanes to speed up scavenging from the working chamber into the annular exhaust outlet. Also, fresh air is injected into the exhaust chamber which lowers the exhaust gas temperature and adds to its volume. The outer rotor, according to the patent claims, is therefore a partial gas turbine! As in the Cooley engine, there is permanent sliding contact between the tip of each lobe on the inner rotor and the working surface. Each lobe apex was provided with a radial groove for a seal strip, spring-loaded and under centrifugal load, against the working surface. The

Elevation of the Sensaud de Lavaud engine. Below, details of the phasing gears and gas seals are shown.

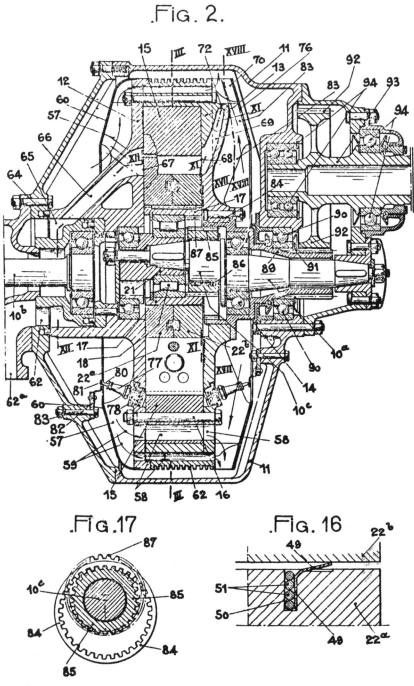

.FIG. 2.

.FIG.17

.FIG. 16

seals are counterweighted. The counterweights are subject to an opposite centrifugal force to that acting on the seals, thereby they limit pressure on the working surface.

Gearing of the Sensaud de Lavaud engine is simple. One side of the outer rotor carries a ring gear with internal teeth, coaxial with the rotor support shaft. It meshes with a smaller planet gear located inside it, mounted on the eccentric part of the support shaft. It is therefore coaxial with the inner rotor. The number of teeth on each must be full multiples of the number of lobes on both rotors.

Ignition is performed by twin spark plugs mounted in the end plates of the outer rotor, directly opposing each other. Plugs are carried on the flanks of the outer rotor. Both plugs have contact rings for high tension current. The ports are so positioned that rotor movement controls their opening and closing. If the plugs are at 6 o'clock, the ports are at 12 o'clock. Ports oppose each other—intake on one side, exhaust on the other.

De Lavaud specified the possibility of diesel operation. Its operation on any cycle would have remained hypothetical if it had only been D. Sensaud de Lavaud's private affair. But one test engine was built by the Ateliers de Batignolles at the instigation of the French Air Ministry, with the support of Citroën and Renault. However, power output never came up to the expected figures, and the project was abandoned in 1941.

MAILLARD

One American tutored in Germany; one Brazilian working in Paris; what next? One citizen of Switzerland, living peaceably in Geneva. Bernard Maillard had no exotic background in any way, but he seems to have come closer to anticipating the Wankel engine than anyone else. He was a mechanical engineer, employed by Adolph Saurer AG of Arbon, builders of high-grade trucks and diesel engines of all sizes. He invented a rotary machine and received a British patent in 1943.

His design shows compressor activity in two phases per rotor revolution, with intake ports at 1 and 7 o'clock, and exhaust ports at 5 and 11 o'clock. This did not allow much time or volume for compression, as the exhaust port was opened only about five degrees after the intake port closed. As a result, the Maillard design is not adaptable to torque-producing engines.

Maillard's basic invention consists of a static two-lobe housing containing a three-lobe rotor, with chamber volumes varying to provide the necessary compression and expansion to operate on a four-stroke cycle. It was intended as an air compressor. Both sides of the minor axis had an intake port and an exhaust port. As an air pump, it has the drawback that the intake period is short and maximum chamber volume is reached after the intake port has been closed. Whatever provision for sealing Maillard envisaged is ·not described in the patent. If Maillard's invention looks like a poor compressor and an unworkable engine, what is the value of his patent? Quite simply this: He showed the first combination of an inner hypotrochoid rotor, eccentrically mounted in an outer hypocycloid working chamber. Not being acquainted with Felix Wankel and his post-war work, Maillard let his patent lapse in 1948.

·   ·   ·

Geometry of the Maillard patent engine.

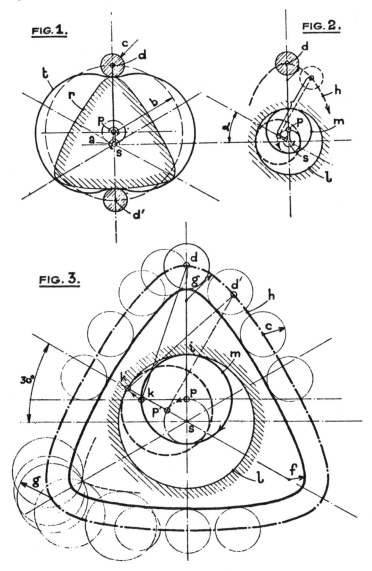

All the rotary engines we have reviewed thus far have had an essential defect when evaluated as an internal combustion engine because primary attention was paid only to modification of a mechanism, or there was no advantage over the reciprocating engine. Another reason why all research work made in the distant past on various types of rotary engines ended in failure was the low level of machining technology and quality of materials which form the background for the development of the engine.

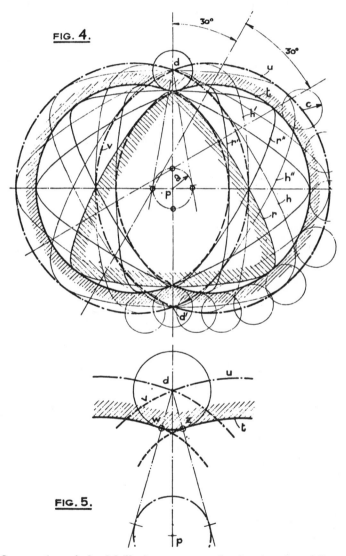

Cross-section of the Maillard compressor, showing its adaptability to both two-lobe and three-lobe inner rotors.

The Wankel engine was the first rotary piston engine on which systematic studies were carried out in full scale with respect to the cooling system, the lubricating methods, the combustion process, port timing, position of spark plugs, effect of machining accuracy and gas sealing. It was unfortunate that no wide-base research such as this was ever made on any previous rotary engine. The arrival, development and success of the Wankel engine has not deterred other inventors from pursuing their dream: a superior rotary engine. Every year new ones come up, and some recent ones are forgotten. But, in common with inventions of the distant past, none of these have had the benefit of a concerted research, test and development program such as brought the Wankel engine to the forefront in record time.

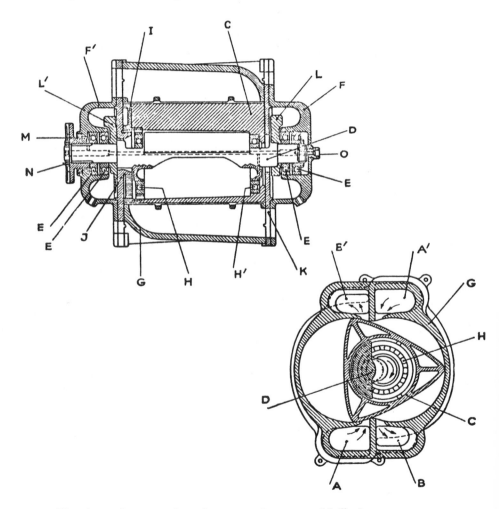

Elevation and cross-section of a proposed prototype Maillard compressor.

At the concept level, the Wankel engine has many rivals. Strange engines with rotating pistons, curved pistons, vane-type pistons, orbiting pistons, and scissor-action pistons. Rotary engines come in three main groups:

1. Eccentric-rotor types, such as the Wankel, Jernaes, and Renault.
2. Scissor-action types with pistons or vanes, such as the Tschudi, Kauertz and Virmel.
3. Revolving-block types, such as the Mercer, Selwood and Porsche.

JERNAES

Finn Jernaes, a Norwegian inventor living in Kristiansand near the southern tip of Norway, has taken the three-lobe rotor and trochoidal envelope from Wankel's design but has come up with a novel type of gearing for his "Planet Motor" patented in the U.S.A. in 1965. The Jernaes engine is built up around a central output shaft fitted with an integral circular plate which works as a rotor hub. Unlike the rotor in the Wankel engine, the Jernaes engine rotor has no internal gearing.

The hub supports three planet gears running on eccentric shafts. These planet gears mesh with a stationary reaction gear mounted concentrically with the output shaft and the hub. Each planet gear has the same eccentricity and they are spaced 120 degrees apart. As the rotor turns, it carries the planet gears along. They are phased to let the output shaft turn at rotor speed (instead of three times rotor speed as in the Wankel). Finn Jernaes

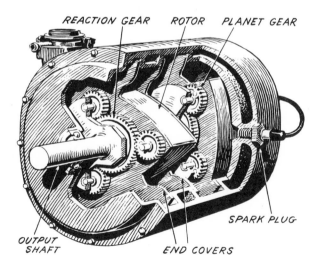

One possible version of the Jernaes engine with twin rotors. (*Drawing: Ray Pioch*)

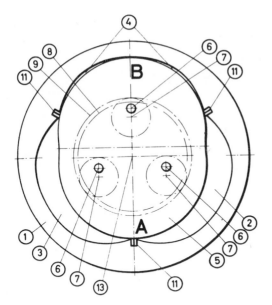

Cross-section of one proposed version of the Jernaes engine, using a three-lobe chamber with an inner epitrochoid. 1 = Stationary casing. 2 = Working chamber. 3 = Working chamber. 4 = Working chamber. 5 = Rotor. 6 = Stub shafts. 7 = Planet wheel. 8 = Disc. 9 = Sun wheel. 11 = Seal strips. 13 = Drive shaft center.

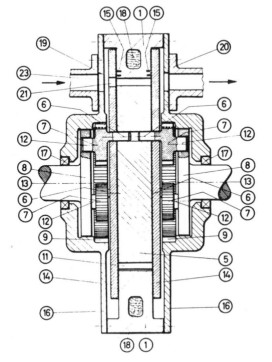

Side section of one proposed form of Jernaes engine, showing one rotor and its chamber. Code numbers as in Fig. 28 with the following additions: 12 = Shafts carried on discs 14. 13 = Drive shaft. 14 = Cover discs. 15 = Sealing rings. 16 = End cover shields. 17 = Mainshaft bearings. 18 = Coolant passages. 19 = Intake manifold. 20 = Exhaust manifold. 21 = Port. 23 = Gas passage.

claims that his gear mechanism makes for a big increase in torque at relatively low r.p.m. In other words, a four-stroke single-rotor Jernaes engine produces the same number of power impulses per output-shaft revolution as a six-cylinder piston engine. The rotor side-sealing is ingenious: the end covers revolve with the rotors and are sealed with two simple rings on each side. Intake and exhaust ports in the casing are opened and closed by the passage of holes in the revolving end covers.

The patent covers a multitude of variations on the engine described. A number of American companies including General Dynamics, Lycoming Division of Avco Corporation, and Rocketdyne Division of North American Rockwell have examined the Jernaes engine, but no real progress has been made.

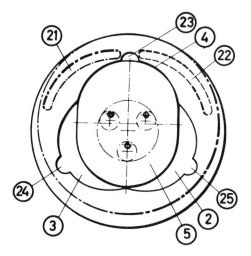

The Jernaes engine discs are provided with suitably shaped openings, which uncover gas passages leading to the working chambers. Intake slot 21 feeds fresh mixture through the gas passage 23, as long as the openings coincide. The same applies to the exhaust slot 22, with respect to gas evacuation, as the rotor and disc turn together.

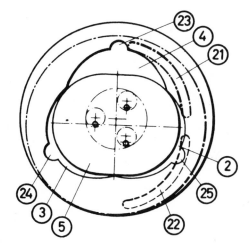

RENAULT

The research department of the Regie Nationale des Usines Renault, France's government-owned automobile company and the largest manufacturer of passenger cars in the country, began to study the Wankel engine in 1958, but centered its own designs and experiments on the Cooley patents. Renault started off with a rotary engine with poppet valves instead of ports as in the Wankel engine. A four-lobe rotor ran in a five-lobe trochoidal

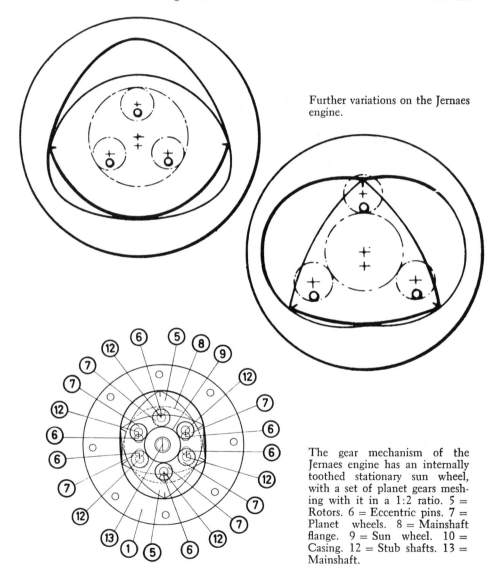

Further variations on the Jernaes engine.

The gear mechanism of the Jernaes engine has an internally toothed stationary sun wheel, with a set of planet gears meshing with it in a 1:2 ratio. 5 = Rotors. 6 = Eccentric pins. 7 = Planet wheels. 8 = Mainshaft flange. 9 = Sun wheel. 10 = Casing. 12 = Stub shafts. 13 = Mainshaft.

chamber, with a separate intake and exhaust valve for each chamber lobe. Another notable feature lies in the fact that the piston seal tips are mounted in the housing and not in the rotating piston, as in the Cooley engine. The patent showed a cross-section of an engine with low compression and a combustion chamber shape of dubious value.

Renault spent considerable time and money on its development, and interested American Motors Corporation in its adaptation to automobiles of American size. A statement made jointly by Roy Abernethy, President of American Motors Corporation, and Pierre Dreyfus, President of the Regie Nationale des Usines Renault, on September 19, 1962, said that the two companies will share engineering and technical resources and costs in the long-range research into lighter, less complex and more efficient powerplants. Ralph Isbrandt, vice president in charge of automotive research and engineering, gave a technical presentation of the rotary engine then under consideration. He described some of its essential features and also drew a comparison with the Wankel engine.

When the rotor in the Renault-Rambler engine turns as a result of gas pressure on 'one or more of its faces, it is carried around its orbit by the

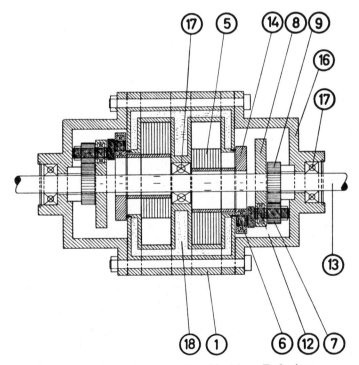

Elevation of the gear system shown on Fig. 32. 14 = End piece. 16 = End housing. 17 = Mainshaft bearing. 18 = Water jacket. 8 = Flywheel flange with starter motor gearing. 9 = Sun wheel. 5 = Rotor.

eccentric shaft. A pair of gears assures that the rotor will turn on the eccentric
in such a manner that clearance is always maintained between the rotor and
the stationary housing. The same set of internal and external gears coordinates
the rotor rotation and crank rotation to keep them in phase. Five radial seals
contact the rotor's outer contour and separate the individual combustion
chambers. Several types of seal can be used to contact the sides of the rotor
and complete the job of sealing off the chambers.

In the Wankel engine, the gases are carried around inside the outer housing, within the moving chamber formed by the rotor as it turns. Fuel-air
mixture is taken in on one side and compressed, fired at the top, and expanded
and exhausted on the other side. In contrast, the Renault-Rambler rotary
engine has stationary combustion chambers. In the Wankel design, the apex
seals move with the rotor and are oil cooled, while in the Renault-Rambler
unit, the apex seals are stationary in the housing and therefore water-cooled.
The Wankel uses ports, while the Renault-Rambler type uses conventional
poppet valves, giving more precise timing and closer control of overlap. But
the engine is soon to be abandoned by both Renault and American Motors,
and neither company is currently engaged in research or development work on
any form of rotary engine. Ralph Isbrandt's closing remarks seem particularly
relevant: "This research engine was designed for convenience of experimental
work, with no provision for installation in a car, and no particular concessions
made to compactness or cost. Since this type of rotary engine allows considerable freedom of combustion chamber shape, the experimental engine

Cross-section of the Renault-Rambler rotary engine.

has been provided with removable heads to permit a thorough, long-range research program with a variety of combustion chambers. While we do not expect to offer this engine in a passenger car in the near future, we feel that it has theoretical possibilities, and we thought you would find it of interest. In the continuing effort to increase the total efficiency of the passenger car, we feel that it is essential to explore in depth developments such as this, which offer the promise of basic gains in simplicity, weight, cost, reliability and comfort."

TSCHUDI

The same year that Felix Wankel was born, another boy was born a few miles to the south, on the Swiss side of Lake Constance, not far from Basel. He was the son of a blacksmith, and his name was Traugott Tschudi. He went to America in the Twenties, settling in New York, and began design work on a toroidal engine about 1925.

The Tschudi engine has four curved pistons running a toroidal track. Each pair of pistons is diametrically opposed and carried on a separate rotor. Each rotor also carries two rollers. These rollers bear against two identical cams fixed to the output shaft. The rollers rotate only because of friction against the cams. Power flow is achieved by having one roller push into a cam groove while the other assures positive cam location. The cam disengages from the rollers to allow one rotor to stop. When two pistons stop and restart, the engine obtains a change in the gas volume between the pistons. Piston travel, controlled by cam and roller action, therefore directs intake, compression, combustion, and exhaust phasing. When one set of pistons stops, the other set continues to turn the output shaft. The shaft is eccentric in relation to the toroids and the rotors, and it makes 1.2 revolutions for each 360 degrees of piston travel.

The type of motion given to the pistons is a hesitating progression. The compression is followed by the two pistons containing compressed air between them and they both move together to the spark plug region where combustion occurs. At this point movement of the lagging piston momentarily ceases while the leading piston proceeds at full speed. Eventually, the leading piston reaches the exhaust port where it comes to a halt while the lagging piston follows it, rapidly completing the scavenge. Induction of a new charge is carried out similarly.

The piston movement in the Tschudi engine is an inversion development of the "Geneva Wheel" motion used in motion picture cameras, but it has a reduction ratio of only 2:1 instead of the more usual 4:1 or 6:1. The drive shaft is mounted eccentrically relative to the center of movement to the rollers which then merely idle round about their own axis. For the rest of the time the cam causes the pistons to rotate at a varying speed, depending upon cam profile. For every revolution of the shaft, one-half a revolution is imparted to the pistons.

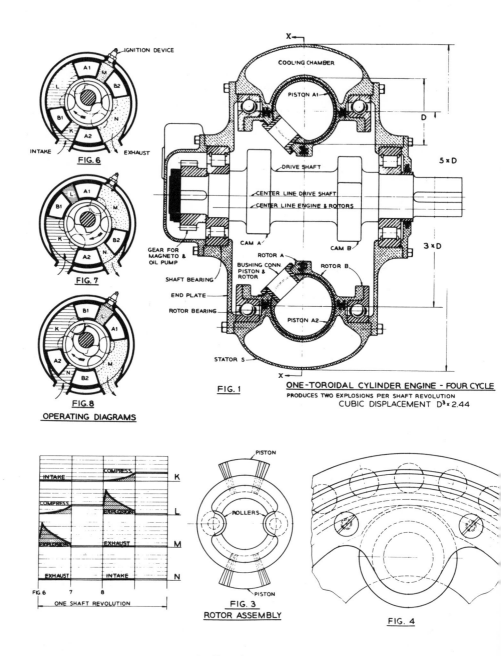

FIG. 6

FIG. 7

FIG. 8

OPERATING DIAGRAMS

FIG. 1

ONE-TOROIDAL CYLINDER ENGINE - FOUR CYCLE
PRODUCES TWO EXPLOSIONS PER SHAFT REVOLUTION
CUBIC DISPLACEMENT $D^3 \times 2.44$

ONE SHAFT REVOLUTION

FIG. 3
ROTOR ASSEMBLY

FIG. 4

Sections and details of the Tschudi engine.

The Tschudi engine works on the four-stroke principle. Each toroid produces only two power impulses per output-shaft revolution. A Tschudi must have two toroids to become a practical torque-producing power plant. Fortunately, the single-toroid unit lends itself to power multiplication simply by

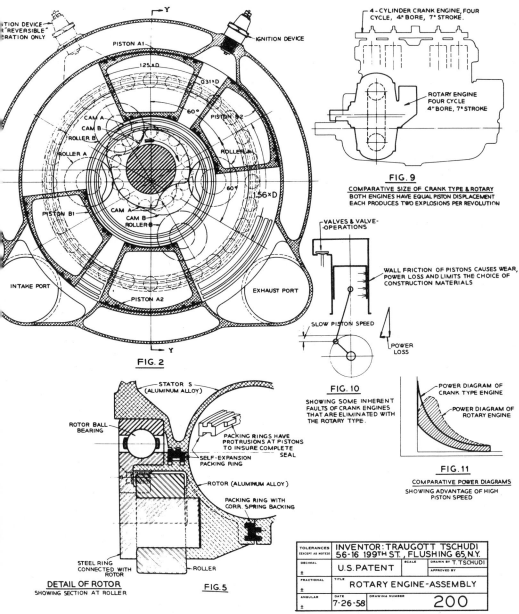

FIG. 2

DETAIL OF ROTOR
SHOWING SECTION AT ROLLER

FIG. 5

FIG. 9

COMPARATIVE SIZE OF CRANK TYPE & ROTARY
BOTH ENGINES HAVE EQUAL PISTON DISPLACEMENT
EACH PRODUCES TWO EXPLOSIONS PER REVOLUTION

FIG. 10

SHOWING SOME INHERENT
FAULTS OF CRANK ENGINES
THAT ARE ELIMINATED WITH
THE ROTARY TYPE.

FIG. 11

COMPARATIVE POWER DIAGRAMS
SHOWING ADVANTAGE OF HIGH
PISTON SPEED

TOLERANCES (EXCEPT AS NOTED)	INVENTOR: TRAUGOTT TSCHUDI 56-16 199TH ST., FLUSHING 65, N.Y.		
DECIMAL ±	U.S. PATENT	SCALE	DRAWN BY T. TSCHUDI
			APPROVED BY
FRACTIONAL ±	TITLE ROTARY ENGINE-ASSEMBLY		
ANGULAR ±	DATE 7-26-58	DRAWING NUMBER	200

Sections and details of the Tschudi engine.

stacking a series of stators and rotor assemblies around the common output
shaft. A double unit would be equivalent to an eight-cylinder reciprocating
piston engine. A 90 degree offset of the cams in contact with the respective
rotors would ensure uninterrupted torque output.

The Tschudi rotary engine can be designed for either spark-ignition or compression-ignition (Diesel) operation. The principal advantage over the reciprocating piston and crank engine is its mechanical simplicity. It scores over other rotary engines by the fact that it is pressure sealed with conventional piston rings. There is also provision for piston cooling. These are important points often neglected on rotary engines.

Tschudi chose the toroidal shape because it simplified pressure sealing. The piston rings are fixed with a pin to prevent their rotation on the pistons. The tangential rings between stator and rotors are the self-expanding type. Since their motion is radial, they are provided with a tongue and groove surface; a feature which is used in the sealing of high pressure steam turbines.

Tschudi's first rotary engine was designed in 1927. It was a scissor-action type, using a ratchet device for power transmission from the pistons to the output shaft. That proved to be a fragile arrangement and by 1935 it had been entirely redesigned, with a gear drive system. Then Tschudi became aware of modern cam roller drives and designed his own, which he patented in 1953. The cam-roller drive is a positive contact mechanism; both rollers of the rotors are in continuous contact with the cam faces. Aircraft engines with cam-roller drives were in use during the First World War. However, the cams of these engines were of the figure 8 shape, with the explosion taking place at the dead point, a disadvantage fully avoided in the present invention. Oval shaped cam-roller drives are now used on the latest European diesel engines as a means to eliminate the crankshaft.

The cam-roller drive system was designed for smooth operation and a minimum of vibration. This type of cam-roller drive was first used on a three-cylinder two-stroke reciprocating piston diesel engine built in 1954 by Svanemölle Verft in Copenhagen, Denmark. It was completely crankless. Tschudi claimed that the pistons were friction-free in his engine and would never wear. They are rigidly mounted on rotors and have clearance in their toroidal chambers. This permits greater choice in the selection of materials. Pistons can be die-cast and need no surface machining. The output shaft is located eccentrically to the engine and contains two cams. Each cam contains a segmental curve serving as the base circle, interrupted by one ovate curve and by one concave curve. Both cams are mathematically identical but are offset 180 degrees on the driveshaft, and axially spaced so as to engage in the respective rotors.

The Tschudi engine has a continuously cool inlet region, ensuring adequate induction of air. It has low pressure loss through elimination of torturous valve passages. Pre-ignition, through hot spots that can occur in a conventional cylinder head, can probably be eliminated. The inventor claims more efficient operation due to a faster piston speed, but other experts consider this doubtful. The Tschudi prototype engine is designed with a jacket for water-cooling but the toroidal cylinder may be provided instead with fins for air-cooling. Several factors indicate that the rotary engine will generate less waste heat than a crank engine: rapid expanding pistons will transform the caloric energy of the fuel into additional power; only one-sixth of the toroidal cylinder is exposed to the explosion. Therefore, pistons and rotor faces will cool while passing through the other zone. Pistons will be splash-cooled with oil contained inside the rotors.

One of the weakest features of the Tschudi engine is the high stress level at the point of contact of the cam and rollers, particularly at the time when the piston is forced in a counter-clockwise direction by combustion pressure. The rapid acceleration and stopping of the pistons and rotors relative to the cam is also bound to result in shock loading of the cam surface. Experts who have examined it express doubts about the lubrication, friction and wear characteristics of the packing rings between the rotors and stator, as well as the use of aluminum for cylinder material. It is not considered likely that the pistons will be frictionless against the stator. The combustion pressures acting to deflect the rotors and seals may cause considerable friction.

Because the pistons are alternately moving and coming to rest, according to the geometry of the cam, high inertia forces will result. The large ball bearing which supports the piston carrier rotates at a mean speed that is half that of the output-shaft. The axial load caused by high gas pressures within the toroid is eccentric and intermittent. The rotational speed is also intermittent. It is therefore, thought that the difficult operating conditions of the bearing may impose a limitation on the engine.

Although rings are provided to stop gas leaking past the pistons, the toroidal passage cannot be perfectly intact due to its construction from three separate parts. Three small fissures run down the length of the toroid past the piston rings and they cannot be completely closed. This is a problem associated with all engines of this type and a full investigation into experience already accumulated would show whether it is a serious matter or not.

A similar sealing problem exists on rotary heat exchangers and is, in fact, the main obstacle to successful development of these units. In view of the considerable development work carried out on this type of sealing problem it is thought that this feature may be the major snag with this type of engine combustion.

KAUERTZ

The Kauertz engine resembles the Tschudi in that it is another scissor-action power unit. It is the invention of Eugen Kauertz, now living in retirement in Hüfingen, Germany, but still active as an inventor. The prototype engine has only 22 parts. A test engine of 7.5 cubic inch displacement has produced 51 horsepower at 4,000 r.p.m. according to the inventor's claims. A larger 61.7 cubic inch unit has delivered 213 horsepower in tests. The Kauertz engine has the same number of firing impulses per shaft revolution as a V8 reciprocating piston engine, and runs with comparable smoothness. The Kauertz engine has vane-type pistons in a circular-section working chamber. Two sets of vanes rotate on the same axis but they continuously change position relative to each other.

This speeding up and slowing down of one set of vanes changes the volume of gas between the two sets. Changes in gas volume produce the pumping action needed for intake, compression, combustion, and exhaust. Four effective combustion chambers give four power phases per output-shaft revolution. Correct phasing of the vane motion is assured by a gear-and-crank system. The output shaft carries a coaxial but stationary sun gear and a planet-gear carrier that revolves with the shaft. A crank fixed to the planet-gear axis describes two complete revolutions for one turn of the planet carrier. The planet-gear crank is linked to a counterbalanced arm that both rotates and oscillates around the output shaft axis. Both movements of this arm are produced by the crank and its linkage. The arm is rigidly fixed to the secondary set of vanes. These vanes alternately catch up with and fall behind the primary vanes, which are fixed to the output shaft and therefore drive the

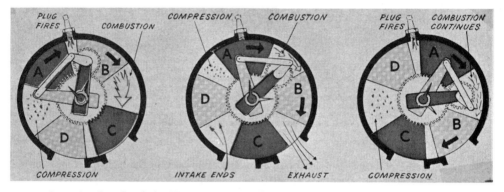

Operational cycle of the Kauertz engine. All piston movement is unidirectional, but it is interrupted in order to produce a scissor-action (with the scissors revolving around the pivot point).

planet carrier. Both faces of all four vanes work full-time, going through all phases of the four-stroke cycle. Vane sealing is obtained by free-riding blades held against the housing by centrifugal force. The same blades also offer some sealing against the end covers. The Kauertz engine has no unbalanced inertia forces, and very high speeds are possible because there is no valve gear. The vanes themselves open and close the ports with accurate timing.

The main drawback in the Kauertz engine seems to be the high inertia loads generated on the phasing vanes and transmitted to the frail-looking linkage. Does it have a future? It may be developed under a contract with the South African government. And a Canadian firm, Hi-Pow'r Roto, which holds Kauertz rights for Canada, Japan and Australia, is reported to be building a factory in Victoria, B.C.

VIRMEL

The Virmel engine is similar to the Kauertz because the vane-type movement is the same, but the power drive is entirely different. The Virmel is the invention of Melvin Rolfsmeyer, an engineer living in Lincoln, Nebraska, who named it for his wife Virginia and himself (Vir(ginia)Mel(vin)). The prototype is undergoing testing and development at the Lynx Corporation in Lincoln, for use in cars and boats.

Displacement is only 50 cubic inches, and claimed power output exceeds 300 horsepower at 3,800 r.p.m. The Virmel engine has two sets of vane-type pistons and a gear-and-crank system that controls piston phasing. It differs from the Kauertz in several respects, but works on a similar cycle. It produces four power impulses for each output shaft revolution. The two vane sets are fixed to two concentric shafts, and there is a stationary sun gear. But in the Virmel engine, the sun gear has two satellites, each one with a crank linked to a lever splined to one of the two main shafts. Rotation of the satellite gears and cranks directs piston acceleration and retardation. The satellites are always diametrically opposed on either side of the sun gear. Satellite rotational speed (and output-shaft speed) is steady, while one set of vanes is momentarily brought to a complete stop twice during each cycle.

In the Kauertz engine, the primary vanes run at steady speed and secondary vane speed varies. In the Virmel engine both sets stop and restart. Unlike the Kauertz, the Virmel engine has no through shaft. The two concentric mainshafts do not continue beyond the sun gear and crank system. Power flow is taken via the satellite gears and crankshafts revolving at satellite speed to another planetary gear train. The satellites in both planetary gear trains simultaneously orbit and rotate. The final gear set reverses rotation in a 1:1 ratio to turn the output shaft in the same direction as the twisting force applied to the engine mass. This practically eliminates torque reaction and simplifies installation.

OMEGA

Another scissor-action engine is the Omega, invented by Granville Bradshaw in 1955. Granville Bradshaw designed Pratt & Whitney's first radial aircraft engine during World War I, and designed the Belsize-Bradshaw car of 1921, using a light V-twin oil-cooled engine. Later he joined BSA (Birmingham Small Arms Company, Ltd.) and designed several flat-twin, air-cooled motorcycle engines. In the early 1950s he was engaged in the design of motorscooter engines.

The Omega engine shares the toroidal chamber and the four pistons with the Tschudi design. The pistons are carried on rocker arms and reciprocate in pairs. The rocker arms are mounted on separate but concentric shafts, each fitted with a crankpin that carries a connecting rod. The connecting rods turn a crankshaft located immediately below. The toroidal engine block revolves around the oscillating pistons, and a 2:1 gear set positioned behind the crankcase assures one revolution of the block for every two crankshaft revolutions. The pistons are double-faced and float between gas pressure phases on opposite sides. Every swing of each piston, in either direction, is a power stroke. Every power stroke is cushioned by a compression stroke at the other end. This drive system eliminates inertia losses caused by crankshaft reversals and sharply reduces bearing loads. Friction losses are claimed to be small, since there is no side thrust on the pistons. Operating on the two-stroke cycle, the Omega engine has only one intake and one exhaust port serving the combustion chambers. This is possible because the block itself rotates. Similarly, there is only one spark plug, mounted on the periphery of the toroidal chamber.

The Omega engine was handed over to Weslake & Co. of Rye, Sussex, for development, in 1966. The Weslake people were unable to start it and had to overhaul the engine before it was put on the test bench. It never ran under its own power, though occasional firings would occur. No actual development was ever accomplished.

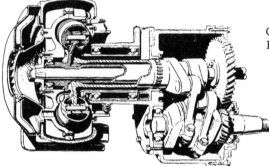

Cutaway drawing of Granville Bradshaw's Omega engine.

MERCER

The Mercer engine begins the third category of rotary engines which have reciprocating pistons and rotary blocks. The first engine of this type was built in 1898 by F. O. Farwell. His design was used in the 1905 Adams-Farwell automobile. Its principles reappeared in the French Gnone-Rhone aircraft engine during World War I, and are now used, with variations, for the experimental Mercer, Selwood, and Porsche rotary engines.

The Mercer engine is a rotary radial with four spoke-like cylinders spinning inside a stationary rim. It was the creation of Austin Mercer, an engineer and inventor, who operates a ten-man machine shop in Bradford, Yorkshire. It was first tested in 1963. Operating on the two-stroke cycle, the Mercer engine was designed for high torque at low revs. Unlike other rotary engines, it has a wide spread of power throughout the speed range.

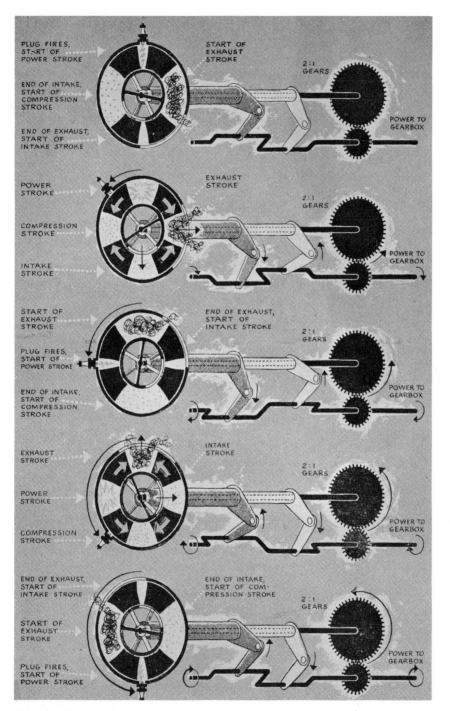

Operational cycle of the Omega engine.

The Mercer engine consists of two opposed pistons in a single cylinder. Each piston pin works as the axis for two rollers. The rollers run on a track that looks trochoidal but is merely the peripheral track of two intersecting circles. Combustion in the center forces the pistons and rollers apart. The rollers can move outward only if the cylinder block turns so the output shaft revolves with the block. The only reciprocating parts are the pistons, and since they move in and out in unison, the engine is inherently balanced. The output shaft does not continue on the other side of the block. A hollow shaft at the other end carries the spark plug, feeds a gas mixture from the carburetor, and delivers lubricating oil. The exhaust ports are in the output shaft. The pistons open and close the ports in the manner of a normal two-stroke piston engine. Since there is no crankcase, the fresh mixture cannot be fed in by crankcase compression as in conventional two-stroke engines. Instead a blower is installed between the carburetor and the induction pipe. This gives the advantage of forcing the gases into the cylinders instead of relying on their haphazard passage through transfer ports uncovered and sealed by sliding pistons. The supercharging arrangement eliminates the common two-stroke complaints: poor scavenging and disposal of exhaust gases, resulting in low efficiency and high fuel consumption.

Intake timing is variable to give peak performance at all speeds. The central induction tube can be twisted axially to alter the point at which fuel is blown into the cylinders. The angular relationship between the inlet valve and the conventional cylinder-wall exhaust port is changed in accordance with speed and load. Scavenging of the Mercer engine is facilitated by the transfer of a fresh charge from the area behind the pistons where it has been compressed. The pistons then move together to compress the new mixture prior to combustion. The fresh charge is admitted to the combustion chamber via a transfer channel similar to two-stroke piston engine practice. Some of the fresh charge is lost to the exhaust during the transfer process. Stresses on roller assembly and cylinder walls are quite high. Each cylinder goes through two complete cycles for each revolution. Four pairs of explosions or eight power strokes accompany one turn of the shaft. At a given speed it makes the engine equal to a 16-cylinder, four-stroke engine of four times the cubic capacity.

The experimental model had a displacement of only 50 cc. (3.0 cubic inches) with a 1⅛ inch bore and ¾ inch stroke, and ran with a 6:1 compression ratio. Mercer claimed a peak output of 10 horsepower at 5,000 r.p.m. The little prototype was ten inches in diameter and weighed 56 pounds including blower and generator.

In 1965 Mercer built a 750 cc. engine (46 cubic inches) that was tested in 1966 and produced 16 horsepower at 1,500 r.p.m. In 1968 and 1969 Mercer revised his invention and built a second-generation rotary engine. The new version is lighter, smaller, and simpler. The cruciform cylinder block has been replaced by one cylinder and the supercharger has been discarded. The cylinder has dual bore and the two pistons have dual diameter. The small bore part is central, including the combustion chamber and the full travel of the piston crown. The large-diameter part of the pistons travels through an intake and partial compression chamber connected to the combustion chamber by a conventional transfer port. In this respect it bears a striking resemblance to the Pescara type of free-piston engine. As in the earlier Mercer engine each piston carries rollers that run along a stationary track consisting of two circular arcs. As the pistons are driven outward by the combustion pressure from the centrally located spark plug, the rollers are pressed against the track, thereby forcing the block to rotate. The outward motion of the pistons also compresses a fresh air-fuel charge that has entered behind the pistons. When the pistons are farthest apart, the exhaust ports are uncovered and vent the burned gases.

This second-generation Mercer engine design is quite versatile because it can be applied to multi-cylinder or banked arrangements. This piston design is intended to give improved charge synchronization at all engine speeds. Detail design for porting and cooling can be varied according to application. A separate system of pressure lubrication to all moving parts was considered necessary.

However, the latest Mercer engine shows no advantage for operation under part load (typical automobile engine conditions). Purging pressure falls off very quickly, in the case of exhaust gas scavenging as well as feeding the fresh charge under compression to the combustion chamber. The high pressure drop is not conducive to high efficiency. The timing of gas intake and exhaust is poorly controlled by piston movement. There is a risk of 20 to 50% loss of fresh charge to the exhaust system, which would make the emission levels unacceptable.

SELWOOD

William R. Selwood of Southampton, England, decided to sponsor development of a rotary engine invented by Cecil Hughes in 1952. Progress was slow, and now Selwood has sold the engine to a new company, Orbital Engineering, in Bristol. Hughes is still working on its development. The Selwood engine is rotary in the sense that the cylinder block revolves, driving an output shaft, while the mainshaft which supports the spider and the pistons, is stationary. The spider is fixed at a 15 degree angle. The Selwood orbital engine has toroidal chambers and turns reciprocating motion into rotary motion by a kinematic inversion of the swash-plate system (now commonly used in hydraulic pumps).

The cylinder block is fixed to the output shaft and rotates as a unit. The motion of the pistons is the key to this engine: as the piston is forced down into the cylinder, its pressure on the wobble plate causes the rotation. The pistons run back and forth in their toroidal tracks. Piston travel is possible only by letting the block turn. But the piston doesn't really reciprocate; it orbits around the mainshaft and goes through 30 degrees of track travel in half an orbit. The next half-orbit, the piston moves 30 degrees back again.

There are 12 cylinders grouped around the output shaft, and the engine is stated to be notably free from vibration. Like the Mercer engine, the Selwood works on the two-stroke principle. When one side of a piston begins a power stroke, the other side starts a compression phase. Selwood's interest in this engine was first publicized in 1961, and little progress seems to have been made since then. Its suitability for automative purposes remains questionable.

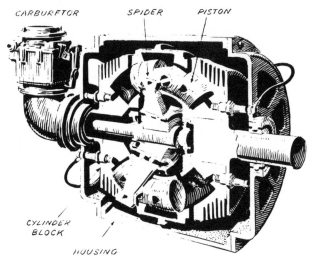

The Selwood rotary engine, invented by Cecil Hughes.

PORSCHE

Porsche (the well-known sports car makers in Stuttgart) patented a crankless engine in 1963. Its four-cylinder cruciform block revolves in a cage, but, unlike the Mercer engine the Porsche gets its power on inward strokes. Each piston has a roller connected by belt-like links to the other rollers. All rollers are in permanent contact with a large two-lobe cam on the central shaft.

The housing has an intake port, an exhaust port and a spark plug. The wrist pins of the four pistons are large rollers interconnected by belt-like links, and are always in contact with a large kidney-shaped cam on the output shaft. The rotor brings each piston through the four strokes with each one making one stroke each way twice per revolution. The patent shows a very short-stroke rotor and sensibly-shaped combustion chambers. Experts have expressed some concern over internal friction problems, and the engine has never been developed. In fact, Porsche has become a Wankel licensee.

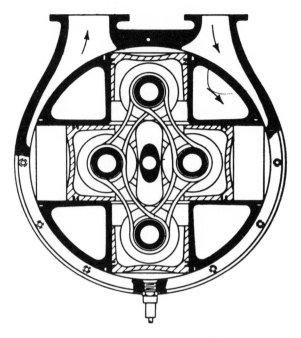

Cross-section of the proposed Porsche crankless rotary engine.

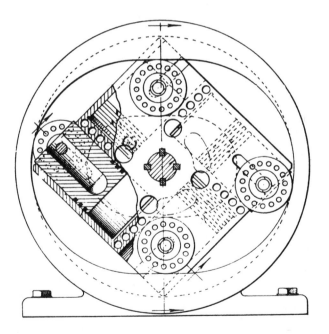

Cross-section of the Leath rotary engine.

LEATH

Another rotary-block engine, invented by American engineer Harry Leath, has four cylinders in a square block which revolves inside a cylindrical housing. Each piston axis runs at 90 degrees to that of its neighbor. The pistons have an axle running diametrically across the bottom of the skirt and the axle carries a pair of rollers that run on an epitrochoidal track inside the housing and control the motion of the block. The block is carried centrally in the housing and the roller track is designed to correspond with the reciprocating motion of the pistons in their cylinders.

The inventors claim to have solved the basic problems of sealing and cooling, but they still know relatively little about the effect of changes in porting, shape of combustion area, or spark plug location and ignition timing and they have no clear idea of production costs, durability, and ease of maintenance. Inventing a rotary engine is the easy part—the difficulty lies in making it a *good* rotary engine. Continued experience with the Wankel engine and improvements on it may facilitate development work on certain other types of rotary engines. As the Wankel engine continues to progress, it will become increasingly accepted, and while we are still thinking of it as an "unconventional" engine, it may become a fairly typical automotive powerplant in the future, while other rotary engines still remain in their embryonic stages.

Bibliography

ARTICLES, SAE, PAPERS, BOOKS

ARTICLES

ANON.: Wankelmut. Der Spiegel, Nr. 31, 1961.

ANON.: Japaner im Nacken. Der Spiegel, Nr. 34, 1963.

ANON.: A History of Rotary Engines and Pumps. The Engineer, January to June, 1939.

ANSDALE, R. F.: NSU-Wankel Engine. Automobile Engineer, May, 1960.

ANSDALE, R. F.: Rotary Combustion Engines. Automobile Engineer, December, 1963, January & February, 1964.

ANSDALE, R. F.: Air-Cooled Wankel Engine. Automobile Engineer, August, 1965.

ANSDALE, R. F.: The Mazda 0813 Wankel Engine. Automobile Engineer, April, 1968.

ANSDALE, R. F.: Feasibility of High-Output Wankel RC Engines. Combustion Engine Progress, 1968.

BENSINGER, DIPL. ING. W. D.: Der heutige Entwicklungsstand des Wankelmotors. Motortechnische Zeitung, January, 1970.

BENTELE, MAX: Rotating Combustion Engine. R. T. Hurley and M. Bentele, SAE Transactions, 1961, based on SAE Paper S-236 presented at a meeting of the Metropolitan Section of the SAE on March 10, 1960, and "Curtiss-Wright Rotating Combustion Engine Delivers 2 HP Per Cu. In.," SAE Journal, June, 1960.

BENTELE, MAX: Curtiss-Wright's Developments on Rotating Combustion Engines, presented at the International Congress and Exposition of the SAE in Detroit in January, 1961, SAE Paper 288B, reprinted in SAE Transactions, 1961.

BENTELE, MAX: Curtiss-Wright's Entwicklungen an Rotationsverbrennungsmotoren. MTZ Motortechnische Zeitschrift, Stuttgart, June, 1961; reprinted by Technische Rundschau, Bern, Switzerland, August 11, 1961.

BENTELE, MAX: Further Developments on Rotating-Combustion Engines at Curtiss-Wright. Paper presented at a meeting of the Twin City Section of the SAE in Minneapolis, Minnesota, on March 21, 1962, SAE Paper S-348.

BENTELE, MAX: Weiterentwicklung von Rotations-Verbrennungsmotoren bei Curtiss-Wright. Dr. Max Bentele. MTZ Motortechnische Zeitschrift, Stuttgart, September, 1962.

BENTELE, MAX: Fortschritte mit Curtiss-Wright's Rotations-Verbrennungsmotoren. Dr. Max Bentele und Charles Jones, MTZ Motortechnische Zeitschrift, Stuttgart, Mai, 1966.

BENTELE, MAX: Curtiss-Wright's New Rotating Combustion Units. M. Bentele and C. Jones, Diesel & Gas Turbine Progress, April, 1966.

BOWLER, MICHAEL: Wankel Progress. Motor, August 24, 1968.

BROCKHAUS, HERBERT AND STROBEL, RUDOLF M.: Ro 80—ein NSU Mittelklassewagen mit Kreiskolbenmotor. Automobiltechnische Zeitung, September, 1967.

BULMER, CHARLES: Serious Contender (NSU Ro 80). Motor, September 9, 1967.

VON FERSEN, OLAF: NSU-Wankel Zwischenbilanz. Automobil Revue, July 24, 1969.

FRANCIS, DEVON: The Engine That's Giving Detroit "Wankel Fever." Popular Science, April, 1966.

FRERE, PAUL: Vierkammermusik. Das Auto, Motor und Sport, Heft 8, 1970.

FROEDE, W.: Entwicklungsarbeiten an Dreh- und Kreiskolben- Verbrennungsmotoren. VDI-Zeitschrift, March 11, 1960.

FROEDE, DR. ING. WALTER: Kreiskolbenmotoren Bauart NSU/Wankel. Motortechnische Zeitschrift, January 10, 1961.

FROEDE, DR. ING. WALTER: Auszüge aus neueren Entwicklungsarbeiten am Kreiskolbenmotor Bauart NSU/Wankel. Motortechnische Zeitschrift, April, 1963.

FROEDE, DR. ING. WALTER: NSU Wankel Engine. Automobile Engineer, July, 1963.

FROEDE, DR. ING. WALTER AND JUNGBLUTH, DIPL. ING. GEORG: Der Kreiskolbenmotor des NSU Spider. Automobiltechnische Zeitung, May and August, 1968.

GROTHMANN, CARL: Japanese Wankel-Powered Car. Popular Science, April, 1967.

HIGGINS, L. R.: The Story of the Rotary Valve. Motor Cycling, July 30, 1953.

HUBER, EUGEN WILHELM: Thermodynamische Untersuchungen an der Kreiskolbenmaschine. VDI-Zeitschrift, March 11, 1960.

HUF, DIPL. ING. FRANZ: Zur Geschichte der Rotationskolbenmaschinen. Automobil Revue, No. 49, 1961.

KORP, DIETER: The Man Who Made Tomorrow Come. Sports Car Illustrated, March, 1960.

KORP, DIETER: Der Wankel-Motor und die Aktien. Das Auto, Motor und Sport, Heft 18, 1960.

KORP, DIETER: So entstand der neue Mercedes-Benz C-111. Das Auto, Motor und Sport, Heft 19, 1969.

LUDVIGSEN, KARL: Technique of Tomorrow's Power. Car and Driver, May, 1961.

LUDVIGSEN, KARL: NSU Ro 80. Car Life, December, 1967.

LUDVIGSEN, KARL: Mazda Cosmo. Toyo Kogyo Joins the Rotary Club. Car Life, December, 1967.

LUDVIGSEN, KARL: Return of the Gullwing. Car, October, 1969.

MOLTER, GÜNTHER: NSU Ro 80. Road & Track, October, 1967.

MUNDY, HARRY: NSU-Wankel Rotary Expansion Engine. Autocar, February 19 and 26, 1960.

MUNDY, HARRY: Progress Report. Autocar, March 2, 1962.

NORBYE, JAN P.: Test Drive of a U.S. Car with a Rotating Combustion Engine. Popular Science, April, 1966.

RIXMANN, W.: Der NSU-Wankelmotor im Fahrzeug. Automobiltechnische Zeitung, March, 1961.

SCHMIDT, ERNST: Die Drehkolben- und Kreiskolbenmaschine, Entstehung einer neuen Bauart des Verbrennungsmotors mit überraschenden Eigenschaften. VDI-Zeitschrift, March 11, 1960.

WANKEL, FELIX AND FROEDE, DR. ING. WALTER: Bauart und gegenwärtiger Entwicklungsstand einer Trochoiden-Rotationskolbenmaschine. Motortechnische Zeitschrift, February, 1960.

WANKEL, FELIX: Rotary Piston Engine Performance Criteria. Automobile Engineer, September, 1964.

WIESELMANN, H. U.: Wankelmut. Das Auto, Motor und Sport, Heft 20, 1969.

YAMAGUCHI, JACK: Wankel twin in production. Motor, July 29, 1967.

YAMAGUCHI, JACK: Wankels for all. Motor, August 17, 1968.

YAMAMOTO, KENICHI: Die Kreiskolbenmotoren der Bauart Wankel-NSU von Toyo Kogyo. Automobiltechnische Zeitschrift, June, 1970.

SAE PAPERS

ANSDALE, R. F.: Rotary Engine Development and its Effect on Transport. Paper #8, SAE of Australasia, presented in Auckland, New Zealand, October 14–18, 1968.

COLE, DAVID E. AND JONES, CHARLES: Reduction of Emissions from the Curtiss-Wright Rotating Combustion Engine with an Exhaust Reactor. SAE paper #700074, presented in Detroit, Michigan, January 12–16, 1970.

FROEDE, WALTER G. DR.: The NSU Wankel Rotating Combustion Engine. SAE paper #288A, presented in Detroit, Michigan, January 9–13, 1961.

FROEDE, WALTER G. DR.: The Rotary Engine of the NSU Spider. SAE paper #650722, presented in Cleveland, Ohio, October 18, 1965.

FROEDE, WALTER G. DR.: NSU's Double Bank Production Rotary Engine. SAE paper #680461, presented in Detroit, Michigan, May 20–24, 1968.

JONES, CHARLES: The Curtiss-Wright Rotating Combustion Engines Today.
SAE paper #886D, presented in San Francisco, Calif., August 17–20,
1964.

JONES, CHARLES: New Rotating Combustion Powerplant Development. SAE
paper #650723, presented in Cleveland, Ohio, October 18–21, 1965.

JONES, CHARLES: The Rotating Combustion Engine—Compact Lightweight
Power for Aircraft. SAE paper #670194, presented in Farmingdale,
N.Y., November 3, 1966.

KELLER, HELMUT: Small Wankel Engines. SAE paper #680572, presented
in Milwaukee, Wisconsin, September 9–12, 1968.

PICARD, FERNAND: The Future of Automobile Technique. SAE paper #980A,
presented in Detroit, Michigan, January 11–15, 1965.

YAMAMOTO, KENICHI AND KURODA, TAKASHI: Toyo Kogyo's Research and
Development on Major Rotary Engine Problems. SAE paper #700079,
presented in Detroit, Michigan, January 12–16, 1970.

BOOKS

ANSDALE, R. F.: The Wankel RC Engine. A. S. Barnes & Company, South
Brunswick & New York, 1968.

BURSTALL, AUBREY F.: A History of Mechanical Engineering. The M.I.T.
Press, Massachusetts Institute of Technology, Cambridge, Mass., 1965.

HÜTTEN, H.: Schnelle Motoren seziert und frisiert. Richard Carl Schmidt &
Co., Braunschweig and Berlin, 1955.

INMAN HUNTER, MARCUS C.: Rotary Valve Engines. Hutchinson's Scientific
& Technical Publications, London, New York, Melbourne, Sydney, &
Cape Town, 1951.

IRVING, CAPT. J. S.: Investigation of the Developments in the German
Automobile Industry during the War Period. British Intelligence Objectives Sub-Committee, London, 1945.

The Author

JAN P. NORBYE is the Automotive
Editor of *Popular Science Monthly* and a Contributing Editor of *Automobile Quarterly*. He is the author of *Sports Car Suspension* and *The New Fiat Guide*. Norbye has been a full-time writer and editor since
he came to the United States in 1961 to take over the position of Technical Editor of *Car and Driver*, where he worked for four years prior
to joining *Popular Science*. He had started writing about cars as a
spare-time occupation while still in high school in his native Oslo,
Norway. His all-absorbing interest in automobiles led him to join the
industry after graduating from Oslo Commercial College in 1951. He
worked for a year with HRG Engineering Company in Tolworth, Surrey, England, and developed his contacts within the British motoring
press. Returning to Oslo, he joined the administration of the Royal
Automobile Club of Norway and handled many assignments for the
Club's journal, *Motorliv*. In 1954 he became service manager for the
importers of MG, Morris and Studebaker in Norway but left in 1956
to take up a position with Esso Touring Service in Paris, France. Two
years later he returned to Scandinavia as a service representative for AB
Volvo, working out of the Gothenburg, Sweden, factory. He helped
establish Volvo's service organization in the United Kingdom and made
frequent visits to France and Belgium. Norbye has driven cars since
1945 and is an experienced, professional test driver. His journalistic
specialities are technical writing and automotive history. Norbye is
thirty-nine, an Affiliate Member of the Society of Automotive Engineers, and a Director of the International Motor Press Association.
He lives in Jackson Heights, New York City.